TIMESPACE

The social sciences and humanities have recently taken a 'spatial turn', with workers drawing upon a range of geographical concepts and metaphors to explore an increasingly complex and differentiated social world. Elsewhere, interest has grown in the role that differing conceptualisations of time play in shaping our understandings of the world. *TimeSpace* is the first book to bring these interests together. Rather than thinking in terms of either time or space, it argues that our accounts of the social world must draw instead upon the more complex notion of TimeSpace.

With contributors drawn from a range of disciplines, including Geography, Sociology, Gender Studies, International Studies and English Literature, *TimeSpace* is wide-ranging in both substantive and theoretical scope. In the first part of the volume contributors explore the 'Making' and 'Living' of TimeSpace. Chapters examine past and present changes in time and time consciousness and the meaning of such changes for the people living through them; changing understandings of Modernisation and Progress and the geographies that underpin them; and the role that understandings of TimeSpace play in projects of national and racial identity and the politics of Belonging. In the second part of the volume, 'Living-Thinking TimeSpace', attention is turned to the ways in which we might most usefully conceptualise TimeSpace itself – whether drawing on the perspectives of a rejuvenated time-geography, some variation of Lefebvre's rhythm analysis, phenomenology or Buddhism.

At the heart of the volume lies a challenge to all those who have uncritically embraced the recent 'spatial turn' and to those working in the field of time studies to think in terms of neither only time or space but a multi-dimensional, partial and uneven TimeSpace.

Jon May is Lecturer in Geography, at Queen Mary, University of London. **Nigel Thrift** is Professor of Geography at the University of Bristol.

CRITICAL GEOGRAPHIES

Edited by **Tracey Skelton**, *Lecturer in International Studies,
Nottingham Trent University,* and **Gill Valentine**, *Professor of
Geography, The University of Sheffield*

This series offers cutting-edge research organised into three themes of concepts,
scale and transformations. It is aimed at upper-level undergraduates, research
students and academics and will facilitate inter-disciplinary engagement between
geography and other social sciences. It provides a forum for the innovative and
vibrant debates which span the broad spectrum of this discipline.

TIMESPACE

Geographies of temporality

Edited by Jon May and Nigel Thrift

London and New York

First published 2001 by Routledge
11 New Fetter Lane, London EC4P 4EE

Simultaneously published in the USA and Canada
by Routledge
29 West 35th Street, New York, NY 10001

Routledge is an imprint of the Taylor & Francis Group

© 2001 Selection and editorial material Jon May and Nigel Thrift;
individual chapters the contributors

Typeset in Perpetua by Taylor & Francis Books Ltd
Printed and bound in Great Britain by Biddles Ltd, Guildford and King's Lynn

British Library Cataloguing in Publication Data
A catalogue record for this book is available from the British Library

Library of Congress Cataloging in Publication Data
TimeSpace : geographies of temporality / [edited by]
Jon May and Nigel Thrift.
p. cm. – (Critical geographies)
Includes bibliographical references and index.
1. Time–Sociological aspects. I. May, Jon.
II. Thrift, N.J. III. Series.
HM656 .T558 2001
304.2'3–dc21   00-045742

ISBN 0–415–18083–X (hbk)
ISBN 0–415–18084–8 (pbk)

CONTENTS

CONTRIBUTORS

Mike Crang is Lecturer in Geography at the University of Durham. He is the co-editor of the journal *Time and Society* (Sage). Recent publications include *Thinking Space* (co-edited with Nigel Thrift, 2000), *Virtual Geographies* (co-edited with Phil Crang and Jon May, 1999) and *Cultural Geography* (Routledge, 1998). He has written widely on issues of historicity, social memory and practice.

Karen Davies is Docent and Senior Lecturer at the Department of Sociology, Lund University, Sweden. Previously Director at the Centre for Women's Studies, Lund University she was one of the founders of Scandinavia's first feminist journals; Kvinnovetenskaplig tidskrift. Her research moves around questions of gender and has focused upon issues of unemployment, labour markets, muncipal childcare and hospital work. Her analysis of time from a feminist perspective has been developed in *Women, Time and the Weaving of the Strands of Everyday Life* (Aldershot, 1990) as well as in numerous articles. She is a member of the international advisory board for *Time and Society*.

Geoffrey DeVerteuil is a Doctoral candidate in Geography at the University of Southern California. His research focuses on public facility location theory, restructuring of the local welfare state and the impacts of welfare reform upon the interactions between service delivery institutions and service-dependent individuals in poor inner-city neighbourhoods.

Lucinda Ferguson was born in-country New South Wales and lived during term times at a School Boarding House in Sydney. She was later employed at this Boarding House while studying sociology at the University of New South Wales.

John Frow is Regius Professor of Rhetoric and English Literature at the Department of English, University of Edinburgh. His most recent books are

Cultural Studies and Cultural Value (Clarendon Press, 1995), *Time and Commodity Culture* (Clarendon Press, 1997), and, with Tony Bennett and Mike Emmison, *Accounting for Tastes: Australian Everyday Cultures* (Cambridge University Press, 1999).

Ann Game is Associate Professor of Sociology at the University of New South Wales. Her publications include *Undoing the Social* (Open University Press, 1991) and, with Andrew Metcalfe, *Passionate Sociology* (Sage, 1996) and the forthcoming *Doing Nothing and Other Ways of Being.*

Martin Gren is Lecturer in Human Geography at the Department of Geography and Tourism, University of Karlstad, Sweden. He is the co-editor (with P.O. Hallin) of *Svensk kulturgeografi: en exkursion inför 2000-talet* (Studentlitteratur, 1998) and (with P.O. Hallin and Irene Molina) of *Place for Culture / Spaces of Power* (Brutus Östlings Förlag / Symposion, forthcoming). He is currently researching the historical geographies of the Swedish 'mad business'.

Kevin Hetherington is Senior Lecturer in Sociology at Brunel University. His books include *The Badlands of Modernity* (Routledge, 1997), *Expressions of Identity* (Sage / TCS, 1998) and *New Age Travellers* (Cassell, 1999). His research is concerned with issues in spatial theory and he has recently co-edited a special issue of *Environment and Planning D: Society and Space* (with John Law) on 'After networks'. He writes on the historical sociology of both museums and factories and is currently researching the spatial and visual politics of access, with particular emphasis upon access to museums

Nuala C. Johnson is a Lecturer in Geography at Queen's University of Belfast. She has published widely on the geographies of national identity, the heritage industry and the cultural geography of public memory. She is currently completing a book on the public commemoration of the First World War in Ireland.

David R. Loy is Professor in the Faculty of International Studies at Bunkyo University, Chigasaki, Japan. His primary interests lie in comparative philosophy and religion and with comparisons of Buddhist and modern Western thought in particular, a subject on which he has published widely. He is the author of *Nonduality: A Study in Comparative Philosophy* (Yale University Press, 1989), *Lack and Transcendence: The Problem of Death and Life in Psychotherapy, Existentialism and Buddhism* (Prometheus Press, 1996), and editor of *Healing Deconstruction: Postmodern Thought in Buddhism and Christianity* (Prometheus Press, 1996).

Andrew Metcalfe is Senior Lecturer in Sociology at the University of New South Wales. He is author, with Ann Game, of *Passionate Sociology* (Sage, 1996) and of the forthcoming *Doing Nothing and Other Ways of Being*.

Elspeth Probyn is Associate Professor and Head of Department of Gender Studies at the University of Sydney. Her publications include *Sexing the Self* (Routledge, 1993), *Outside Belongings* (Routledge, 1996) and *Sexy Bodies* (co-edited with Elizabeth Grosz, 1995). Her forthcoming book is titled *Visceral: Sex, Eating and Ethics* and she is completing a co-edited anthology on the politics of shame.

Jenny Shaw is Senior Lecturer in Sociology at the University of Sussex. Alongside research on various aspects of time and questions relating to the pace of life she has a long-standing interest in gender and the study of consumption. She has recently completed a study of Marks and Spencer and its place in the British High Street (with Janice Winship) and is author of *Education, Gender and Anxiety* (Taylor & Francis, 1995) and co-author of *Making Gender Work* (Open University Press, 1995).

Jeremy Stein is employed in the Business School at the University of Birmingham. He is currently Research Fellow on the 'Dilemmas of a Maturing Technology', a project funded by the Leverhulme Trust. His research interests focus on technological and organisational innovation, corporate strategy and the history of technology.

Jennifer R. Wolch is Professor of Geography at the University of Southern California, where she also co-directs the Sustainable Cities Program. She has published widely on issues of welfare, governance and homelessness and is author of *The Shadow State: Government and Voluntary Sector in Transition* (The Foundation Center, 1990) co-author (with Michael Dear) of *Malign Neglect: Homelessness in an American City* (Jossey-Bass, 1993) and *Landscapes of Despair: From Deinstitutionalization to Homelessness* (Princeton University Press, 1987) and co-editor of *The Power of Geography* (Unwin Hyman, 1989).

ACKNOWLEDGEMENTS

As a number of our contributors have reminded us during the protracted business of putting this collection together, for a book where so much attention is focused on the difficulties associated with an ever increasing pace of life, *TimeSpace* has taken a remarkably long time to reach the shelves. To some extent this is always the way with such collections, but the editors would like to thank the contributors and those at Routledge (Sarah Lloyd, Sarah Carty and Ann Michael) for their unusual patience in what has, by any way of reckoning, been a rather long process. The credit that what emerged from this process is an interesting and innovative collection is theirs. That it took so long can only be down to us.

Jon and Nigel
(Brighton and Bath)

1

INTRODUCTION

Jon May and Nigel Thrift

The impetus for the current collection arises out of a growing sense of dissatisfaction with two recent and related developments in social theory and the social sciences and humanities more broadly. The first of these, evident from the mid-to-late 1980s and of growing significance across an increasing body of work from the early 1990s onwards, concerns the increasing prominence of space and spatiality. Whether relatively simple assertions of the 'difference that space makes' (Sayer, 1985), more grandiose claims as to the inherent spatiality of the postmodern condition (Jameson, 1991), or the growing tendency to draw upon a language of space and place, location and position in writings on subjectivity and identity (Keith and Pile, 1993), as Doreen Massey has remarked, ' "space" is very much on the agenda these days' (Massey, 1994: 249). As geographers we must welcome such developments. But this sudden 'reassertion of space in social theory' as Soja (1989) has described it, also makes us a little uneasy. Certainly, we share the kind of concerns expressed by Smith and Katz (1993), for example, that much of this talk about space is just that; that in the work of cultural theorists especially, there is in fact very little to suggest that the 'spatial turn' has progressed beyond the level of metaphor (see also Cresswell, 1997). More fundamentally, though, our concern is with the basic formulations of space evident within the spatial turn, formulations that appear to us curiously one-dimensional and which, at root, seem premised upon a familiar and unhelpful dualism moving around the foundational categories of Space and Time.

Whilst there is no need to rehearse the details of an argument already cogently expressed elsewhere, suffice it to say that we are then in broad agreement with both the central tenets and general conclusions of Massey's recent critique regarding the limitations of the dualism upon which the spatial turn would seem to be premised (see Massey, 1992a). These are, first, that in the writing of authors otherwise as different as Laclau (1990) and Jameson (1991) the tendency has been to draw a strict distinction between Time and Space. Within such a dualism, where

1

Time is understood as the domain of dynamism and Progress, the spatial is relegated to the realm of stasis and thus excavated of any meaningful politics (see also Harvey, 1993; cf. Massey, 1993; Hetherington, this volume). Second, that this dualism has yet to be seriously challenged by those who would champion a more dynamic conception of space in line with the reconfiguration of the socio-spatial dialectic under the auspices of a radical geography (Soja, 1980). Instead, here too the tendency has been to work within a basic duality, albeit one within which it is space rather than time that is prioritised, such that in place of an earlier and debilitating historicism it may be that social theory is moving towards a creeping – and just as debilitating – 'spatial imperialism' (see also Crang and Thrift, 2000). And third, that rather than continue to see-saw between a prioritisation of either space or time, or attempting to adjudicate as to the political potential of either, we need instead to 'overcome ... the very formulation of space/time in terms of this kind of dichotomy ... [and to recognise instead] that space and time are inextricably interwoven' (Massey, 1994: 260–1) part of a multi-dimensional space-time able to cope with multiplicity (Rodowick, 1997; Assad, 1999).

Though her aim is to move beyond such dualistic thinking, it would be fair to say that Massey's primary concern is with drawing attention to the limitations such dualisms impose upon our theorisation of the spatial. Yet at the same time, and providing for a second source of dissatisfaction, it appears to us that a very similar set of problems have become apparent in recent writings about time. Whilst perhaps less widely acknowledged, the same period that has seen increased attention turned to questions of space and spatiality has also seen renewed interest in questions of time and temporality across a range of disciplines including sociology, anthropology and – to a lesser extent – human geography (see, for example, Bender and Wellbery, 1991; Gell, 1992; Thrift, 1996; Urry, 1999; and the journal *Time and Society*). Whilst, as Massey herself makes clear, recent thinking on time has been strongly influenced by developments in both philosophy and the natural sciences, within the social sciences the majority of such work has been concerned with extending our understandings of the nature and experience of social time rather than with an examination of the nature of time itself (Adam, 1990; Wood, 1990). Much of this work has proved extremely valuable, adding to our understandings of social time in a number of ways. Not least, in contrast to earlier formulations, social time is now recognised as multiple and heterogeneous, varying both within and between societies and individuals and according to social position (see Adam, 1995; Davies, 1990).

The problem is that too much of this work is itself characterised by exactly those limitations noted by Massey in relation to recent writings on space and place. Not least, and especially in the more abstract accounts, attempts to develop what Nowotny (1992) has called a 'social theory of time' have foundered as they have continued to work within the confines of a powerful and persistent dualism.

In other words, and even when considering the work of those for whom the two are clearly inseparable (for example, Bergson, Sorokin or Giddens) rather than seeking to clarify their inter-dependency, such accounts have too often proceeded as though questions of time and space are able to be treated in isolation (see, for example, Adam, 1990; Bergman, 1992). As such, and though certainly enabling a far clearer understanding of the complex timings of social life, such accounts have in the main generally failed to acknowledge the extent to which time is irrecoverably bound up with the spatial constitution of society (and vice versa) or recognised the implications of this for a more developed understanding of social meaning and action. Nor have they yet taken on board the full implications of those studies which point to the spatial variation evident in the making and experience of social time itself (see, for example, Davison, 1992; Howell, 1992; Pawson, 1992). As Glennie and Thrift have argued there is, and always has been, a 'geography of time, timing and time-consciousness' (1996: 280).

And yet from this apparently simple assertion, at least two things follow. First, any search for a singular or universal social theory of time must be doomed to failure as both that which it seeks to account for (the timing of social life) and the frame within which those timings may be set is itself variable across both time and space (and see Adam, 1990). Second, rather than seeking to think in terms of what Massey refers to as a four-dimensional space-time, the more difficult challenge is in fact to think in terms of a multiplicity of space-times or what, in a conscious attempt to move still further away from any separation of the two, we have called TimeSpace. It is this challenge that we have set our contributors.

Practising and imagining TimeSpace: a conceptual framework

Before moving on to a discussion of the essays we want to say a little more about our own understanding of this concept and the logic of the collection. Our starting point here is that just as it has been recognised that the nature and experience of social time is multiple and heterogeneous, so it follows that the manner of its construction – the means by which a particular sense of time comes into being and moves forward to frame our understandings and actions – is in turn both multiple and dynamic. In making sense of its construction we need to pay attention to questions of social practice in four inter-related domains, each of which is spatially constituted (and see Thrift, 1988).

First, a sense of time is still to some considerable extent shaped by our re-sponses to a series of *timetables and rhythms* set according to the inter-relations of Time and Space in the natural universe, ranging from the diurnal cycle to the rhythms of the seasons, the rhythms of the body to the turning of the tides (Parkes and Thrift, 1980; Young, 1988). Though apparently universal, the extent

3

to which a society remains bound up with such rhythms varies across space and over time as the relative import accorded those rhythms shifts and changes in relation to the import accorded to a sense of time moving out of each of the other domains sketched below. So too their effects might be considered socially uneven even as those same rhythms often provide the basis for the regulation of social difference (as with the menstrual cycle, for example). Variation is also apparent across the life course with this too subject to social regulation. For example, at the level of the individual whilst the shift worker must learn to adapt their 'body clock' and the dieter their pangs of hunger, the child works to procure an ever later 'bedtime' (Valentine, 1997). At a broader level of analysis, whilst the calendar first traces then shapes the timing of the harvest (Durkheim, 1915/1965) street lighting moves out from the central districts of the city only gradually, providing for an uneven and ever-changing geography of the night (Schivelbusch, 1988; Schlör, 1998).

Second, a sense of time is thus both shaped by and enacted through various systems of *social discipline* – be they broadly secular or religious. Each such system takes shape within particular settings and achieves purchase according to the spatial arrangements evident within those settings (whether the monastery or factory, office or home). For example, where greater productivity depends upon and (apparently) imposes strict time-discipline so too the worker's use of time can only be properly monitored through an appropriate use of space within the workplace – so as to enable easy surveillance (Stein, 1995). Likewise, just as 'work' time gives shape to 'family' time or 'leisure' time (and vice versa) so such time only acquires full meaning when enacted in the appropriate setting (with feelings of frustration apparent when a person 'brings their work home with them', for example, or when time at the office is disrupted by the demands of family or friends) (Hareven, 1982; Massey, 1995; Shaw, this volume).

Third, a sense of time emerges from our relationships with a variety of *instruments and devices* – ranging from the sun dial to the thermodynamic engine and the video recorder – devised either to mark the passage of time or which work to alter our conception as to the nature and direction of its duration and passing (Adam, 1992). Here, just as many devices which may primarily be thought of as instruments of time work to alter our conceptions of space (as, for example, the advent of the VCR has altered our perceptions as to a shared broadcasting community and hence a wider spatial collectivity) so too devices that appear primarily concerned with space may likewise have significant impact upon our understandings of time (for example, the telephone, telegraph or live satellite broadcast) (Kern, 1983; Urry, 1995). Fourth, a sense of time emerges in relation to various *texts* that may be more properly understood as vehicles of translation (attempts to render social meaning from new conceptualisations of Time itself) and which in setting out particular understandings of time return to regulate that which we would codify (for example, the books of hours).

As a sense of social time is made and re-made according to social practices operating within and across each of these domains so this four-part schema stands in contrast to more familiar accounts of the making of social time which are apt to privilege one domain at the expense of others and so tend towards a certain determinism (see, for example, Harvey, 1989; Thompson, 1967 on labour control and social discipline; Kern, 1983; Urry, 1995 on technology; Young, 1988 on the timetables and rhythms of the natural universe). Further, adding to the various spatialities always already embedded within them, the senses of time associated with developments in each of these latter three domains especially vary according to their impact and reach across space. Thus, the picture that emerges is less that of a singular or uniform social time stretching over a uniform space, than of various (and uneven) networks of time stretching in different and divergent directions across an uneven social field – think, for example, of the uneven dissemination of the mechanical clock through the fourteenth and fifteenth centuries or of railway time in the mid to late nineteenth century (Barrell, 1982; Glennie and Thrift, 1996). Finally, with the impact and reach of developments in different domains varying across space so a further geography is described, as the (already partial and uneven) networks that constitute one domain connect (or fail to connect) with the (partial and uneven) networks constituting another. The result is therefore a radical unevenness in the nature and quality of social time itself, with this spatial variation a constitutive part rather than an added dimension of the multiplicity and heterogeneity of social time or what, for precisely these reasons, we prefer to call TimeSpace. Such unevenness extends, of course, to any broad historical changes in either the nature or experience of TimeSpace, an argument that we develop below.

Before illustrating in more detail precisely why thinking in terms of (a multiple, heterogeneous and uneven) TimeSpace rather than only time and space may be important, and how the conceptual framework sketched above may help us in this task, we want to stress three further points. First, just as any one of these networks of TimeSpace may work to shape numerous and often incompatible if not contradictory senses of time, so we need always to remember that none stands in isolation. Rather, our sense of time is a product of the inter-relationships between each and these relationships are both dynamic and unequal. Second, insofar as it is being constantly reproduced through our material practices, it makes little sense to talk of either the 'making' or the 'living' of TimeSpace but only of what we have (rather clumsily) referred to as Making-Living TimeSpace. And third, whilst in the last of these domains such practices may well involve the production and dissemination of various texts we would not reduce these codifications of TimeSpace to the physical text itself. Rather, whether in oral, printed or electronic form the manner in which we conceptualise TimeSpace has import for the way in which we come to act in TimeSpace (and see Loy, this volume). As such nor would we draw any strict distinction between the 'living' and 'thinking' of TimeSpace and though for the sake of clarity we have set out the remainder of this introductory chapter, and the collection itself, under the headings of Practising (Making-Living) and Imagining (Living-Thinking) TimeSpace we hope that these somewhat artificial divisions are accepted for what they are; a heuristic device designed only to enable the reader to find their way around the collection more easily.

Practising TimeSpace

It would be possible to elaborate upon the conceptual schema outlined above in a number of ways. For example, we could deploy a notion of these multi-dimensional networks of TimeSpace to rework traditional accounts of the making of time consciousness in early modern England – a task already under way elsewhere (Glennie and Thrift, 1996, 1998; Thrift, 1988, 1996; cf. Thompson, 1967). Alternatively, it could be used to develop a more sophisticated under-standing as to the radically uneven development of systems of Standard Time in the latter nineteenth and early twentieth centuries, whether within or between different countries (Davison, 1992; Pawson, 1992). Or it could be used to demonstrate the numerous ways in which questions of time and space interact to provide for radically different experiences of TimeSpace for men and women (Davies, 1989, this volume).

But the example we wish to develop here concerns that radical reworking in the nature and experience of time and space usually referred to by the shorthand of 'time-space compression' (Harvey, 1989). We have chosen to illustrate our thinking around TimeSpace in this way for three main reasons. First, because in its most basic form a thesis of time-space compression moves around precisely that inter-dependency of time and space we would wish to champion – as changes in the nature and experience of one impact upon changes in the nature and experience of the other. Second, because the main elements of such a thesis are by now not only relatively well known but may indeed have assumed the position of something akin to received wisdom (Castells, 1989; Robins, 1991; Urry, 1995). And third, because notwithstanding such widespread acceptance, we would point to a number of quite fundamental problems with the way in which accounts of time-space compression usually proceed. Whilst drawing upon the more developed conceptual schema traced above gives rise to a quite different picture of those changes in the nature and experience of time and space usually associated with a period of time-space compression, we believe that this more nuanced account has implications for how we think about both time-space compression itself and our understandings as to the making and re-making of TimeSpace more generally.

Let us turn first, then, to an outline of the better known elements of such a thesis. The narrative goes like this. From about the middle of the nineteenth century to the outbreak of the First World War, and again towards the end of the twentieth century, there occurred a radical restructuring in the nature and experience of both time and space. Though different commentators identify different processes as underpinning these changes (cf. Harvey, 1989; Kern, 1983) considerable agreement exists as to both the main characteristics of this restructuring and its consequences. In regard to the former, the general consensus seems to be that both periods saw a significant acceleration in the pace of life concomitant with a dissolution or collapse of traditional spatial co-ordinates (changes usually expressed via some kind of discourse on *speed* – or space divided by time). For the latter, the argument is most often that this restructuring was and is profoundly unsettling, as in its midst people must struggle to hold on to more familiar understandings of space and place and negotiate the consequences of radically foreshortened time horizons.

That such a thesis has become so widely accepted is, at first sight at least, hardly surprising. Certainly, when considering the period from the middle of the nineteenth century to the outbreak of the First World War it is difficult to avoid the conclusion that there occurred a radical if not revolutionary change in the nature and experience of both time and space through those years. So too, for both those living through them and those subsequently mapping these changes, the overwhelming impression seems to have been one of a radical compression of spatial and temporal horizons – a notion captured in contemporary accounts of a 'great acceleration' or the progressive 'annihilation of space by time' (Marx, 1987). Indeed, as the century progressed evidence as to this 'great acceleration' was everywhere apparent, discussed most frequently both at the time and since in relation to a series of developments in transport and communication technologies out of which it is in turn usually understood as having arisen.

Consider first developments in transportation. In Britain the beginning of the nineteenth century saw a rapid expansion of the stage coach network and a progressive reduction in journey times such that by 1830 movement between Britain's major towns and cities was some four to five times faster than in 1750. With the development of the railway network journey times were reduced even further whilst the rapid expansion of that network had the effect of opening up these 'technologies of speed' to a considerably enlarged public; by 1870, for example, some 333.6 million journeys had been made by rail, the vast majority of them by third class passengers (Thrift, 1994). Nor were such developments restricted to an increase in the speed of movement between places. Within cities too, the speed of transportation increased dramatically with both a progressive expansion and improvement to road networks and, towards the end of the century, the electrification of the tram system and coming of the underground railway.

Building upon these developments the same period also saw a rapid expansion of the communications network, beginning with improvements made to the postal system which in Britain was carrying approximately 1,706 million letters a year by 1890. So too with the telegraph system. First used in 1830, by 1863 nearly 22,000 miles of wire had been laid across the world, transmitting nearly six million messages a year from 3,381 points (Thrift, 1994). In 1901 the first wireless message was sent across the Atlantic and when in 1912 the *Titanic* sent out its distress signals they were immediately heard by over a dozen ships and picked up at a station in Newfoundland from where they were sent around the world little more than an hour after first being transmitted. Finally, the closing decades of the century witnessed a progressive 'wiring' of a number of the major US and European cities (Stein, 1999). With the laying out of hundreds and then thousands of miles of telephone cable, the telephone network 'allowed people to talk to one another across great distances, to think about what others were feeling and to respond at once without the time to reflect afforded by written communication ... [such that, and for the first time, perhaps, it became] possible, in a sense, to be in two places at once' (Kern, 1983: 69).

Such developments clearly had a significant impact upon people's experience of and relationship to time. For example, the development of a rail network eventually ushered in the need for a Standard Railway Time drawing a variety of communities where the clock would once have been set to different times into a common time frame (Bartky, 1989; Stephens, 1989). More significantly, perhaps, as each successive development increased the speed of travel and communication, it would appear that people began to pay attention to ever smaller fractions of time – as is evident with the increasingly popularity of watches in the last decade of the nineteenth century especially (Zerubavel, 1981). As Kern's discussion of the telephone makes clear, this innovation was also of considerable importance in reshaping people's understandings of space. For example, within cities a still largely localised telephone network greatly altered a sense of privacy, shattering both the geographical and to some extent the social distance between users and raising new anxieties as to its proper use (Marvin, 1988). At the national and regional level, and prefiguring similar debates towards the end of the twentieth century, as places appeared to be moving ever closer together so concerns were raised as to the uniqueness of place (Schivelbusch, 1986; cf. Relph, 1976) whilst internationally the experience seems to have been of both a progressive shrinking of the world and its simultaneous enlargement as people became aware of events in ever more distant parts of the world.

Politically such developments 'made the idea of "empire" practically possible by allowing distant colonies to be controlled from the centre ... even[ing] out commodity markets [and] diminish[ing] the significance of local conditions of supply and demand' (Stein, this volume). They would also appear to have had important consequences for the way in which this 'networked' world was perceived. Not least, developments in telegraphy and wireless technologies in particular changed the nature of news reporting as news became stripped of its local or regional context and organised instead around more systematic modes of thought (Carey, 1989). More generally, by the turn of the century the overriding impression seems to have been of a marked acceleration in the pace of change itself and an increasing sense of insecurity and uncertainty as the world appeared both more interconnected and more unstable (Berman, 1991). For some at least, such feelings can be read as a direct result of the progressive shattering of geographical and social distance and of a speeding up of the pace of life concomitant with those developments sketched above. And their impact was never more apparent than at the outbreak of war in 1914 when for the diplomats of Western Europe events unfolding elsewhere, to which they both felt the need and were able to respond instantaneously, appeared quite literally to 'run away from them' – with disastrous results (Kern, 1983).

A very similar story and with similar if not even more extreme consequences has of course been told with regard to the late twentieth century. Though that story hardly needs repeating here, the work of Harvey has been especially influential and in particular his suggestion that we are currently in the grip of a second great round of time-space compression that is so revolutionising

> the objective qualities of time and space that we are forced to alter ... how we represent the world to ourselves ... [as] space appears to shrink to a 'global village' of telecommunications and a spaceship earth of economic and ecological interdependencies ... and as time horizons shorten to the point where the present is all there is ... so we have [had] to learn to cope with an *overwhelming sense of compression* of our spatial and temporal worlds.
>
> (Harvey, 1989: 240; emphasis in original)

In terms of time Harvey too therefore points to a marked and debilitating acceleration in the pace of life and of a sense of the future as rushing towards us out of control. In terms of space, it is not simply that a sense of geographical distance has been radically compressed, but that more familiar understandings of location and position might have to be abandoned altogether (Shields, 1992). Indeed, in the more extreme accounts recent developments in telecommunications are heralded as bringing with them a fundamental dissolution of space and consequently of the human subject itself (Virilio, 1993) whilst cities too are now not infrequently described as having lost their more familiar form, reduced to mere 'interruptions' in a wider space of global flows (Castells, 1996; Robins, 1991; cf. Amin and Graham, 1997).

And yet, notwithstanding the violence done to such a thesis by the brevity of our summary, there seem to us a number of quite fundamental problems with the way in which accounts of time-space compression of the kind outlined above have more usually proceeded. Not least, where the history of speed in the twentieth century, and hence of any change in perceptions of distance, is better understood as a 'continuation by other means of methods of transport and communication that started in the nineteenth century' (Thrift, 1996: 272) arguments as to the radically disorientating effects of such changes in the contemporary period would seem to us considerably overdone (see also, May, 1994; Thrift, 1997). More generally, we would suggest that rethinking such changes in the light of the conceptual schema traced in the previous section – within which the making and remaking of TimeSpace is understood as a multi-dimensional, uneven and always partial process – provides for a very different picture of these changes. Though this alternative picture could as easily be traced for the contemporary period, for the sake of clarity and because we believe the contemporary experience to be prefigured in numerous ways by earlier developments, we confine ourselves to the nineteenth century. Three elements of this revisionist exercise are of particular importance.

First, we would suggest that in reducing changes in the experience of time and space to simple feelings of acceleration and dissolution the standard accounts of time-space compression are not a little under-developed. To some considerable extent, this under-development is a consequence of too heavy an emphasis being placed upon developments in transport and communication technologies and not enough upon developments elsewhere both in the field of technology (or what we prefer to call the domain of instruments and devices) as well as across a number of the other domains through which the experience of TimeSpace is rendered. When these more numerous developments are considered, and the connections between each traced, the picture is less of any simple acceleration in the pace of life or experience of spatial 'collapse' than of a far more complex restructuring in the nature and experience of time and space reaching through the nineteenth and in to the early decades of the twentieth centuries. With these changes space is seen to both expand and to contract, time horizons to both foreshorten but also to extend, time itself to both speed up but also slow down and even to move in different directions.

Reconsider first, for example, those changes in technology usually ascribed such importance to a changing sense of time and space. Whilst here attention has been focused upon a range of developments in transport and communication technologies it may in fact be that a number of other developments had a far more profound impact upon everyday understandings of time and space than the expansion and consolidation of the rail and road networks, telegraph, telephone and wireless systems. Certainly, rather more account should perhaps be taken of changes in the technologies of light and power. Starting with the widespread diffusion of gas lighting from the end of the eighteenth and developing in pace through the beginning of the nineteenth century there occurred what Melbin (1987) has referred to as a progressive 'colonisation of the night'. Though occurring only slowly, such processes had the effect of blurring the distinctions between night and day, such that by the end of this period they were beginning to displace a sense of time rooted in the more natural rhythms of the diurnal cycle (Schivelbusch, 1988). In terms of space, this extension of the night also had the effect of opening up whole new districts of the city whilst lending the city as a whole a 'phantasmagorical' quality not previously apparent (Clark, 1985).

The progressive development of these various 'technologies of light' (Hillis, 1999) were important in other ways too. Most obviously, towards the end of the nineteenth century the emergence of the cinema in particular had a profound impact upon perceptions of the basic qualities of both time and space, moving us some way beyond a sense of their simple acceleration or collapse. For example, with the showing of Melies's *The Vanishing Lady* in 1896 the impression was one of time first slowing down, stopping and then reversing – as a skeleton suddenly becomes a living woman. With the development of editing procedures the effects were even more dramatic. Here it was possible both to 'chop up' and to reorder a more familiar linear narrative and to show events occurring in different places simultaneously, as in Edwin Porter's *The Life of an American Fireman* (1902) or David Griffith's *A Corner of Wheat* (1909). Where such images had important consequences for any understanding of time as continuous or irreversible, they also therefore lent space a radically discontinuous and fragmentary quality the representation of which was also a concern of more elitist artistic movements (for example, Cubism) (Kern, 1983).

Very similar understandings as to the discontinuity and inter-dependency of time and space then apparent in the world of film and painting were emerging elsewhere in this period too – with a whole series of *texts* attempting both to represent this change in the nature and experience of time and space, and to re-conceptualise the essential nature of both. From about the 1890s, for example, biologists were increasingly making the argument that if different organisms 'lived' at different 'speeds' then time itself was relative, whilst in 1905 Einstein's special theory of relativity destroyed for ever the Newtonian world of an absolute Time and Space in which objects and events simply unfold over time and extend in space. Rather, though in relativity theory (as in biology) time remained unidirectional, the passage of time, or the speed of an object's progress through space, altered according to the position of the observer as time and space became irrevocably tied in a relative and four-dimensional space-time.

Finally, though attention has tended to focus upon what Harvey (1989) refers to as a compression of spatial and temporal horizons, it might at least be argued that through both the eighteenth and nineteenth centuries developments in the natural sciences especially were, if anything, moving in the opposite direction with the discovery of what Urry (1999) has called glacial time. For example, starting first with the Comte de Buffon's assertion that the earth must be at least 168,000 years old and continuing with the work of Charles Lyell and Charles Darwin in the 1830s and 1850s, far from a foreshortening of such horizons the message of geology was of the earth's great antiquity. Though concerned with the past rather than the future such arguments must at least have encouraged people to think in terms of great expanses of time, whilst the broader point is, of course, that this great expansion of time horizons occurred at precisely the moment at which people were also beginning to pay more and more attention to ever smaller fractions of time. Rather than a simple picture of speed and acceleration then, the picture that emerges is one of a growing awareness of living within a *multiplicity of times*, a number of which might be moving at different speeds and even in different directions. In terms of space, whilst developments in transport and communication technologies rendered the world both more extensive and considerably 'smaller' at the same time, in both the biological and natural sciences people were beginning to consider the world rendered visible by the microscope (cf. Amato, 2000) and returning from these experiments in scale to reconsider questions of time (whether the age of the earth or the relative speed of time's passage and the process of ageing).

But we would also suggest that neither this 'great acceleration' nor any more comprehensive restructuring of TimeSpace were anywhere near as disorientating as is usually assumed. This brings us to our second objection to the ways in which accounts of time-space compression usually proceed; that the effects of such changes, even in the nineteenth rather than the twentieth century, have been considerably overdrawn. To some extent this is a problem of sources. With most attempts to codify the new experiences of time and space emergent in this period coming from a few, relatively elitist texts (most notably, the fine arts, experimental literature or the academy) it is difficult to gain a picture of how such changes were experienced by the public at large – and not least in a period before mass literacy. But it is also a problem relating to the more general failure to consider the extent to which the experience of any such changes differed for different people – according in part to *where* a person lived (an issue to which we return below) as well as to a person's social position (and see May, 1994).

Some experience of a great acceleration in the pace of life or of a generalised collapse of spatial horizons might reasonably be assumed to have been relatively widespread, as more and more people took advantage of an expanded rail network, for example, or as the newspapers carried news from ever more distant parts of the world. So too, though initially put forward in the rarefied atmosphere of the academy, ideas as to the great age of the earth emerging in the field of geology are liable to have had considerable reach – especially as those ideas began to challenge accepted religious doctrine. Equally clear, however, is that a number of those other changes traced above, and in a variety of domains, would have formed part of the experiential remit of only a small minority. Certainly, the reach of both theoretical physics and of biology was limited – restricted to small numbers of the middle classes. So too with the attempts to represent this new experience of the world (in Cubism or, in the early years at least, cinema) whilst the telephone network too grew only slowly, remaining for many years a localised service connecting only wealthier households and a small number of businesses (Stein, 1999).

For the vast majority of people it is far more likely that no such sudden or radical disjuncture in the nature or experience of TimeSpace was anything like as apparent, an argument that gains force if we consider the timescale over and the context within which such changes occurred. Exactly this argument is made by Stein in his examination of improvements made to the main communication networks connecting the industrial towns of Ontario, Canada in the 1840s and 1850s (this volume). Drawing upon the various guides to rail and steamship services published annually from the late 1840s, Stein is able to trace a progressive shortening of journey times evident throughout this period. For example, where in 1850 the stretch of the St Lawrence between Montreal and Kingston en route to Cornwall took some twenty-six hours to traverse, by 1853 journey times had been reduced by a little over two hours. As important as this reduction in journey times was the increasing ease with which travellers were able to make their way along the St Lawrence, as the building of a series of locks and canals dispensed with the need for long stage coach journeys between the more treacherous stretches of the river and as improved design made river boat journeys more comfortable. As Stein notes, these improvements to the main transportation routes were of considerable importance to the population as a whole rather than only the wealthy, as even for those residents who did not make the journey upriver Cornwall became more closely tied in to the regional, national and international economy (providing a boost to manufacturing employment and bringing an ever increasing array of goods and commodities to local stores). Whilst more impressive still, and building on improved river transportation, were improvements made to the postal service. Where in 1812, the St Lawrence settlements received post only weekly, by 1830 this had increased to three times a week and by 1856 to twice daily, with these improvements too being felt by the majority of residents.

And yet, as Stein argues, in the case of the rail and steamship network the overwhelming impression is surely less one of any dramatic change in the time taken to complete the average traveller's journey, and thus any radical alteration in people's perceptions of distance, space or place, than of a series of quite small and incremental improvements to the service. Likewise, these improvements to the postal system occurred over a period of some thirty to forty years; a lifetime when viewed from the perspective of individual residents. Though the coming of the telegraph in 1847, and hence of instantaneous communications between Cornwall and the outside world, must indeed have been dramatic for the few able to make use of the network, these people were few indeed. So too with the telephone network, with the town's first telephone directory listing only forty-three subscribers and only a little over a hundred even by the late 1890s. Thus as Stein argues, 'in the case of new communication technologies there was a significant time lag between the introduction of an innovation and its widespread social diffusion [and] this should at least make us suspicious of writers such as Kern whose interpretations of the telephone and similar technologies tend to exaggerate their social and economic effects' (this volume).

Nor should we overestimate the changes wrought in other domains. Following Thompson (1967) a number of commentators have tended to see changes in the workplace, and the imposition of new systems of time-discipline coming with the factory system, as the single most important factor in changing people's basic attitudes towards and experience of social time (Landes, 1983; Hopkins, 1982). Though a number of features of Thompson's thesis have since been challenged – not least, its timing (Whipp, 1981; Harrison, 1986; Glennie and Thrift, 1996) – Thompson himself set these changes at the tail end of the eighteenth century when an orientation towards task was finally replaced by an orientation towards time and the more regular, less humane, less natural domain of clock time (for an alternative reading of task-orientated systems of time, see O'Malley, 1992). But the story of this transformation is central to the standard narratives of time-space compression too. It was in the nineteenth century that the factory system spread significantly beyond its initial base in the English industrial heartland and in the second half of that century when improvements to the manufacturing process enabled a significant intensification of the industrial routine, both necessitating and enabling an accelerated sense of time within the workplace and ever tighter systems of time-discipline within and beyond the factory walls (Harvey, 1989).

Just such an intensification was indeed under way in the textile factories of Ontario, Canada through the 1880s and 1890s. Moreover, there is considerable evidence to suggest that in Cornwall, as elsewhere, with some 20 per cent of the town's labour force directly employed by the factories, and many more dependent upon the spending power of the factory workers, it was the timetables of industrial textile production that shaped the work routines of the town as a whole. But at the same time, and once again, although it was the schedules of the textile factories that set the basic urban routine (the opening hours of shops, the post office or other local services, for example) 'there was in reality no single uniform urban time, but multiple times and multiple routines ... factory work intersected with family, religious and domestic routines' and even, on occasions at least, bent to them (Stein, this volume). Hence 'in December 1876, for example, the new superintendent of the Canada Cotton Mill instituted the arrangement of ringing the factory bells on Sundays, at times that coincided with the start of church services' (Stein, this volume) and suggesting that a more ancient system of time, based upon the religious rather than the productive routine, not only continued to co-exist with these new systems of time-discipline but exerted some influence upon them (on the influence of family routines on industrial working practices in the nineteenth century see Hareven, 1982). Moreover, it would appear that in the sounding of the bells at the beginning and end of each shift the factory owners were in any case simply adapting a pre-existing system initiated by Cornwall's Presbyterian church who had rung for the start of the day at 6 a.m. for the midday break and again to signal curfew at 9 p.m. since at least the 1830s (Pringle, 1972).

Whatever the extent to which people in Cornwall, Ontario struggled to adapt to tighter systems of labour control and time-discipline, then, there is little to suggest a more fundamental reorientation in people's attitudes towards or sense of social time itself through this period – whether emerging from the factory floor, or elsewhere. Certainly, rather than radically new, the routines of the factories built upon a sense of time already shaped by the clock rather than a more 'natural' orientation to task, with the bells of the church rather than the whistles of the factory floor the key instigator in bringing a more routinised clock-time into consciousness.

In fact, the continuing influence of the church and its temporal routines in Cornwall, Ontario offers an important insight into wider arguments around changes in time consciousness more generally (and beyond those famously traced by Weber regarding the intersection of changes in industrial production and the Protestant work ethic). As has been argued elsewhere, both clocks and other instruments and devices for the marking of time (notably bells) were far more ubiquitous even in the Middle Ages than has previously been recognised (Thrift, 1988). Certainly, though originally found mainly in churches and monasteries, as early as the fifteenth century they had already spread to a range of other public buildings – offering an important means by which to regulate not only the religious routine but trade (the opening and closing of the market, for example) as well as a whole host of other activities and ceremonies. Importantly, however, by no means all of these devices were set to the same time – or rather, each would often move to very different rhythms. Thus, whilst in the Middle Ages religious doctrine continued to exert considerable influence over both the daily and other routines (the hours of prayer and devotion, feast days and saints' days, high days and holidays) these routines co-existed with a number of others, many of them moving at a different pace; the calendar of the seasons and of harvest, local and regional market days, the opening hours of the ale house or the sounding of the curfew to name only a few.

Such work challenges the idea that time competence or an orientation towards the clock is a relatively new phenomenon (and certainly not one that first emerges with the factories of the eighteenth or nineteenth century) (Glennie and Thrift, 1996). But it also suggests that for hundreds of years prior to this first great round of time-space compression people had in fact lived according to a multiplicity of times and rhythms, learning to adapt to changes in those rhythms and even to quite significant changes in the 'shape' of the day itself; the more remarkable feature of Cornwall's church bells being that they rang only three times a day, suggesting a change in church time from a more obviously 'religious' to a more obviously 'secular' timetable, perhaps.

Set in this context, that even the coming of Standard Railway Time in 1883 (when the residents of Cornwall, Ontario had to set their clocks back five minutes and forty-five seconds so as to achieve synchronicity with the rest of the eastern seaboard – thus literally 'losing' five minutes) was cause for little public comment is less surprising than it might at first appear (Stein, this volume). Though undoubtedly an important event, not least for continuing social and economic integration, the coming of Standard Time to Ontario Canada certainly does not seem to have precipitated the kind of wonder or unease reported by those who have traced its introduction through the pages of more elite sources (Kern, 1983). In part, and as Stein suggests, this may be because Railway Time was in fact not far out from local time and leading to little readjustment in people's day-to-day routines. But more generally it must surely speak of the ease with which people adapted to new systems of social time – tolerating even a 'change' in time itself – just as they had for centuries, and suggesting that narratives of any sudden or radical disjuncture in people's sense of time in the middle to the end of the nineteenth century are considerably overdrawn.

A more prosaic but none the less significant reason that the introduction of Standard Railway Time passed with so little incident in the Ontario textile towns, as elsewhere, is of course that it had in any case been slowly making its way across the North American continent for several decades (Stephens, 1989). This brings us to the third, and in the context of the current collection, perhaps most important, qualification we would wish to make to the standard narratives of time-space compression. This is, that even as it is apparently premised upon the inter-dependency of time and space, such a thesis would in fact appear to leave little room for basic geography. And yet, if as was suggested at the beginning of this chapter, we take seriously Glennie and Thrift's assertion that there is (and always has been) a 'geography of time, timing and time-consciousness' (1996: 280) the whole nature of the thesis changes.

A very similar argument has of course been made by Massey (1993) and her notion of the 'power-geometries' underlying changes in the nature and experience of time and space in the contemporary period usefully reminds us of the inequalities of opportunity and constraint that are a feature of the uneven geographies of time-space compression. But such unevenness was if anything more extensive in the nineteenth century than today, such that an elaboration of Massey's arguments would seem useful. Consider first, for example, how far we might need to re-conceptualise time-space compression as, quite simply, a metropolitan phenomenon. Certainly, it was the great cities (London, New York, Paris and Berlin) that witnessed the most spectacular effects of the new electric lighting, that saw the most rapid and most extensive development of their transport and communications systems (both throughout their own districts and tying one city to another) and where the academies but also the cinemas and galleries were concentrated. It was these cities too where the implications of any change in either the nature or experience of time and space were most fervently discussed (by an urban intelligentsia and bourgeoisie) and where the most sustained attempts were made to represent those changes (in a new modern aesthetic and sensibility) (Benjamin, 1985; Berman, 1991; Clark, 1985). Whilst we should certainly not conceive of such changes as impacting upon different districts within these cities in equal measure (consider, for example, the uneven geographies of street lighting that even in the closing decades of the nineteenth century rendered certain neighbourhoods the 'city of dreadful night' in both a literal as well as a metaphorical sense), outside of the great metropolises change proceeded only much more slowly – as we have seen (Hall, 1988; Wilson, 1991).

More specifically, we might usefully think of any change in the nature or experience of TimeSpace concomitant with the kind of changes outlined above as spreading from a variety of (mainly though not exclusively) urban nodes so as to describe a series of uneven and partial networks. For example, if by the 1870s one could travel by train from London to Cambridge in a little over an hour, or be in Manchester in a little under three, it might still take almost a day to make one's way fifty miles east of Cambridge across East Anglia and the Fens to a town or village not yet part of the expanding railway network. Such networks did not always follow the patterns we might expect. In terms of a fledgling telephone system, for example, London's telephone network developed only slowly – linking up with Brighton (1884) sometime before the first connection was made with Manchester (1890, via the exchange at Birmingham) (Robson, 1973). More generally, though, they tended to follow existing networks of social, economic and political power. Thus London was connected by telephone to Paris, and hence to the French Stock Exchange (1891), long before it was connected to Bristol. Whilst on the world stage, as Michael Adas (1989) has argued, the result was of course to emphasise the disparities between one part of the world and another – contributing both materially (through improved communications, control and supply routes) and ideologically (assumptions of scientific, technical and social superiority) to Western dominance and colonial expansion.

Even as that great 'annihilation of space by time' described both at the time and subsequently occurred less rapidly and more unevenly than is often suggested, the picture becomes more complex still if we consider the different domains through which a sense of TimeSpace is rendered and the differential development of the various networks associated with each. For example, in the early years of the nineteenth century at least, the time-discipline of an emergent factory system was more likely to be felt in rural than established urban areas (as the factories followed raw materials and sources of water and steam power). But as the needs of industry changed and as the factories moved away, a number of these areas would have found themselves marginalised – locked out of new transportation routes and from the flow of innovations and ideas so important to a changing experience of the world. Nor can we discount the different character-istics of different networks such that there was likely to be considerable geographical variation even regarding the same development. New ideas as to the age of the earth emergent with developments in geology, for example, were very differently received in different areas – depending in part upon the strength and character of the local church. Finally, in the field of new technologies especially, network development often proceeded according to a 'bundling effect' such that any effects of the uneven diffusion of different technologies was magnified. Hence, even by the 1860s, for example, whilst those living in the outlying villages of England might be able to send a message flashing across the wires and around the world in a matter of minutes on reaching the telegraph station (assuming, of course, they could *afford* to send it) it might still take several hours to walk or ride to the nearest post office or railway station so as to access the network.

The radically uneven diffusion of new technologies (of transport and commu-nication, speed and light), of new systems of social regulation and time-discipline, and of new ideas as to the basic nature and qualities of time and space themselves must therefore have rendered the experience of any change in the nature of TimeSpace through this period profoundly uneven. And, though this unevenness is the result of a basic geography, the full implications of that geography are perhaps not always fully understood. Consider, for example, the administrator in London's Colonial Office awaiting reports from outlying districts in southern India or Africa, or the owner of a textile mill in Ontario, Canada, waiting on a new supply of machine parts. Even for people such as these – sitting at the centre of social power, and hence of these networks – the sense must surely have been as much one of the great lack of inter-connectivity then apparent in the world (the huge spaces lying in the interstices of these networks) and of the slowness of things, as much as of any acceleration in the pace of day-to-day affairs or of an ever more integrated social space.

Indeed, in the more usual talk of acceleration or of speed that characterises discussions of time-space compression, the issue of *slowness* has perhaps failed to gain the attention it deserves. Yet if, as is of course the case, acceleration is a relative concept one can not have a speeding-up without also having a slowing down.[1] Thus it is not only that the coming of the railways made travel between one place and another considerably faster, for example, but that in itself this very sense of speed rendered other forms of transportation seem much slower than they had once appeared. The same would hold true when comparisons between the telegraph and the traditional postal service were made or, in the contemporary period, as new and faster computers render the speed of a machine once deemed perfectly adequate apparently obsolete. Even if it was speed that was the dominant experience of the era, then (and we would argue that this is only part of the story and would anyway have been the case only for some people, in some places) accompanying this sense of speed, part always of it, must have been a sense of things getting not faster but slower.

Finally, this sense of the relative speed of things would have been important in other ways too. For example, in understandings as to the different 'pace of life' evident in different parts of the world we can perhaps trace the contours of a second geography, an imaginative geography, emergent out of and giving meaning to those more obviously material geographies outlined above (and see Shaw, this volume). Certainly, as industrialisation and urbanisation continued apace so towards the end of the nineteenth century, for example, Britain's middle classes developed a marked taste for the pastoral and the peripatetic – evident in the growing popularity of the work of John Constable in particular, whose views of a slower moving rural world were nostalgic even at the time of their production some eighty years earlier (Daniels, 1993). Nor was access to these imaginative worlds restricted to representations of an English rural idyll or to the middle classes. Rather, by the middle decades of the nineteenth century the wider contours of such a geography would seem most obviously to have moved around a colonial imaginary (Duncan, 1993) then finding voice in a number of arenas but especially obvious in the new commodities and their display then transforming the world of consumption (May, 1996a).

Here, the emergence of the department store is of particular importance as these stores became a significant social space for women as well as men, rich and poor alike (Leach, 1984). With their fabulous interiors and spectacular displays, rather than simply as a new outlet for the ever increasing range of goods on offer to an expanding urban population, the department store can be

19

better understood as offering their customers entry to a variety of imaginative worlds, or 'dreamscapes', far removed from the everyday world beyond the store windows (Laermans, 1993). In the Parisian department stores of the 1850s and 1860s the fashion was for these displays to take on an Orientalist theme. Though representations of North Africa and its people moved through a variety of forms in this period (each offering their own justification for a renewed French presence in the region) by the time at which Leon Belly produced his *Pilgrims Going to Mecca* in 1861 the dominant trope in such work was one in which the desert appeared as a space of a certain simplicity and spiritual purity. In a period of considerable social upheaval (with Paris itself then being torn apart and reconstructed) such a space thus came to be understood as standing not just apart from Europe but as in some sense 'before' Europe and the spirit of decadence and decay then understood (by some) to be sweeping across the European continent (Heffernan, 1991). Hence it might be suggested that as images of such a space filled the window displays of the department store and galleries, so window shoppers and consumers alike could perhaps have escaped the turmoil that was all around them (if only for a while) with the imaginative geographies emergent out of the uneven geographies of time-space compression themselves holding within them the very means by which people might have negotiated those feelings of acceleration or disorientation more usually associated with the period (on the social upheaval accompanying the reconstruction of Paris see Berman, 1991; Clark, 1985).

Very similar arguments have of course been made with reference to the contemporary period, with a whole number of developments (ranging from a rapidly expanding 'heritage industry' to a new taste for the 'authentically primitive' holiday destination) being understood as both responding to and in some sense enabling some form of 'retreat' from the 'ravages' of time-space compression (May, 1996b; Thrift, 1989; Urry, 1990). We make the argument here, though, simply to draw attention to the extent to which the imaginative realm is itself of considerable importance to our understandings of TimeSpace and hence, in turn, to how we subsequently act in TimeSpace.

Imagining TimeSpace

Even though we necessarily live it, imagining TimeSpace has proved an extremely difficult task, one which often means working at the farther realms of representation in order to sketch out new orders of experience. In tracing out the contemporary ways of theorising TimeSpace it is, of course, possible to travel far back into the historical record and there find all kinds of fragments of the present. But, given our introductory remit, we want to begin more or less where we left off in the previous section – starting again at the turn of the twentieth century. By this point in history, recognisably 'modern' ways of thinking about time and space had already arisen in the West, the result of the demands of new kinds of time and space competences (for example, those arising from the advent of the telegram and telephone), and the impacts of geological discoveries and the theory of evolution on the historical time frame within which humanity imagined itself set (Toulmin, 1996).

But by the time of turn of the century a second wave of thinking about Time-Space, one based upon notions of energy, motion, dynamism and industry, was reshaping the intellectual landscape (Kern, 1983). Whilst it would be very easy to tell a story of this new thinking as unproblematically occurring across a whole series of registers, we should be wary of this kind of narrative continuity. More accurately we can say that a certain style of thinking about TimeSpace becomes prevalent which, like the case of painting in the twentieth century, is not so much

> a devising of a new description of the world – one in which, to take the most widely touted example, the terms of space and time were recast in a way that responded to changes out there in physics or philosophy. It was a counterfeit of such a description – an imagining of what kinds of things might happen ... if such a new description arose. And a thriving on that imagination; thriving here simply meaning an immense, unstoppable relish at putting the means of illusionism through their paces, making them generate impossible objects, pressing them on to further and further feats of intimation and nuance – all for the purpose of showing the ways in which they might form a different constellation.
>
> (Clark, 1999: 213–55)

Of course, such a style had many determinants. We can list a few of them. There are developments in the natural sciences – this is the time of the special theory of relativity (1905) and the general theory of relativity (1916) which relate the construction of space and time to movement. There are developments in engineering. This is the time of the rise of time-shifting technologies like the automobile and the airplane. There is the rise of moving pictures which provide new forms of locomotive magic which intensify action (Moore, 2000). There are developments in the human sciences. This is the time of abandonment by psychologists of 'elemental' models of human consciousness in favour of operational and functional models which characterise awareness as a stream of activity allowing only fleeting subject effects (Kern, 1983; Crary, 1999).[2] We could go on.[3] But the point is that these developments should not be read as simple reflections of one another, adding up to a kind of intellectual earthquake in which TimeSpace takes on an entirely new form but rather as something more hesitant, less brash.

21

However, if one is looking for a representative thinker of the time, then Henri Bergson (1859–1941) more than fits the bill.[4] For Bergson shows both the strengths and the weaknesses of this style of thinking. Bergson is, of course, known for his attention to time – but a different kind of time from the succession of instants supposedly characteristic of scientific knowledges. (Bergson calls such time 'cinematographic' because movement is an illusion created by a succession of static frames.) This is the 'real time', of duration, able to be captured by intuition, a time which is continuous as opposed to the discontinuous time of the instant of science. Intuition (the organ of philosophical knowledge) grasps reality directly and truly through a kind of heightened attention to life, a kind of 'rapt concentration' or 'absolute absorption' (Crary, 1999) made possible by memory which allows us to convert the brute reflexes of any moment – and Bergson contends that 'the majority of daily acts have many points of resemblance with reflex acts' (Bergson, 1991: 168) – into something new, reworking the object of perception. 'By allowing us to grasp in a single intuition multiple moments of duration, (memory) frees us from the moment of the flow of things, that is to say, from the rhythm of necessity' (Bergson, 1991: 95). The richest and most creative forms of living occur, then, in the 'zone of indetermination' which the human subject has the potential to occupy – so-called because, by drawing on memory, she or he 'has the capacity to recreate the present, that is to escape from a relationship of constraint and necessity with one's lived milieu' (Crary, 1999: 318–19). In other words, following an evolutionist schema[5] in which the most developed organisms (human beings) have the most independence from slavish reactions to the stimuli provided by the environment; Bergson is able to argue that

> the fact that our nervous systems not only can delay response to a stimulus but have the possibility of 'variable' responses is a precondition of a free and autonomous subject. But what made this variability important were the ways in which perception was penetrated by 'a thousand details out of our past experience' and Bergson provides an extended commentary on what goes into determining the particular quality of this 'mingling' of memory and perception. He indicates that the interaction can happen in ways that are either creative or reactive and habitual, but makes it clear that the latter is what occurs most often.
>
> (Crary, 1999: 319)

For Bergson 'real time' then is the privileged dimension, since it is the dimension of the new, of expansiveness, of opening up.

Thus the living being essentially has duration; it has duration precisely because it is continually elaborating what is new and because there is no elaboration without searching, no searching without grasping. Time is this very hesitation.

(Bergson, 1991: 93)

And time is a constant melding of past, present and future, a 'mode of stretching' (Grosz, 1999: 25) which produces a kind of simultaneity in difference:

[T]ime is something. Therefore it acts. [T]ime is what hides everything from being given at once. It retards or rather is retardations. It must, therefore, be elaboration. Would it not then be a vehicle of creation and of choice? Wouldn't the existence of time prove that there is indetermination in things? Would not this be that indetermination itself?

(Bergson, 1991: 93)

It follows that for Bergson space is a subsidiary domain since mental life is not extended in space but in time. But it is a domain which, since Kant modelled time on space, has been allowed to run riot. There is an 'obsession of Space'.[6] For example, Bergson was incensed by the way that contemporary thought (and especially science) tended to distort the real experience of duration by representing it spatially, as on a clock (Kern, 1983). Real duration, unlike clock time, cannot be spatialised without being deferred, it is quantitative, dynamic, irreversible.

At the time, Bergson's thinking created something of a sensation and he became probably the most influential philosopher of the first half of the twentieth century. But in time, his work with its emphasis on flux and 'pure mobility' began to fall into disuse, to be replaced by new ways of thinking TimeSpace.

In particular, at mid-century, perhaps the work with the most pressing claim to having an account of TimeSpace was the 'slow-motion phenomenology' of writers like Bachelard, Heidegger and Merleau-Ponty. Of course, such writers have very great differences but they share some important similarities. For example, they are all against Cartesian rationalism and are intent on finding a new relation to Being which dissolves the abstract subject of cognition and liquefies the rigidified spirit. Again, each of them produces a conception of intentionality which is much more dynamic than that of Husserl, one which allows the 'other' to challenge 'me' through a much greater sense of interaction with the world. And each of them locates the truth of the world in a poetics of 'true sensation' which through a proper being-relation lets be that which lets one be.

But what, perhaps most of all, binds these writers together is their attention to not just time, but also space. This attention, which only strengthens during the course of each of their careers, shows up in four ways. First, space figures as a key element of the showing-up of the world, most especially through the primacy of lived space, whose most obvious operator is the body. Merleau-Ponty's account is fairly typical:

> practical space is bodily space: it is oriented around both the physical structure of the body and the projects undertaken to fulfil the needs of the body. Thus near and far, accessible and inaccessible, within reach and out of reach, can be described in terms of bodily motility, which includes both the body's physical capacities and limitations, and the body's self-fulfilment or frustration in pursuit of its goals. The crucial point here is the claim that the body has its own internationality, one which is prior to and independent of any symbolic function, categorical attitude or intelligible condition of consciousness conceived as representation. Within this framework we can understand bodily mobility as consisting of movements that are both immediate (in the sense that they are not moderated by thematic or reflective conscious acts of deliberation or decision) and at the same time, purposive (hence intentional). For example, the lightning movement wherein the lid closes to protect the eye is performed long before the threatening particle is taken as a conscious theme. In short, Merleau-Ponty's claim is that 'the body is an expressive space'. Nor is it merely one space among many: rather it is the origin of the rest, that which projects significations beyond itself and gives them a place.
>
> (Dillon, 1997: 135)

Second, this orientation to space as well as time qualifies Bergson's predilection for mobility. This is a world of stops as well as starts. 'Any discontinuity in psychic life is taken by Bergson to be a fixing and an indication of stasis. He does not allow for moments – moments of stasis – in movement, or acknowledge that separate structures imply a simultaneous connection' (Game, 1997: 120). Third and relatedly, TimeSpace, can be conceived as a set of dwellings in the world, 'intricate immensities' (to use a phrase from Bachelard) that can provide tranquil moments (cf. May, 1991). Fourth, there is then a kind of romanticism that underlines these writers' thoughts on TimeSpace. Georges Canguilhem once – rather dismissively – described Bachelard as having a 'rural philosophical style' (cited in Jones, 1991: 14) and, though the judgement is unfair, there is a certain 'downhomeness' to these writers' work which can be equally lambasted (as evidence of a small-minded provincialism) and praised (as evidence of a nascent ecological consciousness) (e.g. Abram, 1997). But what we can say is that most of the time the air of the city is not one that they seem to be breathing.

24

Why, then, has there been a rediscovery of dynamism, flow and mobility at the turn of this century, a rediscovery which often goes under the title of 'the new Bergsonism' (Watson, 1998), when so much of the debate had already been played out earlier? There are, of course, many reasons. To begin with, there are the developments in the natural sciences, many of them associated with discoveries in biology, which have reworked the natural world as a seething mass of life, meanwhile expanding what counts as life based upon notions of information (cf. Ansell Pearson, 1999a; Grosz, 1999). Which leads to the second reason: the development of new kinds of technologies, especially so-called information technologies. Such technologies allow authors to emphasise mobility and to take the world to be in transit to a generalised state of circulation which can continually accelerate (Urry, 1999). Then, there are developments in the human sciences. Most especially, there is a move to a general emphasis on performativity which takes each strip of interaction to be an opportunity to act out of (Abercrombie and Longhurst, 1998). Hence the rise of an emphasis on embodied creativity from karaoke to various forms of therapy; cultures are intent on creating fleeting narrativities which can hold the moment. We could continue as before, but hopefully these few examples begin to demonstrate the changing cultural background against which intellectual assumptions are made.

But, though the times may fit Bergson again, like his concept of duration, there can be no simple reprise of his way of thinking TimeSpace. This time round, there are at least four differences to take into account. First, in certain senses, his thought is now seen as very much of its time. For example, his separation of intuition – the organ of philosophical knowledge – from the instant – a scientific concept which exists precisely because intuition is not available to science – hypostasises both philosophy and science. Similarly his separation of the mode of comprehension of life (intuition) from the mode of comprehension of inert matter (intellect) is at variance with much of how we now view the world (Chimisso, 2000).[7] Not least, this is because the boundary between the organic and inorganic – in a world of viruses – is increasingly difficult to trace. 'How can any clear line be drawn, such that material objects are characterised by inertia and by temporal self-containment (i.e. by being) that the organic world enlivens (through becoming)' (Grosz, 1999: 23).

Second, Bergson's idea that time is modelled on space and equated to it, so being drained of its essential dynamism, seems odd to us now.

> Spatiality [is] just as susceptible to the movement of difference as dura-
> tion. If time has numericised and mathematised duration, then so too,
> mathematisation has rendered space itself a kind of abstraction of place
> or locus. Just as time is amenable to both flow and discontinuity (to de-
> and re-territorialisation), so too is space. Space is no more inherently

material than duration and is no more the privileged domain of objects than memory is subjective and to be denied to spatial events; each is as amenable as the other to being disconcerted by difference, which in any case refuses such a clear-cut distinction between them.

(Grosz, 1999: 22)

Third, much more care is taken to understand the general social and cultural practices which construct senses of time and space. Bergson's sense of social critique was limited – which is not to say it was non-existent. For example, Crary (1999) argues that much of his writing on duration was a response to the general standardisation and automation of experience that took place at the turn of the century: 'The way in which new arrangements of spectacular consumption within a mass society seemed to be fundamentally productive of redundancy and habit obviously disturbed Bergson. The more conditioned and predictable behaviour became, the fewer openings he saw for memory to play any inventive role in it' (Crary, 1999: 318). Seemingly remote from social and cultural polemic, Bergson did therefore – as Benjamin realised – mount a social critique upon which Benjamin would subsequently construct some of his thoughts concerning the dominance of the mechanical. Later in his career, in his most explicitly sociological study, Bergson (1932/1977) would begin to allow not just sense of time but also sense of space a greater social role. In his constant crusade to open up opportunities for difference (and to argue against closed societies which would slow difference down) he begins to allow space a greater role (Grosz, 1999; Mullarkey, 1999).

How are these three differences being operationalised in rethinking Time-Space? In what follows, we will take each difference in turn and point to the new conceptual flurries that are now occurring.

Take, then, first of all the distinction between organic and inorganic – and its relationship to rethinking TimeSpace. There have been two major reactions. One is that of latter-day Bergsonians like Gilles Deleuze and Keith Ansell Pearson (Deleuze, 1991; Ansell Pearson, 1997, 1999a). They have both used Bergson as a way of rethinking evolutionary time. For them, Bergson's great achievement in thinking about creative evolution is his '*élan vital*', the general life force, as an example of literary production, that is, 'the constant elaboration of novel forms without end' (Ansell Pearson, 1999b:157). What both Deleuze and Ansell Pearson take from Bergson is therefore the notion that life forms are the vehicles through which the '*élan vital*' 'charges its energy and reorganises itself for further invention and creation' (Ansell Pearson, 1999b: 157). However, for both Deleuze and Ansell Pearson, Bergson's conception of evolutionary time remains touched by a residual anthropocentrism and perfectionism, which allows the form of man to be taken too seriously. But this 'organicism' can be suspended by emphasising that the organism is a complex assemblage. They therefore propose a rhizomatic model of evolutionary time.

> A rhizome is an assembly of heterogeneous components, a multiplicity which functions beyond the opposition of the one and many. It is neither a One which becomes Two (which would be to presuppose evolution as involving little more than a linear accumulation), nor a multiple that is either derived from a One or to which one ought be added. It is composed not of units but of dimensions, or rather directions in motions. The rhizome is 'anti-genealogy', since it operates not through filiation or descent but via 'variation, expansion, conquest, capture, offshoots'. It is the rhizome that Deleuze and Guattari wish to privilege as the most inventive domain of creative evolution.
>
> (Ansell Pearson, 1999b: 160)

Such a rhizomatic conception applies not only to time, but also to space. Against the grain, Deleuze needs Bergson as not only attempting to complicate time, but also attempting to conceive of space as more than simply 'a sort of screen that denotes duration' (Deleuze and Guattari, 1998: 485).

Thinking of TimeSpace in this way leads to a second means of blurring the organic and inorganic. True Space. That is actor-network theory or, as it is often described by Latour, with a nod to Deleuze, 'actant-rhizome' theory. Actor network blurs the organic and inorganic in a similar way to that found in the work of Deleuze and Ansell Pearson, by emphasising an order of ceaseless connection and reconnection which refuses the border between the organic and inorganic by emphasising actor networks which are circulations, rather than entities or essences.

> To have transferred the solid from what was a substance, a territory, a province of reality, into a circulation is what I think has been the most useful contribution ... of ANT. It is, I agree, a largely negative contribution, because it has simply rendered us sensitive to a consequence which is also the most bizarre: if there is no zoom going from macro structure to micro interactions, if both micro and macro are local effects of hooking up to circulating entities, if contents flow inside narrow conduits, it means there is plenty of space in between the tiny trajectories of what could be called the local productions of 'phusigenics', 'sociogenics' and 'psychogenics'.
>
> 'Nature', 'Society', 'Subjectivity' do not define what the world is like, but what circulates locally and to which one subscribes ... as we subscribe to Cable TV and several – including of course, the subscription

that allows us to say 'we' and 'one'. This existing space 'in between' the networks, those terra incognita, are the most exciting aspects of ANT because they show the extent of our ignorance and the immense reserve that is open for change.

(Latour, 1999: 19)

Thus actor-network theory sketches out a conception of TimeSpace which traces out the evolution of correspondences/correspondances based upon four principles. First, that time is not in itself a prime determinant of change: a 'time passes or not depending on the alignment of other entities' (Latour, 1997a: 178). Second, that the event cannot be split into spatial and temporal components: 'if a place counts as a no-place, it counts as a non-event. Place is not a feature easier to understand than time. When a place counts as a topos, it also counts as a kairos' (Latour, 1997a: 178). Third, in any account there is going to be – simultaneously – a shift in space, a shift in time and a shift in action: 'we should not speak of time, space, and action but rather of temperisation, spatialisation, actualization (the words are horrible), put, more elegantly, as timing, spacing, acting' (Latour, 1997a: 179). And, fourth, that the question of timing, spacing and acting should be combined with a consideration of their intensity. As Latour (1997a: 179) understandably points out:

What has occurred, an event or a non-event? Process isn't in itself associated with time more than space. Process is not the fourth dimension, but a fifth. This is well known, as far as time is concerned, since we have used (at least since Husserl) the notion of 'historicity' in order to differentiate process from the simple passage of time. But the same should be true for space, though there is no spatial term as widely accepted as 'historicity' is for time. To differentiate the understanding of being in a space, a topos-kairos, from simply being located on a map, we would need a term as clear cut as 'historicity'.

Latour therefore wants a notion of what he calls 'spacificity' in other words, and, so far as he is concerned, the stakes are high. As he goes on,

Writers like Bergson with his distinction between spatialisation and duration, Peguy, with his contrast between the history of historians and the history of events, Whitehead with his insistence on process, Deleuze with his early work on difference and repetition, were obsessed by the question of the intensity of time as opposed to its expansion. The difficulty in using their writing to trace the fifth dimension of a process is that they were engaged in a battle that they saw as a scientific delineation of time and space. The difficulty also comes from their unfairly favouring time ... over space in order to evade the evident spatialisation produced by Science, as if process were in any way more easily connected into time than with space.

(179)

Coming from a background in the sociology of science, Latour does not, of course, see scientific practice – or scientific time and space – as necessarily objective and therefore does not want to have the battle. In explaining process instead, he wants to start up the comprehensive study of timing, spacing and acting but, as he himself says, 'Why is the fifth dimension of time-space so difficult to register?' (Latour, 1997a: 187).

Take, second, the distinction between a dynamic time and an inert space which Latour necessarily addresses in his search after a language of process. Here we can take a hint from Bergson's own work on rhythmic extensity. By using the concept of rhythm, Bergson tried to overcome some of the problems he faced:

In Bergson's durational cosmology, mind is continuous with matter, so that matter is simply the lowest state of mind and mind the highest state of matter. To separate matter wholly from mind would be to deny that distinction has extensity; to rob matter of mind would make it wholly homogeneous and discontinuous, in short, devoid of duration. The permeability of mind and matter, duration and extension, lies in the fact that duration has a rhythm, it is a synthesis of the temporal and of spatial. Thus in his 1911 lectures, Bergson could claim that there is an 'enormous difference between the rhythm of our own duration and the rhythm of duration of matter,' while in *Matter and Memory* he speculated that 'it is possible to imagine many rhythms which, slower or faster, measure the degree of tension of different kinds of consciousness'. For this reason, Bergson compares our consciousness to a 'melody'; similarly sensations such as colour are the 'qualitative' milestones of the absorption of vibratory matter into the rhythmic tension of our duration. As the penultimate symbol of pure change, colour was an intermediary between the qualitative and the quantitative, vibrating matter and the rhythmic pulse of our own consciousness. Colour as well as melody could signify rhythmic duration, for a state of intensity reportedly permeated our 'psychic elements, tingeing them, so to speak, with its own colour'. To perceive such extensive rhythms required an element of intuition, for the intellect could not discern such inner duration, but focussed instead on an object's surface appearances, rather than plumbing its durational depths. If 'the faculty of seeing should be made one

29

JON MAY AND NIGEL THRIFT

with the act of writing', wrote Bergson, 'our vision would transcend its habitual function and, through an element of intuition, discern the melody of "inner" duration'.

(Antliff, 1999: 188)

But Bergson was hardly alone in his reaching out to rhythm as a means of interpreting time and space. The quest was of its time. In psychology, rhythm was a crucial term. And across the humanities rhythm was being used as a key to unlock awareness. For example, the notion of rhythm became a part of the practice of painting – as in the work of Matisse (Antliff, 1999) – and cinema (Buck-Morss, 2000). Equally it was crucial to the study of music and dance. For example, the Swiss composer and pedagogue Émile Jacques Dalcroze (1865–1950) developed the practice of eurythmics, a technique designed to heighten musical sensitivity by relating music and movement. But before long eurythmics was adapted to choreographic improvisation and Dalcroze became something of a guru for many dancers. At the peak of its fame, before Dalcroze left for Switzerland at the outbreak of the First World War, there were 500 students enrolled at the Educational Institute for Music and Rhythm in Hellerau in Germany, with branches of the Institute founded in several other European cities, including Kiev, Prague, St Petersburg and Warsaw (Segel, 1998; Armstrong, 1998; Bachmann, 1993).

This interest in rhythm has continued to surface.[8] For example in *The Dialectic of Duration* (1936/2000), *Psychoanalysis of Fire* (1938) and *The Poetics of Space* (1957/1964). Bachelard (1884–1962) dedicated a considerable amount of time to the notion of a rhythmanalysis. Drawing on *La Rhythmanalyse* (1931), the work of the Brazilian philosopher Lucio Alberto Pinheiro dos Santos, Bachelard intended to produce an alternative to Bergson's '*élan vital*' which, at the same time, was a model for a kind of psychoanalysis of reason.

> A fundamental part of Bachelard's criticisms is that Bergson's philosophy prevents human beings from judging, choosing and acting to a plan. Bergson's creative evolution proceeds thanks to an internal, 'natural', motor: the vital impulse. By contrast, for Bachelard, change (rather than 'evolution') is brought about by work, commitment and critical assessment of past ideas and practices. That the *Dialectic of Duration* was intended not only to provide a metaphysical justification of discontinuity, but also an ethical reason for it is clearly stated. The concluding chapter of the book, 'Rythmanalysis', is an articulated proposal of how we should give a 'rhythm' to our life. This rhythm is analagous to the rhythm of intellectual knowledge, that is to say, to its dialectic processes. Just as knowledge proceeds by alternating negation and creation, our life

should be an alternation of the effort of intellectual knowledge and the rest from its demands. Indeed Bachelard presents this book as an 'introduction to the philosophy of repose' (p17). This repose, during which imagination can express itself, should also receive a 'rhythm': continuity for Bachelard is just not creative. He argues that poetic rhythm has been achieved by Surrealism, an avant-garde movement that he greatly admired. In Chapter 7 of Dialectic of Duration, Bachelard compares the rhythm of surrealist poetry with the rhythm he proposes for rational knowledge. He suggests that the dialectic between rational plot and dream devised by Tristan Tzara for poetry could represent a model for rational knowledge, so that 'rationalism' would turn into 'surrationalism'.

(Chismisso, 2000: 11)

More recently, this interest in rhythm has surfaced again, chiefly from two sources. One is Henri Lefebvre. Lefebvre's long life (1901–1991) provides a bridge between the interest in rhythm as exemplified by those like Bachelard writing in the 1930s and contemporary writers. Near the end of his life Lefebvre transformed his analysis of everyday life into a 'rhythmanalysis', drawing at least in part on the work of dos Santos and Bachelard (who are both cited in Lefebvre, 1991a). For Lefebvre rhythmanalysis must attend to the ways in which, through the lived, time and space are folded into rhythms. Famously, 'every rhythm implies the relation of a time with a space, a localised time, or if one wishes, a temporalised place. Rhythm is always linked to such and such a place, to its place, whether it be to the heart, the fluttering of the eyelids, the movements of the street, or the tempo of a waltz. This does not prevent it from being a time, that is an aspect of a movement and a becoming' (Lefebvre, 1996: 230). Since a rhythmanalyst was closer to the lived, she or he will be 'more aware of times than of spaces, of moods than of images, of the atmosphere than of particular spectacles. He not only observes human activities, but he hears (in the double meaning of the word of noting and understanding), the temporalities in which these activities take place' (Lefebvre, 1996: 229). Lefebvre, then, wants to think TimeSpace in new ways, ways which will provide a kind of psychoanalysis of the intricate space-time of the everyday lived by keeping his 'ears open' to rhythm and texture which are the modes of existence that systems or networks 'assume at those times when they are not being actualised through practice, when they enter into representational spaces' (Lefebvre, 1991a: 118). Unactualised relationships awaiting their moment. A spectral haze of the undone and yet to be done. In other words, rhythmanalysis allows us to understand everydayness – the name set aside for those modes of existence as they come to precede (and recode and exceed) their actualisation into representational spaces: real without being present, ideal without being abstract. That is, while it serves as the third term in Lefebvre's 'everyday life' triad (gathering together and opening up the first two terms: daily life and the everyday), everydayness also extends its thirdness across and into two other triads (the perceived, the conceived, the lived and spatial practices, representations of space, spaces of representation): at once all-embracing and opening them up as a single plane. Everydayness is 'the space' of all spaces, the 'life' of all the lived. Indeed, everydayness is not just a plane of immanence but the very plane of immanence of Lefebvre's entire philosophical project.

(Seigworth, 2000: 251)

The second source, one already signalled in the previous quotation, is again Gilles Deleuze. For Deleuze, too, rhythm is a crucial operator in that it allows him to link time, space and ordering, as Brown and Capdevila (1999: 36–7) make clear.

Consider the notion of repetition. What is repeated becomes a basic element, a rhythm which is discernible as such and not as a noise. Rhythm marks out the time through a simple ordering achieved by a spacing between elements. In so doing it becomes located in a rudimentary space. Something like a 'territory' is formed. The rhythm then serves as the basis for more complex operations: signals, speech, music. We are to be here to what Deleuze calls a 'refrain'.

A child in the dark, gripped with fear, comforts himself by singing under his breath. He walks and halts to his song. ... The song is like a rough sketch of calming and stabilising. ... Now we are at home. But home did not pre-exist. It was necessary to draw a circle around that uncertain and fragile centre, to organise a limited space. ... Finally, one opens the circle a crack, opens it all the way, lets someone in, calls someone or else goes out oneself. ... These are not three successive movements in an evolution. They are three aspects of a single thing, the Refrain (Deleuze and Guattoni, 1988: 11–12). A refrain here is a rhythmic series – the child's song – that creates, by its very repetition, a sense of the familiar, a sense of place. Refrains circulate around this 'uncertain and fragile centre', creating a limited packet of organisation. For Deleuze and Guattari refrains as they are not intrinsically territorial, they are the basic means by which ordered space is marked out from disorder: 'the refrain is essentially territorial, territorialising and re-territorialising' (1988: 300). As the territory becomes secured, so the refrain is 'picked up' or reiterated by others who come to occupy the same space, much like the bird songs or, they argue, cultural myths. Each time the refrain is picked up, it is articulated anew, yet it still remains recognisably the same repetitive series.

In turn, Brown and Capdevila (1999: 37) link these thoughts to actor-network theory by arguing that repetition is what holds refrains together; the here, there and back again of the network is like Deleuze and Guittari's refrain. And this circulating movement

> is as much figural as discursive. It is about the repetition of certain forms as much as the repetition of particular significations. It shifts the border-line between what cannot be presented and what can be made familiar. It allows us to say that networks do not have to be conceived as massive programmes of world ordering (although they can indeed resemble this) but can also, like the territories and 'songlines' of the Australian Abo-riginal, be sung into being.

Then, third, there is the question of different cultural practices of time and space. Here we can draw on a whole series of studies of non-Western peoples, one which takes in most of the classical figures of western anthropology; Kroeber, Mass, Evans-Pritchard, Lévi-Strauss, Leach and more recently, writers like Bordieu, Gell, Fabian and Weiss (for a review see Munn, 1992). Notwithstanding the considerable differences in approach of these and other authors – and most particularly the tendency of some identified by Gell (1992) to transfer Western metaphysics lock, stock and barrel on to non-Western senses of time and space compared with the tendency of others to fix on the density of practices with perhaps too little regard to times beyond the 'shallow' time of everyday life – what surely cannot be disputed is that they have built up a formidable and formidably rich archive of ethnographic work with which to think TimeSpace.

What also cannot be disputed is a strange lacuna. The same archive simply does not exist for Western peoples and especially urban peoples. There are, no doubt, various reasons for this state of affairs. Here we will note just two. To begin with, there is no doubt the colonial legacy. Western time and space, at least in a Euclidean form, has been seen as a norm which simply does not require investigation. Then, there is surely the very great influence of just a few socio-cultural accounts of TimeSpace in the urban West, notably of late those of Baudelaire, Simmel and Benjamin. Continually replayed by cultural theorists, these have taken on iconic status in much the same manner that sociological accounts such as those of Sorokin and Moore held sway in the 1960s and 1970s. Yet they are, to put it but mildly, unproven. This point is worth expanding.

Take for example, Benjamin's notion of time. On one level, Benjamin is trying to make a decisive break with a vision of history as linear progress through the method he calls 'dialectic at a standstill' which 'brings together historical materialism and the mystical concept of the nunc stans concentrating its energies on those moments in history which are laden with now time, for example the manner in which the French revolution viewed itself as ancient Rome incarnate; thus to Robespierre ancient Rome was a past charged with the time of the now which he blasted out of the continuum of history' (Dillon, 1997: 14; my emphasis). The 'state of emergency' would be a perpetual interruption of chronology, a kind of timeshift which never ceases to take place. As Agamben (1993: 91) puts it, this is

> a revolution from which there springs not a new chronology, but a quali-
> tative alteration of time [which] would have the weightier consequence
> and would alone be immune to absorption into the reflux of restoration.
> He who, in the epoche of pleasure, has remembered history as he would
> remember his original home, will bring this memory to everything, will
> exact this promise from each instant: he is the true revolutionary and the
> true seer, released from time not at the millennium but now.

This emphasis on temporal (and spatial) disjuncture continues into Benjamin's analysis of urban TimeSpace which in part centres on 'shock'. Benjamin was interested in investigating Freud's hypothesis,

> that consciousness parries shock by preventing it from penetrating deep
> enough to leave a permanent trace on memory, by applying it to 'situa-
> tions far removed from those which Freud had in mind'. Freud was con-
> cerned with war neurosis, the trauma of shell shock and catastrophic
> accident that plagued soldiers in World War I. Benjamin claimed that the
> experience had become 'the norm' in modern life. Perceptions that
> once occasioned conscious reflection were now the source of shock im-
> pulses which consciousness must parry.
>
> Under conditions of modern technology, the aesthetic system under-
> goes a dialectical reversal. The human sensorium changes from a means
> of being 'in touch' with reality into a means of blocking out reality.
> Aesthetics – sensing perception becomes anaesthetics, a numbing of the
> senses' cognitive capacity that destroys the human organism's power to
> respond politically even when self-preservation is at stake. Someone who
> is 'past experiencing' writes Benjamin is 'no longer capable of telling ...
> proven friend ... from mortal enemy'.
>
> (Buck-Morss, 2000: 104)

This is the dulled city. But in modern culture theory cultural hypotheses like these often come near to the status of a report from reality, to be built on rather than questioned through ethnographic encounter: theory shall speak unto theory.[9]

The problem is that we know remarkably little about the everyday TimeSpace of the city that is not surmise: we have a battery of results from time budget studies and the like which are indicative, but rarely more (Gershuny, 2000), we have some phenomenological interventions, and we have one or two ethnographic studies. Yet, even a short period of reflection suggests that it is unlikely that the TimeSpace of the Western city can be encompassed by just one account. It is not so much one big screen but many tiny windows. Take just one example, from Benjamin's avatar, Freud. As Crary (1999: 365) describes it, in 1907 Freud wrote home from Rome describing the life of the crowded Piazza Colonna on a warm autumn evening. He evoked an urban scene

> in which an individual and a collective affect take shape in a multiplicity of images, sounds, crowds, vectors, pathways, and information, and his letter documents one particular attempt at cognitively managing and or- ganising that overloaded field. It provides a leisurely image outside of the Baudelaire/Simmel [/Bergson] tradition of shock-ridden inner life with its 'swift and continuous shift of stimuli'. Freud's genial evocation of couples and groups, strolling or lounging in wicker chairs in the soft summer night air, seems far removed from the neurasthenic edge of Simmel's 1903 essay on the metropolitan life. Nor does this Rome seem to be prime arena for the post-Haussmann flaneur, the isolated dis- tracted spectator, moving along new urban routes, yielding to a continu- ally changing stream of excitations.

The scene, a palimpsest of different historical moments (from military bands, to lantern slides, and short cinematic performances, from shouting newsboys to electrically lit advertisements), half remembered landmarks, and sociable bustle does not add up to one thing. It is irreducibly plural, 'a multi directional field of stimuli' (Crary, 1999: 367) which casts a kind of spell on Freud. And as Crary (1999: 370) argues, stepping back a little from some of his previous utterances (Crary, 1990), it is a different kind of TimeSpace,

> the plural, hybrid space of this Roman square ... foretells how spec- tacular society is not irrevocably destined to become a seamless regime of separation or an autonomous collective mobilisation; instead it will be a patchwork of fluctuating effects in which individuals and groups con- tinually reconstitute themselves – either creatively or reactively. But even if the latter adverb may be applicable to the most catastrophic social

re-constitutions in the twentieth century, on this particular September 22 the crowd of several thousand, even when spellbound, does not in the least resemble the regressive, docile masses of le Bon or others. In the 'delicious' evening air, the entrancing repetition of faintly glimmering images copied on a makeshift screen does not impede the spontaneous play of such aggregation within this enduring arena of conviviality and life.

In other words, there are all kinds of TimeSpaces which are there to be practices and thought – once we stop using theory as a declamatory tool, once we put aside an oracular mode of analysis, once we understand that, even within capitalist social relations, diversity of experience is not some kind of mirage or unfulfilled yearning. There is a lot to do.

Perhaps we can summarise what is at stake by trying to use one more litera-ture, that of modern feminism. Since at least Irigaray's now classic (1973) paper 'Women and time', feminists have been attempting to refigure gender and sexuality by reworking the time and space of the subject. At one level, this has simply involved a heightened understanding of the direct biological, social and cultural bases of gender and sexuality (e.g. Valentine, 1989), what we might call, following Gell (1992), an ecological understanding. At another level it has involved what might best be described as a sustained programme of experimen-tation in writing[10] and otherwise re-engaging time and space in ways which question the limits of representation in order to represent time and space. For example, in *Rootprints* Cixous (1997) uses her own position as a French-Algerian, an outsider 'abroad at home' (1997: 10), to sketch out 'an aerial, detached, uncatchable' (4) TimeSpace, less grounded in and by representation. As Fulford (1997: 151) summarises it,

> Drawing on an image from William Blake, Cixous imagines a sublime space, ensuring that a 'poetic imagination does not fall into dust'. There is a tension in Rootprints between writing from a particular place – as a woman who grew up in Algeria, while refusing the gravitational pull of roots: 'going off on a voyage, not knowing where' (p. 8). Swinging out on her own, Cixous launches herself and her readers 'into a space-time whose coordinates are all different from those we have been accus-tomed to' (p. 9) with the effect of making strange, seeing things anew, 'dehierarchising everything' (p. 10) … attempting to chart unmapped territories, writing is beyond us, always going forward (p. 102).

In an even more ambitious vein, one of Cixous' occasional collaborators, Clément (1994), attempts to rewrite philosophy by dislocating time through the figure of the syncope, a kind of stop time, 'a vanishment, a gap, or a swoon of meaning' (Conley, 1994) which is 'the queen of rhythm, ... the mother of dissonance, ... the source ... ... in short, of a harmonious and productive discord ... Attach and haven collision, a fragment of and the beat disappears, and of this disappearance, rhythm is born' (Clément 1994: 5). This syncopation, the result of practices like dance, music and poetry which 'traffic in time' (1994: 5), acts both as a simulation of death and an affirmation of life. Surprise! This was an attempt to produce a model of philosophy which can denature and renature.

> Nothing, however, seems natural in the practice of syncope: on the contrary, the signs overflow with excesses against nature. It is not natural to twist the body into catalepsy, it is not natural to lead the singer to voice into singing exercises that are unbearable for the body and sublime in their beauty. It is not natural to dance the waltz, the Bharat-Natyan, or to whirl around, 'Sama, Sama, Sama.' Nothing in all that is natural, and contortion is de rigeur. And it is natural for human beings to couple, it is not natural to 'make love'. When man is in action, he wards off his animal side, even if it means meeting it again in his dreams. That is what the word ecstasy means: a displacement, a deviation. In this sense, we are all deviants.
>
> (Clément 1994: 257)

> Significantly, Clément turns to the task of the Indian syncopated renonçant for inspiration to allow her to move from a male philosophy based on activity – on models of movement, speed, time-space compression – to a certain 'feminine passivity', understood not as absence of action 'but as a different and more valuable form of non-action'.
>
> (Conley, 1944: xiii)

In other words, feminist theory is concerned with providing richer scenes of now. In this, though it is so different in many ways, there is something in common with Bergson. For in many ways the quest of all modern thinking on TimeSpace has been to be filled by and to amplify the presence of the now, to make the present habitable and visible by remaking what counts as past and future, here and there. How can we inhabit the present as if it were a place, a home rather than something we pass in a mad scramble to realise the future? Somewhere here there is a politics, part feminist, part ecological, part visionary which can help us to stop and ponder what we are doing.

The essays

In line with these opening remarks the essays in **Part I** of the volume are mainly concerned with the Practising (Making-Living) of TimeSpace, whether in the contemporary era or in the past, those in Part II with the Imagining (Living-Thinking) of TimeSpace. The opening essays focus upon the making of social time and the politics that accrue to its different forms and circulations. Thus, Hetherington explores early ideas of Modernisation and Progress, Frow some of the consequences of imposing these ideas upon other, quite different understandings of time (those held by the indigenous peoples of Australia, for example). Johnson's focus is on the various ways in which these ideas have been used in narratives of nation and national identity. Following Hetherington's concern with 'Industrial Time' (the substantive focus of his essay being the early factories of eighteenth-century England) the remaining essays in Part I (Stein, Shaw, Davies, Wolch and DeVerteuil) are each concerned with some notion of TimeSpace emerging out of the intersecting domains of instruments and devices and systems of social discipline. Stein is exercised with charting a more sophisticated account of those changes usually described as part of a thesis of time-space compression; Shaw, Davies and Wolch and DeVerteuil the various ways in which people in differing positions of social power negotiate those changes.

In **Part II** contributors turn their attention to the Imagining (Living-Thinking) of TimeSpace. Each of the chapters is very different, though each is committed in their different ways to producing versions of TimeSpace that emphasise process and intensity (and, perhaps, extensity). They have other things in common too, not least the notion of proximity – not a proximity seen in simple metric terms, but a proximity of affect and reflect. This notion is explored in quite different ways by different contributors with some (Game, Metcalfe and Ferguson) turning to phenomenology, some (Gren) to a reworking of time-geography. In his chapter, Crang reviews the attempts of those working from a variety of perspectives to unpack the lived experience of the city and its rhythms, including Torsten Hägerstrand, Henri Lefebvre, Merleau-Ponty and Martin Heidegger. In the final essay, David Loy returns to debates around time-space compression seeking an answer to the profound disorientation such processes are widely assumed to engender by stepping outside of the canons of Western philosophy and social theory altogether and turning instead to the insights of Buddhism.

In more detail, the opening essay by **Kevin Hetherington** traces a genealogy of early modern ideas of Progress and Modernisation. In doing so, he turns to the ideas of Louis Marin (1984) and the concept of utopics; social practices within which future orientated ideas of order and improvement are given expression through new spatial forms. As Hetherington notes, utopics are now evident almost everywhere in modern societies (whether in the design of prisons, asylums, hospitals or schools) but their earliest expression took shape in writings around new forms of manufacture and production centred upon the then novel spaces of the factory. Chief amongst these new spaces was the manufacturing complex laid out by Josiah Wedgwood at his pottery works at Etruria in Stoke-on-Trent, England in the mid to late eighteenth century. As Hetherington argues, Wedgwood's factory soon came to assume the position of what he terms an 'obligatory point of passage' in a wider network of manufacturing production – providing for a new form of industrial manufacture against which other systems of manufacture came to be judged.

Yet the importance of Wedgwood's factory is not simply that it provided for a new utopics of capitalist production. Rather it is that in following changes in contemporary accounts of Etruria through the latter seventeenth to eighteenth centuries it becomes possible to trace a fundamental shift in the discourse of improvement framing those accounts. In the latter seventeenth century such a discourse was essentially spatialised, such that manufacturing activities at Etruria might be compared to systems of production found elsewhere. But by the end of the eighteenth century their orientation had changed to one more concerned with questions of time, with contemporaries recording the improvements made to Wedgwood's works and the surrounding area over the past hundred years or so. In tracing changes in the writings of contemporary authors concerned with describing a new system of manufacturing production Hetherington thus in fact traces the emergence of the first explicit articulation of ideas of Modernisation and Progress that were to prove central to social thought for the next hundred and fifty years. And, far from contained within the realm of Time as is more commonly assumed (see above) such ideas first emerged through a complex geography of TimeSpace moving around the design, construction and broader conceptualisation of new spaces of manufacture and production.

In the two chapters that follow the concern is with tracing the various politics attached to different constructions of these ideas as they are mobilised within debates around race and nation, 'the' national past and national identity. They show that the past is never over, is always done with; rather, pasts continue to act in the present. In particular, both papers stress that nations and peoples are amalgams of different temporalities, proceeding at different speeds and intensities in different places. The paper by **John Frow** examines the extent to which a nation's history can be witnessed in such a way that all peoples figure, taking as his focus the case of the indigenous peoples of Australia. One of the problems he draws our attention to is that in Australia historical discourse mimics the distancing from past events which it is the whole purpose of Aboriginal aims to being close again. In other words, a new model of witnessing is needed which can produce new kinds of incentive act (Grossbery, 2000). Agamben (1999: 163–8) has put the dilemma rather well in another context:

Just as the remnant of Israel signals neither the whole nor a part of a people, but rather, the non-coincidence of the whole and the part, and just as messianic time is neither historical time nor eternity but, rather, the disjunction that divides them, so the remnants of Auschwitz – the witnesses – are neither the dead nor the survivors, neither the drowned nor the saved. They are what remains between them.

Frow explores this ambiguous testimony and how it can be given force by opening up new spaces of history.

Nuala Johnson's chapter also focuses upon the varying constructions of time evident in discourses surrounding the development of the nation and the nation state. Such discourses, she suggests, take four essential forms each of which relates to a different conceptualisation as to the 'shape' of the past, identified by Graham (1997) as Progress, Decline, Recurrence and Providence. Developing Graham's thesis, Johnson's aim is to demonstrate the extent to which these questions of time are inextricably bound up with questions of space – as each such shape is in turn dependent upon a particular rendering of the key spaces of national identity.

Taking each of these shapes in turn, Johnson starts with the ideas of evolutionary Progress to be found in Frederick Jackson Turner's frontier thesis. Here, Jackson argued that: 'The history of our [the north American] political institutions, our democracy, is not a history of imitation, of simple borrowing ... [but] a history of the evolution and adaptation of organs in response to a changed environment' (1896/1962: 205–6). Hence, Johnson suggests that though we tend to associate ideas of evolution and Progress with questions of time, as for Hetherington so too for Turner (evolutionary) change was necessarily and inevitably bound up with questions of space. More specifically, as the American national character was understood as emerging out of the unique challenges posed by a movement across space in line with a shifting frontier, so a whole series of spatial categories – the West, the frontier, the wilderness – worked as synecdoches of temporal progress.

In contrast, the activities of Ireland's Gaelic League in the latter part of the nineteenth century first emerged in response to concerns over national decline. As Johnson shows, both this decline and the nationalist discourse it gave rise to moved around a particular geography. More particularly, as it became possible to trace a geography of linguistic decline moving from the East to the West of Ireland so a marked nostalgia for a period before Anglicisation came to be mapped on to Ireland's western seaboard. Hence the West of Ireland emerged as both the purest space of Gaelic identity (then under threat) and one out of which the nation might revive itself (and see Shaw, this volume). This notion of revival leads Johnson to consider questions of Resurgence. Here she turns her attention to the concerns of National Socialism, considering the links forged between a resurgence of a Volk identity rooted in the land and the conservation and expansion of Greater Germany's 'primordial homeland'. Finally, Johnson turns to ideas of Providential time – the idea that Divine Providence plays a central part in shaping the past and a sense of time which, unlike each of the previous three, may view time as either progression, decline or as cycle. Central to understandings of Providential time is an understanding of the sanctity of place, whether those places within which God is thought to dwell or in which He is held to have revealed Himself. In the case of national Zionism both concepts have been brought in to play, with Jerusalem emerging as a key site of sanctuary and as the settlement and expansion of the Israeli nation state has come to be constructed in line with an understanding of a return to God's chosen land.

In his chapter **Jeremy Stein** returns us to a discussion of 'Industrial time' which provided the context for Kevin Hetherington's essay. In doing so Stein also provides the context for the remaining chapters in this part of the collection, each of which is in one way or another concerned with the ways in which those in different positions of social power negotiate contemporary restructurings of TimeSpace. Stein's specific concern is with the changing experience of time and space under conditions of time-space compression (and see above). Considering what is usually understood as the first great round of time-space compression in the middle to late nineteenth century, Stein begins by asking a number of apparently simple questions. First, given that the better known interpretations of such processes typically rely upon the accounts of a few relatively privileged observers, is it right to assume that the experiences of time and space documented by these observers were the same as the experiences of the general population? Second, to what extent are such interpretations reliant upon an implicit technological determinism? Third, were all of the new technologies emergent in this period necessarily faster and more efficient than earlier ones? And finally, what happens when these changes in technology are embedded in a more careful analysis of the social practices surrounding their uptake and use?

To answer these questions Stein turns not – as is more often the case – to changing conditions in the great metropolises but to contemporary accounts of urbanisation and industrial change in a small Canadian textile town on the St Lawrence river then undergoing significant expansion. Focusing upon the period 1830 to 1880, his conclusions are significant. First, examining changes in the experience of transport and communication he finds that rather than any radical or generalised shift in the experience of time and space brought about by developments in the railway and canal networks, postal, telegraph and telephone systems, such changes were instead typically experienced as cumulative and gradual whilst also being profoundly socially uneven. Second, turning to a consideration of changes in the workplace here too he finds that though there is evidence of a heightened system of time-discipline and of the growing importance of a standardised industrial routine in the latter decades of the nineteenth century especially, such changes were far less wide reaching than is usually assumed. Not least, as factory owners found to their cost, those who would attempt to impose a new system of time-discipline were confronted by the continuing importance of a sense of time attached to those other routines around which the day-to-day lives of the factory workers moved; notably, the rhythms of the family and of religion. The result was a complex mosaic of multiple times, structured according to a sense of time emergent across a multitude of local spaces and providing for a far more complex picture of the changing experience of time and space in this period than is usually apparent.

In her chapter **Jenny Shaw** is likewise exercised with the need to construct a more nuanced account of processes of time-space compression. And, though her focus is a contemporary one, like Stein she also seeks to build this picture with regard to the complex geographies shaping the experience of such processes. Shaw's starting point is the longstanding association between pace and place – the popular belief that in different parts of the world and in different kinds of places, life proceeds at different speeds. Though widely acknowledged, the significance of this relationship is not always appreciated. Not least, it would appear to provide the means by which people negotiate an otherwise rapid acceleration in the general pace of their day-to-day lives, as they are able to assume a change of pace by means of a change of place whether for a little while (when going on holiday, for example) or on a more permanent basis (for example, when retiring).

One of the strengths of Shaw's piece is that, like Stein, she develops her argument with recourse to a grounded analysis of popular experiences of such changes rather than via more abstract or elitist accounts. And, drawing upon material from the Mass Observation Archive, one of the more striking features of her respondents' accounts is the extent to which efforts to negotiate this change in the pace of life differ for men and women. Not least, as Shaw observes, part of the slowness which is sought on holiday and from places where they are taken stems from a definition of holidays as family time – the maintenance of which is in turn usually understood as being the responsibility of women. Thus, even as there is now growing evidence to suggest that this acceleration in the pace of life is more keenly experienced by women than by men, so too it would appear that the ability to respond to such changes are also powerfully differentiated according to one's gender.

Questions as to the fundamental role that gender plays in differentiating the experience of TimeSpace lies at the heart of **Karen Davies'** contribution. The chapter starts with an attempt to show why gender needs to be incorporated in discussions of TimeSpace with a sophisticated critique of the time-geography project (cf. Gren, this volume). Elaborating upon her earlier work in this area,

Davies argues that the approach of time-geography is dependent upon an atomistic and resource-based notion of time quite different from the more relational understandings characteristic of women's experience of time and temporality. Central to this more relational experience is what Davies refers to as a 'rationality' or 'ethics' of care. Such an ethics appears as an important component of most women's lives and works to severely delimit the possibilities for 'time out'; time away from others, when one does not have to take the needs of others into account. Such restrictions are in turn, of course, as much a product of the gendering of space as of time and hence it becomes more useful to speak of TimeSpace rather than only time or space. For example, with regard to leisure time where men's time out tends to take them away from the home or from those for whom one has a responsibility of care, for women any such time is far more likely to be taken in close proximity to those towards whom one has responsibilities and rendering such activities less time out in the traditional sense so much as the experience of 'waiting'.

To elaborate upon such differences, the mainstay of Davies' chapter moves around a detailed analysis of the working practices of women employed in the 'caring professions' (nurses, nursery workers and care assistants) in Sweden. Examining their experiences, Davies charts the difficulties such women have in carving a space for the 'pause': moments at home or at work when the usual flow of work is halted so as to allow for a period of reflection. As Snow and Brissett (1986) have argued, far from simple inactivity, such reflection is an essential part of the development of the self or what Giddens' refers to as 'self-actualisation' (Giddens, 1991). Having illustrated the difficulties women find in carving a space for this reflection, the chapter therefore ends by considering the implications of such difficulties for those who would position the 'reflexive turn' as a central feature of late or postmodernity. Denied the reflexivity essential to self-actualisation, Davies concludes, the relationships of dependency that characterise the care relationship may also deny the possibility of the more equal relationships that are the aim of both Giddens (1991) and Bauman (1995) and that would in turn help create that space.

Though **Jennifer Wolch** and **Geoffrey DeVerteuil** also turn to the approaches of time-geography the experiences they trace using its framework differ markedly from the experiences of either Shaw's or Davies' respondents. Their concern is with the changing experiences of homelessness and with the increasing fragmentation and instability evident in the time-space paths of homeless women and the homeless mentally ill in the cities of the United States. Whilst imposing severe constraints upon the life chances of those whose 'homeless careers' are now increasingly characterised by a continuous 'churning' between the spaces of the homeless shelter, the clinic, city jail and the street, Wolch and DeVerteuil propose that to properly understand this fragmentation we need to turn our attention to the production of a new landscape of urban poverty.

Such a landscape, they argue, is the product of the interaction of a divergent set of time-space cycles; ranging from the longer Kondratieff waves of the global economy, through the shorter cycles of state policy and institutional practices to the local pathways and routes of Hägerstrand's time-geography. Whilst such cycles operate at different scales they also therefore tend to operate at different speeds. When macro-economic restructuring (resulting, as in the current case, in considerable social dislocation) coincides with shifts in the national welfare state (currently undergoing processes of devolution, privatisation and dismantlement) the result can be a dramatic change in the circumstances of marginalised groups, as a growing population struggle to access a shrinking and increasingly disparate set of welfare institutions spread across the urban landscape. Beyond what it adds to our understanding of recent changes in the production and negotiation of urban poverty regimes, the value of Wolch and DeVerteuil's analysis therefore lies in the manner in which it seeks to provide an explicitly temporal dimension to a divergent set of approaches (regulation theory, theories of the capitalist welfare state and institutional studies) more usually addressed only at different scales of analysis.

In **Part II** the concern is with the different ways in which we might imagine TimeSpace and the implications of our various conceptualisations. The section starts with a chapter by **Elspeth Probyn** who uses the presentation of TimeSpace to reinterrogate some familiar concepts: gender, sexuality, class, race. Probyn argues that these concepts need to be relocated by crossing over from the solitary Timespaces engendered by individual categories to more complex Timespaces which can produce categories which can cope with crossover and proximity. The force of the argument is underlined by the new tangibility to be found in relations of proximity in the modern world which enable us to understand the far as close in to ourselves – a relationship rendered visible through mundane practices like eating. In a sense, Probyn is reworking a tradition which questions the degree to which compassion can be felt at a distance, all the way from Adam Smith's discussion of sympathy to Luc Boltanski's (1999) discussion of speech about suffering. A vital part of this tradition has always been the telling of new spatial stories that can redefine who is nearby and who is distant (cf. Thrift, 1998; Gilroy, 2000).

The two chapters which follow document the difficulties in defining these new TimeSpaces. **Mike Crang**'s chapter struggles with how to understand the TimeSpace of the city. In this ambition, it follows a tradition of work which has attempted to interrogate everyday life in the city as mundane but not humdrum (see, most recently, Seigworth, 2000; Thrift, 2000). Crang starts with the use of time in the city. Then with each succeeding visit to the city he adds in more temporal and spatial complexity. Thus, the paths and projects of time-geography constitute the next visit, and serve as a means of prefiguring (in its limited sense of motility) the next visit in which the phenomenological sense of experience of the TimeSpace of the city is added in. In the third visit, Crang wrestles with phenomenological critiques by producing an even more dynamic sense of city TimeSpace which can take in flow. He ends by insisting that what we need is a much more dynamic notion of space which can equal the dynamism usually attributed to time.

Martin Gren's chapter considers time-geography, the attempt by Torsten Hägerstrand to produce a new understanding of the choreography of existence by producing new kinds of notation. Proceeding from his remarkable book (Gren, 1994), Gren shows, first, that time-geography has much more complex consequences than many geographers have understood, but second, that time-geography needs to consider anew the TimeSpaces in which the braidings of time-geography are placed. Not least, and in an innovative turn, Gren inverts the usual criticisms of time-geography which tend to see the approach as 'too physical' (see Davies, this volume) to argue instead that it is not physical enough. Only when the physical ontology is carried all the way through, Gren argues, will time-geography fulfil its full potential.

And so to the last three chapters which all emphasise, though in different ways, the realm of myth and religion. These begin with a chapter by **Ann Game** who over at least the last ten years has been attempting to understand the time of lived experience through a reworked phenomenology (see Game and Metcalfe, 1996). In a seminal paper, Game (1997) considered what might be called the tension of tense through a series of contemplations. In particular, she was intent on establishing the richness of temporal experience, in part attempting to overcome Bergson's over-emphasis on continuity by emphasising forms of discontinuity and rupture. In this chapter, Game extends the argument by focusing on the mythic realm as a means of reworking ways of being. Like Probyn, she is clear that this involves our sense of now and here, so allowing a new sense of belonging to come into existence, one that, like Probyn's, consists of a constellation of ethical relations.

In some senses, Ann Game's long-time collaborator, **Andrew Metcalfe** provides a similar argument in his chapter, co-written with **Lucinda Ferguson**. Here the emphasis is on betweenness, those TimeSpaces of experience which confound inside and outside. The chapter proceeds to describe these TimeSpaces through the consideration of a variety of situations which allow in-between to show up. Finally, **David Loy**'s paper follows on from his important work on Buddhist time (see Loy, 1996, 1997). His aim is to show the cultural peculiarity of Western time with its individualised orientation dependent on always making something happen, on accomplishing something. In contrast, he outlines an

45

orientation which is concerned instead with an affirmation of the now rather than a sense of a future which because it strives so hard to accomplish will always find the present lacking. The case for such a politics becomes ever more pressing as the latest model of the individualised orientation to the future, the entrepreneur, has become the focus of contemporary projects of governmentality (du Gay, 2000; Thrift, 2000). Such a variant poses the real possibility that the Western world will win all and lose everything. That so much of contemporary business seems to yearn for something like the sacred – in work, in the construction of the managerial self, or in the construction of a business – suggests both how close and how far we are from an effective critique.

NOTES

1 With thanks to Marcus Doel for encouraging us to think through our position on this point more clearly.
2 Pioneered by writers like William James
3 For example, to the rise of 'stream of consciousness' novels.
4 No doubt, we could have started earlier, with Nietzsche, or even before.
5 In Bergson's durational cosmology, we can say that:

> Mind is in a continuum with matter, so that matter is simply the lowest state of mind. To separate matter wholly from mind would be to deny that duration has extensity; to rob matter of mind would make it wholly homogenous and discontinuous, in short, devoid of duration.
>
> (Antliff, 1999: 188)

6 There is no space here to go into the work of McTaggart, which is obviously relevant both historically and substantially. But see the excellent discussion of his work in Gell (1992).
7 Part of the problem may lie in Bergson's characterisation of ordering and scientific knowledge as 'cinematographic'. As in the cinema, movement is an illusion created out of a succession of static frames. 'In Bergson's eyes, philosophers have not managed to go beyond this: the ancient philosophers denied the importance of time, and sought truth outside it. Modern philosophers have considered time, how they have shared the world-view of the sciences, and have reduced it to a succession of snapshots' (Chimisso, 2000: 3).
8 Thus, my own work in the 1970s and 1980s with Don Parkes was concerned to operationalise the notion of space-time as rhythm across a wide variety of phenomenon, from the geography of circadian rhythms to the differential distribution of the building cycle (see Parkes and Thrift, 1980).
9 This account is a little unfair to Benjamin. See, in particular, Caygill (1999).
10 The same kind of experimentation is also going on in art (see Sudenberg, 2000).

Part I

MAKING-LIVING TIMESPACE

2

MODERNS AS ANCIENTS

Time, space and the discourse of improvement

Kevin Hetherington

The greatest improvement in the productive powers of labour, and the greater part of the skill, dexterity, and judgement with which it is any-where directed or applied seems to have been the effects of the division of labour.

(Adam Smith)[1]

The factory and the future

Utopian experiments are inherently spatial in character. The issue of time emerges when their orientation turns from the present to the future (see Hetherington, forthcoming). We often think of such experiments in terms of distinct communities, either imaginary ones on paper, or lived ones attempted in practice. As particular spaces that set themselves apart as islands of ordered perfection, their aim is to establish a source of comparison against which the seemingly disordered and troubled mess that makes up society might be compared and judged. The hope of the narrator is that outsiders will judge their own society as not living up to the utopian community. Not only will it offer an example of an alternative moral community that better addresses the problems of the day but it will also suggest practical everyday examples of social and economic relations through which society might come to be ordered. In practice, of course, the comparison can work in the opposite direction too as the dystopian tradition, as well as the failure of many utopian experiments, would suggest.

The history of the early factory played an important role in the expression of utopian ideas in the eighteenth and nineteenth centuries. Given the utopian tradition that I have identified, we might, then, turn to established utopian communities like Robert Owen's New Lanark in Scotland (Owen, 1970 [1813]) or to imaginary ones like Charles Fourier's early-nineteenth-century Phalanstery (Fourier, 1971), to see how the factory came to be seen as a means through which an ideal moral community might be developed. These ideas of the working community developed against a very different reality of factory work with exploitative, unregulated working conditions and a harsh, dangerous, alien environment that de-skilled and exploited its workforce (see Engels, 1987; Ward, 1970; Tann, 1970). However, it is in that reality rather than in the alternatives of Owen and Fourier that we find the utopics of the factory in the nineteenth century.

In the hands of nineteenth-century writers like Andrew Ure (1967 [1835]) and Charles Knight (1845), even in those of Marx, who recognised the achievements of the factory just as much as its detrimental consequences (1938 [1867]), the factory was idealised as the source of progress within society. And in the hands of writers like Samuel Smiles, the industrialists and engineers who ran them became the great heroes of England in its great age of progress and empire (1968 [1867]). The experiments in social engineering that sought to contest this progressive view of the factory were instead alternative visions to that of the emerging factory system.

Certainly the romantic tradition that socialists like Owen promoted, challenging the role of the machine and the factory that housed it, was to become a vehicle of another kind of Victorian utopian experimentation in the hands of writers and designers like Augustus Pugin, John Ruskin and William Morris in later decades. Their craft medievalism, with its celebration of the principles of revealed construction (craftsmanship) and truth to materials as well as the studio or workshop that was not organised upon the lines of a specialised division of labour, was to have a lasting influence on late-Victorian and later-twentieth-century sensibilities (notably in art and craft manufacture). However, their romantic outlook was a direct challenge to the ideas of progress that had come to be associated with industrial manufacture and designs epitomised in the optimism of the Great Exhibition of 1851.

My concern here, then, is a genealogical concern (see Foucault, 1977a) with the question of how the factory in the broader sense came to be seen as a source of utopian ideals within modern societies. I am concerned with how a *utopics* of capitalist (and later socialist) modernisation, in which an idea of an imagined future could be engineered, came to be associated with that type of productive space which we call a factory.

In the twentieth century, supporters of the factory system like F. W. Taylor and Henry Ford sought to engineer this social space such that it might better achieve the promise of modernisation that it offered. From an entirely different perspective, Marxists also saw the factory as a utopian space. As a site of class consciousness, class struggle, organised labour in the form of either trade unions or workers' councils (soviets), the factory was seen as a space out of which a process of conflict would lead to the generation of the conditions for a new stage in the modernising process – socialism. The factory, never a model utopia in itself, came to express instead the idea of utopia through a notion of a modernising process that would help create a better future, capitalist or socialist, for society. It expressed a utopics not only concerned with the spatiality of social engineering but its temporal aspect too.

Utopics, a term coined by Louis Marin (1984), does not refer, then, to the establishment of small utopian communities but to broader social practices whereby the utopian ideals that a society has are expressed spatially. He calls this utopics a spatial play and we see examples of it in such things as town planning, modernist architecture, landscape gardening, prison design and civic building. In these forms of spatial play we witness a modernising utopics put into practice in which future-oriented ideas about order, improvement and development were expressed through certain kinds of planned social space and the discourse surrounding them.

This modernising spatial play is central to the idea of modernity and is not limited to the factory alone. Prisons, asylums, hospitals, schools, libraries, museums, housing estates and so on – all sites of some kind of social engineering – are all testament to this modern utopic expression of social ordering through new forms of spatial arrangement (see Foucault, 1977a; Scull, 1982; Markus, 1993; Hetherington, 1997a). Utopics turn up almost everywhere in modern societies – canals, roads, sewers, public squares – anywhere where social aspirations that focus on creating a better society in the future can be spatially engineered in the here and now.

There is nothing ontological about such spaces that makes them a contender for this utopic practice, rather it is a site's importance in relation to other sites against which it can be compared that marks out its significance. Within the process of the utopic engineering of social space, certain sites will be more amenable to this utopic practice than others. They will become nodes in a network of social spaces that have a degree of centrality and influence within that set of relations. Elsewhere, I have described these sites, following Foucault (1986a), as heterotopia, 'other places' which I define as a site of alternate ordering (1997a: 9). In other words, within a society and the social order through which it represents itself, certain new sites, or newly interpreted sites, will emerge that offer an alternative expression of social ordering to that which currently prevails. Within modern societies, that alternate ordering is often a utopic one that looks to how society might be improved in the future. There is, therefore, a distinct temporal frame that is associated with such sites – they are often oriented to the future, or to some sense of the new, rather than to the present or past. It is this temporal frame – newness, and the uncertainty that that brings – that gives it its alternate 'Other' status.

It follows that there is also nothing intrinsic about any site in itself that might lead us to describe it as heterotopic. Rather, this definition of heterotopia is relational. It is about how we define one site in relation to others surrounding it and how the future or the new comes to be represented against the present in these surrounding sites. Of course, any site might be defined as heterotopic in this way. However, such sites only attain this conceptual status, I would argue, when viewed as part of a distinct spatial practice – in this case that of utopics (see Hetherington, 1997a). Within the practice of utopics, only certain sites, certain nodes within the network, will attain such a status. Following Callon (1986) and Latour (1988) we can say that it is only those that come to act as obligatory points of passage within a network of social spaces that should be viewed as successful heterotopic sites through which this utopic practice is expressed and realised.

An obligatory point of passage is an important site within a particular network. In particular, it is a site that attains a legitimating status for that network. The laboratory in the network of science, the council chamber in the network of local politics, the gallery in the network of art are all examples of obligatory points of passage within their particular fields. What passes for science, politics and art in each of these fields only attains a legitimate and recognisable status if the practices and materials of those fields pass through these ordering and legitimating sites. But laboratories, council chambers and art galleries are not examples of heterotopia – at least not any more. Only when new sites, alternatives to these ones, emerge within their respective fields, offering alternative and new orderings of science, politics or art, do we see examples of heterotopia. The utopic aim is to turn these new sites into obligatory points of passage.

In the field of production, the factory in our time is the established site in ordering the production process. It cannot be described as heterotopic in the sense that I have defined it here. Heterotopia would be alternative production sites offering something new or different that the factory does not offer. But in the eighteenth and early nineteenth centuries factories were novel and indeed quite uncommon spaces (see Tann, 1970; Berg, 1985), offering a utopic alternative to the existing ways in which work was then organised, notably through the domestic system of production known as 'putting-out'. Over time they became obligatory points of passage defining the kind of space best suited to the needs of capitalist production and to the disciplines of work (Hetherington, 1997a: chapter 6).

In this chapter, I want to offer an example of one particular heterotopic factory site that became an obligatory point of passage for certain utopic ideas about production – Josiah Wedgwood's factory at Etruria in Stoke-on-Trent, England (an area also known as The Potteries). This particular space became an important site in which a new utopics of capitalist production emerged, one based upon the principles of the division of labour, time-management, sophisticated accounting techniques and the marketing of well-made, fashionable products. Its success, not only in financial terms, but in terms of defining thereafter how pottery would be produced – at least until the resurgence of craft and studio pottery a century later – is testament to its status as an obligatory point of passage within its particular network of production.

However, the factory is also significant in another way too. It also helped to order something else – something much more fundamental to the idea of modernity. As well as providing a sense of how production should be organised, it also helped to express a new sense of social time that was then just coming into acceptance. I do not refer here to the process of instilling time-discipline within the factory (Thompson, 1967) although it certainly achieved this in practice. Rather, I refer to its role in becoming an obligatory point of passage for a discourse of improvement that led to the idea that society progressed as it moved forward in time. This discourse was expressed through the idea of the improvement of place and of that improved place moving forward in a linear social time. My argument is that over the course of the eighteenth century we see that ideas of improvement, which started out as part of a discourse of space, changed to become part of a discourse of time. That shift occurred through utterances around the space of the factory and the utopics attached to it.

In a world after Hegel, Marx and Darwin and after the progressivism of avant garde movements within modernism, it is perhaps difficult for us to think about social time in anything other than a narrative of progress and stages. It is also clear that this sense of modernisation has not gone unchallenged, by nineteenth-century romantic and conservative ideas about the past or more recently by contemporary postmodern ideas of stasis and the challenging idea of the end of history.[2] But in the mid-eighteenth century, when Wedgwood was setting up his model factory, the discourse of improvement that it came to express was only just emerging and beginning to give shape to an emerging sense of social time.

Debates about the status of the moderns in relation to the ancients were just under way. Writers like Adam Ferguson were only just beginning at that time to suggest the idea that society might develop and progress in stages (1966 [1767]). Similarly, sciences like geology were just beginning to offer views on the earth's temporal development that challenged established creationist ones. In these approaches the idea of temporal societal development was present. But the idea of temporal progress was a novel and a radical one to modern Enlightenment thinking that had not yet become established as a way of thinking about social time.[3]

What I offer here, then, is an aspect of the genealogy of the ideas of progress and modernisation. In particular, the factory came to be seen as a way in which society, represented in a particular place, could be improved not only in terms of the use of natural resources or in comparison with other places but also in comparison with its own past. To illustrate this argument I look at economic surveys and what might be described as accounts by industrial tourists written between the 1680s and 1800s. I work back from the end of the eighteenth century towards the latter years of the seventeenth in order to show how ideas about improvement and social time developed together. In the writings of contemporary eighteenth-century authors, as well as in the production practices and in the products that were being made, we witness how this discourse of improvement emerged and the way in which it placed the factory at the centre of utopian ideas of progress.

The discourse of improvement

I begin with the point nearest to the contemporary in this genealogy, with an account of a particular ending, the death of a subject, encountered in an obituary that appeared in *The Gentleman's Magazine* in January 1795:

> At Etruria, in Staffordshire, aged 64, Josiah Wedgwood, Esq. F.R. and A.S.S.; to whose indefatigable labours is owing the establishment of a manufacture that has opened a new scene of extensive commerce, before unknown to this or any other country. It is unnecessary to say that this alludes to the pottery of Staffordshire, which, by the united efforts of Mr. Wedgwood and his late partner Mr. Bentley, has been carried to a degree of perfection, both in the line of utility and ornament, that leaves all works, ancient or modern, far behind ... it remained for Mr. W. to propose such measures for uniting the Duke of Bridgewater's Canal with the navigable part of the River Trent ... as first fully carried the great plan into execution, and thus enabled the manufacturers of the inland part of that county and its neighbourhood to obtain, from the different shores of Devonshire, Dorsetshire, and Kent, those materials of which the Staffordshire ware is composed; affording, at the same time, a ready conveyance of the manufacture to distant countries; and thus not only to rival , but undersell, at foreign markets, a commodity which has proved, and must continue to prove; of infinite advantage to these kingdoms. ... Mr. W. planned, and carried into execution, a turnpike-road, ten miles in length, through that part of Staffordshire called The Pottery; thus opening another source of traffic, if by frost or some other impediment, the carriage by water should be interrupted.
>
> (anon., 1795: 84)

Wedgwood died but an improved place was born as his legacy. As one would expect with an obituary, the deceased, the famous pottery manufacturer Josiah

Wedgwood, is the centre of attention. He is, however, also placed in this description in the centre of a place that is called 'The Pottery' and its main industry, pottery manufacture. That network consists of Wedgwood himself, pottery manufacture, his partner Mr Bentley, pots of high quality, canals, materials (such as clay and flint) from other parts of the country, overseas markets and disadvantaged competitors, and turnpikes (by implication both canal boats and carriages). This network is arranged in this text so that they centre, or are given meaning, by Wedgwood who is synonymous with his factory. The account implies that this was the agent of change and improvement that made a place called 'The Pottery' a centre for industrial activity then the envy of the world. On his death the place in which he had worked had been transformed by his labours, his invention and his entrepreneurial skills.

In the same year that the obituary was published, a physician called John Aikin published *A Description of the Country from Thirty to Forty Miles round Manchester* (1795). In it we find another account of The Potteries and of the place of Wedgwood and his factory in it. In its section on places of interest within Staffordshire, pottery figures strongly:

> The principal object for which we have included part of this county in our design having been its potteries, we shall almost solely confine our account of particulars to those parts connected with them.
>
> (1795: 513)

The discussion then moves on to describe Newcastle under Lyme, then the principal market town in North Staffordshire, mentioning its corporate status, religious composition, alms houses, appearance in relation to coal burning in the area, markets, theatres, schools, parks and hat manufacture, pottery making and the planned canal through the town. The account then suggests: 'The markets of Newcastle have declined, since several have been established in the potteries; yet they are still considerable' (1795: 515). Further on in the text, after this brief description of Newcastle and its surroundings, there is a lengthy section under the heading 'The Potteries' (516–37). The account begins by describing the region of North Staffordshire known as The Potteries, suggesting that it was made up of a series of villages, which, due to recent industrial expansion over a period of about twenty years have now become an extensive urban conurbation. Aikin then goes on to describe each of the villages that makes up The Potteries, focusing on the nature of the manufacture in each, principal manufacturers, the environs, types of housing, turnpikes, the influence of Methodism in the area, the speed and unplanned nature of the urban development, improvements linked with the canals, and the use of the steam engine in pottery manufacture. Given the importance of Wedgwood that Aikin goes on to discuss (528ff.), I quote here what Aikin has to say about the area of The Potteries specifically associated with him:

> Etruria belongs solely to Josiah Wedgwood, Esq. who has a very exten-
> sive earthen ware manufactory here, a considerable village, a handsome
> seat, and complete grounds. In his pottery pursuits he has most deserv-
> edly acquired a great fortune with an equal share of reputation. The
> name of this place was given to it by Mr. Wedgwood after an ancient
> place in Italy, celebrated for the exquisite taste of his pottery, the re-
> maining specimens of which have served greatly to improve the beauty
> of the modern articles. The Staffordshire canal goes through Etruria
> grounds, which of course renders it a good manufacturing situation; but
> the whole belonging to one individual will most likely operate an in-
> crease of manufactories.
>
> (520)

Aikin then talks at length about how the pottery is manufactured. He does this by way of a description of the history of pot making in the area since the late seventeenth century. He begins by saying that the area was not particularly favourable to animal husbandry but that the plentiful supply of clay, found on the surface coal fields in the area, made it a suitable area for pot manufacture.[4] His account orders the subsequent development of The Potteries into six stages in time starting with the earliest and moving forward to the latest: (1) the manufac-ture of coarse earthenware butter pots and tobacco pipes, (2) the manufacture of 'rude' slip-ware in the late seventeenth century, (3) the introduction by a Dutch potter Elers of salt-glazing into the area around 1690, (4) the use by Astbury of ground flint mixed with clay to produce a finer, whiter stoneware pot, (5) the development of a fine creamware by Josiah Wedgwood, known as Queensware because it was favoured by the Queen who commissioned pieces by Wedgwood and made him the royal potter. Wedgwood is also credited in bringing in finer clay to the local marls from Dorset and Devon to be used in the manufacture of this new, more refined tableware, (6) a range of further developments in the clay body and design of the product, again attributed to Wedgwood. The description ends by noting the recent death of Wedgwood and quotes from the same obituary mentioned above (535–7).

The text orders the space known as The Potteries through a meaningful temporal series of changes or improvements. This description of pottery manufacture and of the place in which it occurred emerges in the account through a discourse of the improvement of both production process and place. It sees The Potteries as a place of urban growth due to industrial improvement, begun in the late seventeenth century by a handful of innovative potters and later brought to full fruition by Wedgwood and his factory after the 1760s.

If we look at other accounts from around this time we begin to see a similar, though still unsure, discourse of temporal improvement emerging. In 1793, William Pitt from Wolverhampton was commissioned by the Board of Agriculture to produce a survey of the agriculture of Staffordshire and means by which it might be improved (1808). As it is principally about agriculture and not manufacturing it has little to say about pottery (1808: 232–8). There is, however, a paragraph on The Potteries:

> The manufacture of potters' ware in the north of the county is very extensive and important, the value of manufactured articles being, as it were, a creation of the manufacturer, from a raw material of low value. The Potteries consist of a number of scattered villages, occupying in extent about ten miles; and may contain about twenty thousand inhabitants, including those who depend upon them for employment and subsistence. They have not been so flourishing since the war. – Mr. Wedgwood (1808: 235).

The name, the importance of the industry as an employer and the extent of the area of production are recognised and Wedgwood is credited as a source of information on the slow down in prosperity due to the war. Improvement was not without its reversals. John Byng's account of 1792 gives Wedgwood a prominent role in a more positive story (1936). His diaries of his tours through England and Wales, published as *The Torrington Diaries* (1936) provide us with another brief account of The Potteries. In volume 3 of these diaries he gives a description of a brief visit to the area:

> I now descended to the village of Stoke-upon-Trent, around which are numberless new buildings, and many pleasant villas for the principal merchants; and there is likewise a good inn building in this place: here I cross'd the Trent; and soon, many branches of navigation. These intersecting canals, with their passing boats, their bridges, the population, the pottery ovens, and the bustle of business, remind me of a Chinese picture, where the angler is momentarily interrupted by a boat.
>
> The late village of Handley, now a great town, upon the hill above, cuts a flaming figure from its new church, and newly built houses. The village of Skelton, likewise, is swell'd into great bulk.
>
> Now I enquired for Etruria, the grand pottery establish'd by Mr. Wedgwood; and putting up my horse at the adjacent inn, sent up my name and compts to Mr W., with a desire to view his manufactory: in

the mean time, as the workmen were at dinner, and would be for about an hour, I saunter'd about Mr W.'s grounds; which are green, and pleasant, with some pretty plantations, views of navigation – The house seems to be good, and is built of staring red brick; as are many in the vicinage, belonging to the principal trader. I was now sheun about the several workshops of this great potter, wherein are employ'd 300 men; but this is a dull observation for any person who has seen China manufactories.

(1936: 126–7)

Byng, writing as a tourist, treats The Potteries as a spectacle of industrial and urban growth. He then goes on to view the key site in this spectacle, Wedgwood's large manufactory at Etruria, the reputation of which clearly preceded Byng's visit there. Even if he was not much taken by his visit to the factory he still felt it important that he had to see it.

Built in 1769 after Wedgwood purchased the land (the lease on his smaller, earlier factory having expired) his pottery centre was modelled on a country estate with parklands. Wedgwood successfully petitioned parliament to have a section of the Trent and Mersey canal built across his newly acquired land and next to this he built his factory. Inspired by Matthew Boulton's manufactory at Soho in Handsworth, Birmingham, he was well aware of the potential economies of scale of large workshops and the opportunity they afforded for the rational layout of space and the opportunity for developing an effective, specialised division of labour. He trained his workforce in specialised tasks, provided some of them with housing and insisted on a high degree of time-discipline, then a rare thing in manufacturing, notably by introducing an early clocking-in system and fines for lateness. He also introduced the then still uncommon system of double-entry book keeping in order to manage his accounts and insisted on high quality workmanship throughout his factory.

In it he manufactured both useful (table)wares and ornamental wares in the new and fashionable neo-classical taste, separating out the spaces in which they were made and the tasks involved in their manufacture. He would often visit the homes of aristocratic friends after they had returned from a grand tour in order to see their recently acquired ancient Roman pots, all of which were thought to be Etruscan at the time. He would often copy them and use their designs as inspiration for his own. After conversations with his friend Erasmus Darwin, he also named his factory estate Etruria after these Etruscan pots that he thought he was copying and manufacturing. Such an association with the past established what he was doing in a European and classical tradition of pot manufacture that had a long history. It was also used as a good marketing tool. Indeed, Wedgwood's success is often put down to his marriage of accomplished production processes with marketing skills (see McKendrick, 1982). He counted the Queen, Catherine the Great and many aristocrats as patrons of expensive dinner services. These 'loss leaders' as we would call them today helped him to promote his more ordinary, but more profitable tablewares and ornamental vases, to his main market – the growing middle class.

The introduction of the neo-classical styles was also important to Wedgwood's success. Certainly, he was a stylistic innovator though the adoption of this style also tells us something else – something interesting about the relationship then between ideas of improvement and social time. Put simply, the neo-classical style, which developed across all of the arts in the late eighteenth century, honoured the works and times of the classical past of Greece and especially Rome (see Honour, 1977). In ceramics it was defined against other styles that had been dominant in the earlier decades of the eighteenth century: established peasant styles, chinoiserie (a European adaptation of Chinese styles found on the porcelain that was being imported into Europe) and the style of the French rococo – fluid, asymmetrical and ornate in character. The geometric symmetry and formalism found in neo-classicism suggested one thing: order, with ancient Rome taken as a model for social order that might be adopted by a newly emerging modern society organised along capitalist and democratic principles.

The discourse of improvement was new to this time but it came to be articulated within the context of neo-classicism. Thus the heroic certainties – the utopics – that such a society developed were not grounded in a view of its own present and the projection of its ideals into the future (as they were to become in the nineteenth and twentieth centuries) but through trying to model that society along the lines of its classical European past. The formalism of neo-classicism, that Wedgwood is often credited with introducing into this branch of the arts, was very much in keeping with this expression of a utopics of order and improvement that had strong historical antecedents. These moderns draped themselves in the clothes of the ancients and in their emphasis on the formal, the monumental and the classically ordered. The commercial success of this enterprise helped to reorientate this idea of improvement from a focus on the classical past on to the as yet unwritten future. The style created a fold in time that linked the ancients and the moderns together in their desire for order that was then projected into the future. It helped to fix an idea of improvement and an orientation towards the future in the context of a solid and dependable present imagined as a replica of the ancient past and all its achievements.

Contemporary descriptions of Wedgwood's famous factory itself are rare. Manufacturers at that time were very concerned about industrial espionage, fearing that descriptions of their production process might lead to them losing secrets about both their manufacturing process and product design (see Pollard, 1965). Skilled workers were not plentiful and owners were very fearful too of losing them to competitors. They became very protective about who they let into their factories. This added to their mystique. They were, however, very eager to obtain the services of skilled workers from other factories in order to gain a possible advantage over their competitors. One contemporary account of Wedgwood's factory is provided by one such person who became something of a double agent. In 1786, Louis-Victor Gervenot arrived in Britain from France. Having been trained by the porcelain factory at Sèvres, he then spent most of his early career travelling around the pottery and porcelain factories of Europe selling and stealing secrets relating to the manufacture of porcelain (see Hillier, 1966). His intent in coming to Britain was to sell the secret of the manufacture of porcelain to a Staffordshire potter and to set up in a profitable partnership with him in return. He advertised in newspapers in London and was contacted by Wedgwood, who eventually declined to go into partnership with him (Hillier, 1966: 91). Gervenot ended up instead in partnership with one of Wedgwood's nearby rivals, John Turner at his pottery at Lane End. Some time during his contact with Wedgwood he was, however, able to get to see round the Etruria manufactory and he gives us an account and an evaluation of it in a letter:

> Wedgwood lives in a mansion which he has built for himself, situated in a beautiful, large park with a canal running round it linked to one running past the factory. The latter is an enormous building, practically a small town, which Wedgwood has called Etruria. The factory is a marvel of organization and Wedgwood himself is its first workman.
>
> There is only one entrance for the workmen through which everybody must pass, and a second entrance for customers. Between these two entrances lie the office from which one can see and control everything. All other factories in Staffordshire have modelled themselves on Wedgwood's factory. There are about 100 of these employing 6000–8000 workmen, and all factory owners have become prosperous within a few years.
>
> (Hillier, 1966: 93)[5]

The account goes on to discuss process and volume of manufacture, Wedgwood's personal quality control practices, wages, working hours, local customs associated with the factory, working regulations and factory discipline and sickness benefits provided by Wedgwood. (1966: 93–5). It is a short, succinct account that praises the virtues of the factory system and sees Wedgwood's enterprise as a model factory of its time. Improvement and growth are linked with the factory system and the development of The Potteries is seen to culminate in the Etruria factory.

The growth and improvement of the area by the 1780s also impressed the Methodist preacher John Wesley, who had been coming to the area to preach since the 1760s. In March 1781 he noted:

I returned to Burslem. How is the whole face of this country changed in about twenty years! Since which, inhabitants have continually flowed in from every side. Hence the wilderness is literally become a fruitful field. Houses, villages, towns, have sprung up: and the country is not more improved than the people.

(Quoted in Ward, 1984: 33)

Around the same time Wedgwood leaves us an account of his own. Perhaps more overtly than all of these other accounts, Wedgwood very clearly locates his own enterprise within a discourse of improvement that has a strong temporal element to it. After the suppression of food riots at Etruria in 1783 (see Ward, 1984: 445–6), Wedgwood produced a pamphlet: *An Address to the Young Inhabitants of The Pottery*. In it he makes comparison between the area in 1783 and 1750 in order to try and convince the young people who have recently rioted that they had never had it so good:

Before I take my leave, I would request you to ask your parents for a description of the country we inhabit when they first knew it: and they will tell you, that the inhabitants bore all the marks of poverty to a much greater degree than they do now. Their houses were miserable huts; the lands poorly cultivated, and yielded little of value for the food of man or beast, and these disadvantages, with roads almost impassable, might be said to have cut off our part of the country from the rest of the world besides rendering it not very comfortable to ourselves. Compare this picture, which I know to be a true one, with the present state of the same country. The workmen earning near double their former wages – their houses mostly new and comfortable, and the lands, roads, and every other convenience bearing evident marks of the most pleasing and rapid improvements. From whence and from what cause has this happy change taken place? – You will be beforehand with me in acknowledging a truth too evident to be denied by any one. Industry has been the parent of this happy change – A well directed and long continued series of industrious exertions, both in masters and servants, has so changed, for the better, the face of our country, its buildings, lands roads, and notwithstanding the present unfavourable appearances, I must say the manners and deportment of its inhabitants too, as to attract the notice and admiration of countries which had scarcely heard of us before.

(Wedgwood, 1969: 17)

In these accounts of the 1780s and 1790s Wedgwood and his factory are located at the centre of the development and improvements of The Potteries that have

61

occurred over a period of time. His famous factory at Etruria, the site and model of this industry, emerges in these accounts as an obligatory point of passage in this discourse of manufacturing improvement. The utopics of production that are expressed through this discourse also illustrate, albeit in a still tentative way compared with later-nineteenth-century accounts (see below), an idea of society as something that gets better as it moves forward in time. In these accounts, and in the idea of improvement, we see a discourse of a place moving through social time and becoming a better place as it does so because of the development of a system of factory production. As a proto-discourse of modernisation, this was a novel view for the time. The cause of that progress is taken to be industrial improvement organised through a factory.

All the same, Wedgwood's fame predates the establishment of his model factory. In an account of the pottery industry by Arthur Young, dating from around the same time as the opening of the Etruria factory, Wedgwood's fame and the association of the development of the area with him was already an acknowledged view:

> From *Newcastle-under-line* I had the pleasure of viewing the *Staffordshire* potteries at *Burslem*, and the neighbouring villages, which have of late been carried on with such amazing success.
>
> (1770: 306)

He goes on to say how many houses there were and how many were employed in the industry before moving on to talk about Wedgwood:

> It dates its great demand from Mr. Wedgwood (the principal manufac- turer) introducing, about four years ago, the cream-coloured ware, and since then the increase has been very rapid. Large quantities are ex- ported to *Germany, Ireland, Holland, Russia, Spain,* the *East Indies,* and much to *America*: Some of the finest sorts to *France.*
>
> (1770: 306)

He concludes his description by saying:

> In general we owe the possession of this most flourishing manufacture to the inventive genius of Mr. Wedgwood; who not only originally intro- duced the present cream coloured ware, but has since been the inventor of every improvement, the other manufacturers being little more than mere imitators; which is not a fortunate circumstance, as it is unlucky to have the fate of so important a manufacture depend upon the thread of one man's life: However, he has lately entered into a partnership with a

man of sense and spirit, who will have taste enough to continue in the inventing plan, and not suffer, in case of accidents, the manufacture to decline.

(1770: 309)

We are beginning to get the picture of The Potteries as a place developing in time: growth industry Wedgwood improvement The Potteries. Wedgwood's factory gives that idea of improvement a focus and a cause. It can also be described as heterotopic in the sense that the ordering it brought into play was novel and different. The manufacturing process, the working conditions, new styles and types of clay body (Wedgwood improved existing variations of earthenware, creamware and basaltware, invented jasperware and was amongst the first to adopt the neo-classical style in pottery manufacture), and indeed of the ideas about place and time, all emerged out of this new kind of social space. It existed as a site of Otherness to the other workshops and potbanks of the time that were still principally organised around non-industrial forms of domestic production within households that had a low degree of division of labour and often made pieces for merchants or travelling crate-men who would pay piece rates for them and then sell them at market (on domestic production see Berg, 1985; see also Weatherill, 1971). Its success in becoming the model that all other aspiring potters had to follow, and follow they did, is testament to its success in attaining the status of an obligatory point of passage within the network of ceramic production. It was not until the 1870s and 1880s, a century later, that studio potters like William de Morgan and the Martin brothers were to react against the factory as the site for making ceramics and when they did that reaction was against the Staffordshire model that had been started by Wedgwood.

If we move back further in time we begin to see a change in this discourse of improvement. In the 1750s we begin to lose Josiah Wedgwood. He was born in 1730 and was apprenticed to his brother in 1744 but did not go into partnership (with John Harrison) until 1752. Later he more famously went into partnership with Thomas Whieldon but he did not set up on his own until 1759 and did not begin the Etruria factory until 1769. In 1750 Wedgwood would have been an unknown apprentice. What, then, does an account by Dr Robert Pococke in 1750 tell us of The Potteries?

On the 6th I went to see the Pottery villages, and first rid two miles to the east to Stoke, where they make mostly the white stone. I then went a mile north to Shefly where they are famous for the red china; thence to Audley Green a mile further north, where they make all sorts, and then a mile west to Bozlam, where they make the best white and many other sorts, and lastly a mile further west to Tonstall, where they make all

sorts too, and are famous for the best bricks and tiles; all this is an uneven, most beautiful, well improved country, and this manufacture
brings great wealth to it; and there is such a face of industry in all ages
and degrees of people, and so much civility and obliging behaviour, as
they look on all that comes among them as customers, that it makes it
one of the most agreeable scenes I ever saw, and made me think that
probably it resembles that part of China where they make their famous
ware.

(1888: 8)

Here we also have an account, not of 'The Potteries', but The Pottery villages.
That account too is one of industry, improvements and industrial diversity but no
Wedgwood, and of villages rather than an industrial conurbation. The idea of
improvement is still there but it is unfocused, piecemeal and local and not
concerned with a comparison with its earlier past.

The sense in which Pococke speaks of improvement is different. For him it has
more to do with the human improvement of a natural environment into one of
human enterprise and economic advantage. Moving back further still we find
another account written in 1695 by the traveller and journal writer Celia Fiennes
who gives us a similar brief description of the region and pottery manufacture.
Like many others mentioned above she visited the market town of Newcastle
under Lyme and while there went on to see pottery being made in the vicinity:

I went to this Newcastle in Staffordshire to see the making the fine tea-
potts cups and saucers of the fine red earth, in imitation and as curious
as that which comes from China, but was defeated in my design, they
comeing to an end of their clay they made use off for that sort of ware
and therefore was remov'd to some other place where they were not
settled at their work, so could not see it.

(Fiennes, 1949: 177)

Here we have another traveller going to view something she had heard about
elsewhere. The manufacture of redware was in fact carried out by Elers who is
mentioned in the later account by Aikin. There is no mention of The Potteries, no
mention of improvement and industry, but there is, by implication at least,
mention of the fame of the area associated with its novel use of clay in the
manufacture of tableware which must have been seen at the time as an item of
taste that one might want to own. There is one other short account, from 1693,
which also refers to the manufacture of these types of tea pot, in *Philosophical
Transaction*, that would tend to support this view (see Lewis, 1969: 4):

The teapots now to be sold at the potters in the Poultry in Cheapside which not only for art, but for beautiful colour too, are far beyond any we have from China; these are made from the English Haematites in Staffordshire, as I take it, by two Dutchmen, incomparable artists.

(1969: 4)

Wares of fine quality were being manufactured in the region later to become known as The Potteries and were being sold in London. Celia Fiennes and those of her class able to afford such wares most likely encountered them there and talk no doubt got round to where they were made. The fact that they were from England and were seen as of better quality than wares coming in via the East India Company from China may well have done something to excite some curiosity as to the region the pots came from. But one gets a sense from such accounts that it is motivated by ideas of curiosity rather than a desire to see the great improvements going on in Britain at that time.

Of course, the most famous account of this type is Daniel Defoe's *Tour Through the Whole Island of Great Britain* (1971 [1724–6]). The purpose of his journey was to discover the economic improvements being carried out in Britain (he was paid as a spy by the government to undertake such a survey). One has the sense from his account that no-one really knew what was going on in the economic activity of a nation and to what extent improvements were taking place. When improvements were discussed it was much more through a discourse about space, a discourse that was concerned with the improvements that could be made to a place when the natural resources were used to their full extent, than a discourse about time.

Finally, then, we arrive at 1686 and one of the most famous accounts of the manufacture of pottery in North Staffordshire, that by the Oxford Professor of Chymistry [sic] and keeper of the Ashmolean Museum, Robert Plot (1686). While there are known references to pot manufacture in the area in historical documents dating back to 1348 (see J.C. Wedgwood, 1913: 5ff.), Plot's *The Natural History of Staffordshire* contains the earliest known published account of the manufacture of pottery in the region and was based upon his own first-hand observations in the late 1670s. This account can be found in chapter 3, *Of the Earths*. Plot gives an account of all things that have to do with the earth in this chapter by ordering them through a series of geological layers through which he moves vertically downwards: things that graze on the surface, the turf, cultivation of the soils, the clays beneath and coal deposits below that. This is quite different to Aikin's temporal ordering a century later. He describes the related forms of agriculture and industry that correspond to each of these geological layers. Not surprisingly his detailed account of pot making is associated with the geological layer of the clays (1686: 121–4). He begins by discussing the making of tobacco pipes all over the county before focusing on the area around Newcastle: 'And Charles Riggs of Newcastle makes very good pipes of three sorts of clay, a white and a blew, which He has from between Shelton and Hanley green' (1686: 121). He then goes on to talk in detail about the manufacture of pots:

> But the greatest Pottery they have in this County is carried on at Burslem near Newcastle under Lyme, where for making their several sorts of Pots, they have many different sorts of Clay, which they dig round about the Towne, all within half a miles distance, the best being found nearest the coale.
>
> (1686: 122)

He then describes the four different types of clay being used: bottle clay, hard-fire clay, red blending clay, white clay (122) before going on to talk about the use of slips, or liquid clays (clay mixed with water) which were used to decorate the pots. Plot also discusses the manufacture of slip-ware, lead glazing, the firing of pots and the manufacture of 'crude' butter pots which he has referred to earlier in the chapter in his brief description of Uttoxeter market:

> The butter they buy by the Pot, of a long cylindrical form, made at Burslem in this County of a certain size, so as not to weigh above six pounds at most, and yet to contain at least 14 pounds of butter, according to an Act of Parliament made about 14 or 16 years agoe, for regulating the abuses of this trade.
>
> (1686: 108)

Finally, Plot mentions the sale of the pots, 'chiefly to the poor Crate-men, who carry them on their backs all over the Countrey' (1686: 124). We can discern from this account that pottery manufacture was already known in the region, or more specifically in the village of Burslem (The Potteries, as a conurbation, did not exist in the 1670s) and that there was already some degree of industry in the area with crude products like butter pots, not dissimilar in style from the centuries of medieval earthenware being made and sold locally, while slip-ware and lead-glazed ware was hawked and sold all over the country.

I have ended with a description of pre-industrial domestic manufacture, making relatively simple products and selling them at markets in a fairly unsophisticated way. Plot's writing on pottery manufacture was certainly known to some of the later writers discussed above. He is discussed in Aikin as follows in his section on Burslem:

Doctor Plot, in his history of Staffordshire, written in 1686, makes particular mention of the potteries of this place, and points them out as the greates of their kind. He also gives admirable detail, describing most minutely the process and manner of making earthen ware in those days. But as the wares of the present time are of a different kind, and very different also in the composition and manufacture, from that described by Dr. Plot, we shall, before we quit this neighbourhood, describe the present mode of manufacturing earthen ware, from clay to its completion.

(1795: 519)

He is positioned as describing an area before improvement, an area at a certain point in time from which improvement then begins to take place. Some sort of comparison with Plot 110 years earlier is intended. After all, as we have seen, since Plot's time we have had improvement, industry and Wedgwood.

Time and the survey

I have given here an account of improvement based on texts contemporary to the period 1800–1686. Working backwards has made visible the developments made during this time while also revealing how The Potteries was constituted in the later accounts through a discourse of improvement that involved an ordering of industry and growth through the agency of Wedgwood and his new kind of factory. This discourse was to continue after the period under discussion here. In the nineteenth and twentieth centuries the idea of improvement as progress in time becomes stronger and more certain. The first history of the Staffordshire potteries was published by Simeon Shaw in 1829 (1970). Using the oral testimonies of old potters, his own observations, as well as some of the sources discussed above, Shaw constructed a history of The Potteries as one of improvement and growth over the period from the late seventeenth century to the beginning of the nineteenth:

Were a person to place himself, in succession, on the hills at Green Lane, Wolstanton, Basford, Hartshill, and Fenton Park, and take a bird's-eye view of the different parts, he would be much gratified with the many indications of the utility of well-directed industry, and its results, a vast increase of population; numerous and extensive manufactories, with beautiful mansions; maintenance for the employed, and opulence for the employers.

(1970: 7)

In contrast to the seventeenth century:

> Little more than a century ago, its existence was scarcely noticed; it was more then a barran [sic] aspect, and was a mere range of straggling and detached hamlets, with few inhabitants and little trade.
>
> (1970: 9)

Wedgwood's factory, of course, is prominent in the account of these improvements:

> When it is considered, that nearly the whole of the Materials used are native productions, and that five parts in six of the manufactured articles are exported, few Branches of manufacture have greater claim to the gratitude and admiration of their countrymen, than these valuable establishments, and the persons who have founded, fostered and advanced them. The late Mr. Wedgwood, in his day was a principal promoter of this advancement; and since his time additional and great improvements have been made by the united genius of the present Potters, Spode, Wood, Ridgways, Minton, Turner, &c, and it is a fair presumption that specimens of their production will be found, not only in the cabinets of princes and opulent persons of taste, but in the markets of every state where British commerce extends.
>
> (Shaw, 1970: 16)

Soon after, a series of biographies of Josiah Wedgwood began to appear and through them he was turned into one of the leading heroes of the Industrial Revolution. The first was by Eliza Meteyard (1865, 1866), followed by Llewelyn Jewitt (1865), then an account of Wedgwood the man by Samuel Smiles (1905) and another in the same vein by Julia Wedgwood (1915). There are also a succession of minor biographies up until the most recent significant one by Robin Reilly (1992). In these accounts the format is much the same: a description of pottery manufacture in The Potteries prior to Wedgwood in a couple of early chapters followed by a series of chapters on the various ways in which Wedgwood was important to the improvement of both pottery manufacture and the place in which it was made. The personal biographies also seek to canonise the man as the great industrialist who put The Potteries on the map and did so much for British trade and the arts in the eighteenth century. His establishment as a name that became associated with a factory suggests how he, as the representation of that factory, acted as an obligatory point of passage within this branch of manufacture. This is apparent in these nineteenth-century and twentieth-century accounts just as it is in the organisation of the narrative in the display of pot in the City

Museum and Art Gallery in Hanley, Stoke-on-Trent (see Hetherington, 1997b). We have already seen this picture emerging around 1770 and it is established by the end of the eighteenth century. The Potteries are improved by Wedgwood's factory. The Potteries come to exist as a place because of the improving effects of Wedgwood's manufacturing agency. This view continues through these accounts and was fixed in stone, literally, when in 1863 the Prime Minister, William Gladstone, laid a foundation stone for Burslem's town hall in order to commemorate Wedgwood's heroic achievements. Gladstone's speech at this occasion (1863), reinforces the idea of Wedgwood as hero and genius, a national asset, able to unite both art and industry and able to enhance the countryside by his enterprise rather than ruin it with urban squalor (1863: 263–4).

When we move forward into the twentieth century and more general accounts of the Industrial Revolution in Britain, Wedgwood, along with contemporaries like Matthew Boulton and Richard Arkwright, becomes a figure of a great divide; initiators of the factory system that transformed Britain into the first industrial nation (for example Mantoux, 1961; Landes, 1969). Wedgwood becomes the pioneer of factory discipline, double-entry book keeping, the rational organisation of factory space, the commercialiser of pottery manufacture and orderer of time within the factory (McKendrick, 1961, 1982; Thompson, 1967). His factory becomes a point of passage through which the idea of societal improvement (now expressed more confidently as modernisation and progress) is expressed.

It was not until 1971, and the publication of Lorna Weatherill's account of pottery manufacture in The Potteries between 1660 and 1760 (1971), that Wedgwood received something of a reversal. Using contemporary sources, Weatherill was able to show that many of the innovations and developments attributed to Wedgwood in all of these accounts were already under way before he began his work as a manufacturer. She saw his success as little more than being the best able to organise production, product design, patrons and markets. She is no doubt right. However, Wedgwood's real achievement was to turn himself into an obligatory point of passage. That these changes are all attributed to him, rather then other potters in the area, is testament to his success in this regard. That success is largely down to the novelty and alternate status of his factory with which he is identified. Others may by 1769 have begun to introduce a specialised division of labour in the potbanks but none did it so effectively and on such a scale as Wedgwood.

Looking at these contemporary accounts we see how the discourse of improvement changes during the course of the eighteenth century. In Plot's account, human beings are making use of the materials around them to fashion, albeit crude, objects of art. By the time we reach Wedgwood's obituary the process of manufacture has been transformed by factory production, the art object has been refined and – as a consequence of its market success – the space in which it has been manufactured has been improved. The discourse of improvement is already there by 1750, prior to Wedgwood, but he reinforces it and gives it a causal focus. Having done so, the discourse of improvement no longer makes sense without him as the embodiment of the factory system.

Whilst the authority of previous texts has not been refuted (not until 1971 and then only partially) it is changes to the form through which this discourse of improvement is expressed that is perhaps the most significant argument in this chapter; an argument concerning what such a discourse tells us about the ordering of time. This argument can be most easily seen in the changing status of the surveys discussed earlier.

All surveys offer spatial descriptions. But it is only the later ones that begin to operate through a discourse of social time. Robert Plot was very much a man of the classical age (Foucault, 1989a). His survey fits with the seventeenth-century model of the classifying table that seeks to establish a universal picture of everything associated with a particular place. The world is ordered in his account by earthly strata and human activity associated with the strata of the earth. Improvement in an account like this is about what one does to nature. Plot's aim, like others of the time (John Aubrey being the most famous in his survey of Wiltshire) was to survey the natural history of a county and what human activity was doing with the resources provided.

The accounts by Fiennes, Pococke, Young and most famously by Defoe, are less concerned with discussing the issue of natural resources (although it is still there in their accounts) and more with surveying the extent of human economic activity in particular places as compared with others. Doing this allows them to build up a picture of human economic and social life within the nation and of the kinds of improvements that are being made locally. The intended contrast is still more spatial rather then temporal. It is concerned with providing a comparison with economic achievements in other nations as well as in comparing parts of Britain with each other. It is only in the accounts of the latter part of the eighteenth and especially the nineteenth century that the survey takes on a comparative temporal perspective. A particular place is no longer surveyed only in terms of what humans are doing with the natural resources, or with how one place compares with another but with how the improvements to a place compare with its own past. It is only in these later accounts that we begin to see a future-oriented utopics emerge.

Conclusion

The modern figure for representing utopia is that of the temporal horizon (Marin, 1993; Hetherington, 1997a, forthcoming; Lee, 1998). In the contemporary world utopias have ceased to be concerned, as they had been in earlier times, with making comparisons in the here and now between an existing society and an ideal elsewhere. Instead, they have shifted that ideal form to one found not elsewhere but in another time – the future. The survey plays an important part in this discursive shift from the spatial to the temporal around the notion of societal improvement. It is in the eighteenth-century discourse of improvement that the idea is established that society can be improved and that given the right conditions this will inevitably happen over a course of time. This idea of improvement is the antecedent of progress and modernisation.

We know from the utopic status that the factory has attained within this idea of modernisation that it is an important site in the articulation of this discourse. The beginnings of such a discourse can be seen in examples of early factories like that of Wedgwood. That heterotopic space, different to the norm – other, radical, challenging and successful – was able to attain the status of an obligatory point of passage. Because of this it became the focus of the discourse of improvement and in particular in its shift from a spatial to a temporal frame of reference.

As with other genealogical accounts (see Foucault, 1977a) there is nothing inevitable in this process. Any sense of inevitability, and the discourses of progress and modernisation are all about inevitability, is a product of the discourse and not an antecedent to it. The idea of modernisation was able to take hold because of the status that the factory system was able to achieve. Of course, it later came to be coupled to the political ideals of a democratic and pluralist society (Rostow, 1960) and was used to try and counter the socialist model that tried to mobilise the idea of the factory in other ways for different political ends. No doubt the political aspect of the modernisation discourse could also be subject to a genealogy, but that is beyond the scope of this chapter.

The argument that I want to end with is that in order to understand how discourses emerge and become effective we have to pay attention to the spaces in which the utterances, of which discourses are constituted, are made (see Foucault, 1989b). The model that I have suggested here is that new discourses emerge in places that are Other; heterotopic, sites that suggest a novel or different mode of social ordering that contrasts with the established sense of order within a particular social field. That mode of social ordering is articulated discursively. We have seen that in the texts written but also in the kind of pots being produced. Once a space attains the status of a point of passage for the statements that make up this discourse, then that discourse is able to attain the status of something natural, legitimate and seemingly inevitable. In effect, that discourse becomes invisible. Statements made about improvement in time came out of the new modes of social ordering established by the factory. Over time, the factory became an obligatory point of passage for the discourse of improve-ment which in time became a discourse of social time as one of progress and modernisation. We see the beginnings of such a discourse in Wedgwood's factory.

NOTES

1   These are the opening lines to Smith's *The Wealth of Nations* (1922 [1776]: 5).

2   Even in the idea of postmodernity, the idea of temporal stages is still present. The postmodern is linked to developments in capitalism in particular (there are many examples but see in particular Lyotard, 1984; Jameson, 1984; Lash and Urry, 1987; Harvey, 1989).

3   The *Oxford English Dictionary* records a number of examples of the use of the word 'improvement' in relation to a sense of temporal progress from the latter half of the seventeenth century. Nearly all of these early examples refer to aspects of Christianity or to moral education. I would argue that a more secular usage applied to societies as a whole is much later in date (*OED*, Volume VII: 751).

4   It was the coal that was more significant than the clay. A ready supply of cheap fuel to fire the kilns was essential to the development of this form of manufacture. There are many areas that have large supplies of clay but which did not develop as centres for the manufacture of pottery. Those that did invariably have ready supplies of fuel – notably coal.

5   The suggestion of panopticism (Foucault, 1977a) in this contemporary quote is clearly evident. In many factories of this time it was not a central 'watchtower-like' office that became the space of the disciplinary gaze but the entrance or gate through which all workers had to pass and be observed (see Biernacki, 1995). All the same, we should not forget that Jeremy Bentham's original inspiration for the panopticon came from a factory, notably a naval workshop run by Bentham's father that manufactured ship's biscuits (see Semple, 1993).

3

A POLITICS OF STOLEN TIME

John Frow

I

Indigenous peoples inhabit modernity not as an archaic remnant but as a fold, a complication of its singular but fractured and internally disparate time. This complication is repeated spatially. In Australia, as in Canada and New Zealand, the destiny of non-urban Aborigines is closely tied to the struggle over possible forms of belonging to land, which represents at once the chance of economic autonomy and, on a quite different plane, the atemporal locus of Law and personhood. Legal ownership is in one sense peripheral to the more ancient rights in country given by that other Law which the Law does not recognize (or rather, since *Mabo vs Queensland* (1992), now both does and does not recognize); in another sense the politics of land rights has been at the very centre of the disputation of nationhood in the three settler colonies.

One way of understanding the play of conflicting – and radically unequal – claims to rights in land is as a knotting of diversely structured temporalities. Within Australian society's 'mosaic of pieces in different developmental stages, and of different ages' Kubler's (1962: 28) metaphor for the non-synchronicity of the social, which he contrasts to the historicist metaphor of a 'radial design conferring its meaning upon all the pieces', quite incommensurate structures of value and of historicity are asserted in relation to the normative rhythm of a modernization process which has marginalized rural capital and landless Aboriginal peoples alike, and sets them against each other. This clash of social forces draws together a heterogeneous cluster of times: that of indigenous movements seeking to form and assert a cultural and political identity against the forces of territorial and cultural dispossession; that of several generations of indigenous children stolen from their kin groups and their culture; that of a traditional indigenous social order, rooted in the non-time of the Law, which shadows the previous two; that of a modernization process (a process of macro- and micro-economic 'reform') actively espoused by the current Liberal government as it was by the previous Labor government; and that of those social classes which, feeling themselves left behind by history, have underwritten a right-wing backlash both against the modernization process which has devastated rural Australia, and against the politics of Aboriginality, including the overdetermined issue of native land rights. This knot of forces, with their quite different speeds and rhythms and trajectories, is at the same time a complex of voices, of stories, of arguments and claims which intersect and clash, are heard and not heard, and assert conflicting versions of historical memory and of the just use of the past. Drawing on Michel de Certeau's reflections on the disruptive articulation of place (including the place of knowledge) with time, this chapter examines the recovered voices of the 'stolen generation' of Aboriginal people and the question of whether and how discursive injustice can be repaired.

II

Someone must have been telling lies about Millicent D. because when she was 4 she was taken away from her family and made a ward of the state. Until the age of 18 she was kept in Sister Kate's Home in Perth, Western Australia, where she was forbidden to see any of her family or to know where they were. She was told that her family didn't care about her or want her, but that in exchange she would be brought up as a white girl 'in a good religious environment' (*Bringing Them Home*, 1997: 115).[1] That's what she was told by the Protector of Aborigines and the Child Welfare Department; but Millicent tells the Inquiry from whose Report I am quoting that 'all they contributed to our upbringing and future was an unrepairable scar of loneliness, mistrust, hatred and bitterness' (115).

When she was in her first year of high school Millicent was sent to work on a farm as a domestic servant; she went back there in the next school holidays, and this time 'it was a terrifying experience, the man of the house used to come into my room at night and force me to have sex. I tried to fight him off but he was too strong' (117). Back at the Home she reported this to the Matron; but the Matron 'washed my mouth out with soap and boxed my ears and told me that awful things would happen to me if I told any of the other kids. I was so scared and wanted to die. When the next school holidays came I begged not to be sent to that farm again. But they would not listen and said I had to' (117).

Millicent ran away from the Home in order to try to return to her family, but she was recaptured, punished, and sent back to the farm to work. This time, she says, 'I was raped, bashed and slashed with a razor blade on both of my arms and legs because I would not stop struggling and screaming. The farmer and one of his workers raped me several times. I wanted to die, I wanted my mother to take me home where I would be safe and wanted' (117). Instead, she was returned to the

Home. Again Millicent reported the rape to the Matron, and again she was punished: 'I got a belting with a wet ironing cord, my mouth washed out with soap and put in a cottage by myself away from everyone so I couldn't talk to the other girls. They constantly told me that I was bad and a disgrace and if anyone knew it would bring shame to Sister Kate's Home.' She ate rat poison to try to kill herself, but 'became very sick and vomited. This meant another belting' (117).

Some weeks later Millicent was examined by a doctor who told her that she was pregnant; again she was blamed and punished. She gave birth to a baby girl who was taken away from her, and she was told that she could have the child back when she left Sister Kate's. Some time later she asked the Matron for her daughter's address, and she was told first that it was not Government policy to give out this information, and subsequently that the child's whereabouts were not known. She then rang the hospital and was told that they had no record of her or of the birth of her daughter; and when she wrote to the Native Welfare Department they told her that they had no record of her family since the records had been destroyed by fire. Ten years after her daughter's birth she returned to Western Australia and again asked the Matron of Sister Kate's about her family and child; this time, Millicent says, 'she told me that my daughter was dead and it would be in my best interest to go back to South Australia and forget about my past and my family' (118). A footnote to this story says that Millicent was reunited with her child when the daughter was 36.

Listen to the actions that are reported here:

- a child is told a story about her family not wanting her;
- she is made untrue promises about the kind of care she will be given;
- when she reports being raped she is punished and her mouth is washed out with soap to signify the uncleanness of the words she has spoken;
- she begs not to be sent back to the farm, but the authorities in the Home refuse to listen;
- reporting a second set of rapes, she is punished, her mouth is washed out with soap, and she is put into solitary confinement so that she cannot speak and the other girls cannot listen;
- she is told that she is bad and shameful; she takes and then vomits out poison;
- she is told a story about the child that has been taken from her;
- she is later told that it is not known where the child is; then she is told that there is no record of the child's birth; then she is told that there is no record of her own or her family's existence; then she is told that her child is dead.

This is a story about an injustice that is partly enacted in language. It is about acts of telling that are true and acts that are false; it is about being told things and not being heard; it is about the relation between telling stories and existing, or about being made not to exist.

III

Millicent's story is a part of the Report of the National Inquiry into the Separation of Aboriginal and Torres Strait Islander Children from Their Families, entitled *Bringing Them Home*. Delivered to the Australian Federal Government in 1997, the Report is a record of the history of forcible removal of indigenous children, usually of mixed descent, from their families and communities, and it makes recommendations about current laws, practices and policies, about compensation for the victims of past laws, practices and policies, and about the services that are or should be available for those victims. I say that Millicent's story is a part of the Report, not that it is told in the report, because the difference is important. In citing Millicent's story the Report is allowing her words to describe a system in which speech receives no answer, or in which it is shameful, or in which it is met with lies. As a bureaucratic document the Report is striking for its attempt to give a voice to those who have not been listened to, or who have had the language in which to tell a story taken away from them. It does this by embedding in its text fragments or extended passages of stories told in the first person by witnesses to the Commission of Inquiry, usually in confidence and sometimes 'with great difficulty and much personal distress' (3). A preliminary passage says that 'throughout this report we have remained faithful to the language used by the witnesses quoted' (20); and later, speaking of the 535 pieces of evidence that the Inquiry heard from indigenous people, it claims to 'relay as many of those individual stories as possible' (21). 'Faithful' and 'relay' suggest a transparency of transcription and passing-on that suppresses the processes of writing and of the insertion of these stories into a narrative context; much of my attention in this chapter will be to the politics of writing that is worked out in those processes of citation, and in particular to the tension between the political need to speak *on behalf of* indigenous people, to lend the authority of the Human Rights and Equal Opportunity Commission to those who are unauthorized in the public sphere, and the desire directly to restore a voice to them within and as a part of the Report.

'To report' is to carry a story back from one place to another. This may be an act of commenting on something observed, or it may be the repeating to another person of something that has been said, or it may be the naming of a person to an authority as having offended in some way, or it may be the making of an official act of judgment by a teacher or an investigative body. Millicent reports to the Matron the farmer who rapes her, but her act of reporting is turned back against her. She gives witness, but she is not believed. The Report of the National Inquiry listens to witnesses and reports their words in two ways: by repeating them directly, and by turning them into the material of a larger and more comprehensive narrative.

As a speech act this 'Report of an Inquiry' weaves together a double set of enunciative relations: that which pertains to the witnesses, splitting them between an I who speaks in the present and an I who once suffered; and that pertaining to the Commission, split between a present order of exposition and a past order of inquiry. These two relations are hooked together by the temporal coincidence of the order of inquiry and the speech of the witnesses: to inquire is to listen to that speech, and the Report is the secondary articulation of those prior acts of speaking and listening. In making this double division between a present of speaking and writing and a past of reality the Report is doing what Michel de Certeau (1988: 2–3) says all writing of history does: instituting a reality by establishing a division between past and present such that the past functions as the other to the time of writing and is made intelligible by this relation. History, he writes, 'constitutes something real to the extent that it pretends to be the representation of a past reality. It takes on authority by passing itself off as the witness of what is or of what has been. It seduces, and it imposes itself, under a title of events; which it pretends to interpret' (de Certeau 1986: 203). This interpretation is characterized above all by its 'avoidance in the unifying representation of all traces of the division which organizes its production' (de Certeau 1986: 205).

In the same way, the Report deploys the citational strategy that produces, according to de Certeau, the characteristically 'laminated' text of historiography, split between a singular, coherent, continuous writing and a plural and 'disseminated' set of languages which are quoted, interrogated and judged as though they were the primary matter of the real itself. It is from the citation of this language of otherness – the chronicle, the archive, the document – that knowledge of historical reality is achieved. History is constituted by a play of languages in which

> the role of quoted language is ... one of accrediting discourse. With its referential function, it introduces into the text an effect of reality; and through its crumbling, it discreetly refers to a locus of authority. From this angle, the split structure of discourse functions like a machinery that extracts from the citation a verisimilitude of narrative and a validation of knowledge. It produces a sense of reliability.
>
> (de Certeau 1988: 94)

Citation thus works as an authorizing strategy, in which the cited texts are strictly subordinated to the text of knowledge. De Certeau explicitly distinguishes this stratification of discourse from the formal structures of 'dialogue' and 'collage'.

The 'Stolen Generations' Report displays a formal ambiguity in its commitment on the one hand to a performance of knowledge grounded in citational reference, and on the other to the attempted performance of a discursive mixing without superordination. The collaging strategy which episodically punctuates third-person reporting and analysis with fragments of stories is designed to allow these voices to have a space of effectivity, of answerability. It makes a claim for justice in relation to the voices of those who were removed from their families, who have lost their language and their traditional country and knowledges and even perhaps the knowledge that they have lost these things, and who have been shamed when they tried to report the wrongs done to them; it is a claim for but also an attempted enactment of a discursive justice. The Report both gives its witnesses a hearing (which it relays), and takes their words up into a counter-speech which is *reporting-on* a system of government to which this Government is the legal successor. But this reporting-on is not being done for the sake of shaming; it is done as a claim that a kind of listening – a response, a taking-on of responsibility – must take place. This is why the recommendation for an official apology is so central to the Report, and why the Federal Government's refusal of an apology, its refusal to assume responsibility for that earlier refusal to listen to Millicent D. and many others like her, is so shameful. The politics of this Report, then, is, like all historiographic politics, that of the fabrication and the authorization of a domain of facticity; but the Report both separates a past in which reality resides from a neutral time of writing, and at the same time refuses that separation by insisting, against the argument made in the Commonwealth's submission, that the past is not over and done with, and the present is not a pure space of presence. It is thus, necessarily, an ambiguous document: formally it has the structure of the reporting systems, the systems of close bureaucratic surveillance, that were used to record and control the lives of indigenous people, and its citational practice is not distinct from that of other historiographic or bureaucratic documents; but it seeks to turn this formal isomorphism to different ends.

IV

Millicent is told to forget about her past and her family. The speech act that governs most of the narratives of witnesses to the Inquiry is that of remembering. An excerpt from a confidential submission which acts as a preamble to the report begins: 'so the next thing I remember was …' What this 'I' remembers is not knowing: 'I was all upset', says this witness, 'and I didn't know what to do and I didn't know where we were going. I just thought: well, they're police, they must know what they're doing. I suppose I've got to go with them, they're taking me to see Mum. You know this is what I honestly thought. They kept us in hospital for three days and I kept asking "When are we going to see Mum?" And no-one told us at this time' (2). The witness's not-knowing is matched by a supposition that 'they' (the police, and perhaps also the welfare authorities) 'must know'. But what 'they' tell the witness is that the children are going to see their mother, and this is a lie. The Report cites this lie in order to deprive it of its power; it produces a knowing which is official but which is not that of the police and the welfare authorities. The witness will know the truth by giving his or her words and having them taken up into a knowledge which is a counter to the lying knowledge of the officials. The Report will offer a judgment on the basis of this knowledge; this judgment, too, will reverse the judgments made about indigenous people, and taken up by them as the truth about themselves. Listen to the memories of another witness:

> I remember my Aunty, it was her daughter that got taken. She used to carry these letters around with her. They were reference letters from the white fellas in town. ... Those letters said she was a good, respectable woman. ... She judged herself and she felt the community judged her for letting the welfare get her child. ... She carried those letters with her, folded up, as proof, until the day she died.
>
> (213)

The Canadian Government, apologizing recently to the indigenous nations of Canada for the forcible and systematic removal of their children to residential schools, said: 'We wish to emphasize that what you experienced was not your fault. It should not have happened' (*Weekend Australian*, 10–11 January 1998). Like that apology, the Report replaces a language of lies and blame, not with praise but with release; it breaks the hold of a false language, indeed a false reality, a reality which 'should not have happened'. Its function is thus performative as well as descriptive, or rather it uses a description as the basis for a performative act. It is a more public version of those letters carried around by the witness's Aunty, a more public and official letter of reference.

Within the system of systematically distorted communication into which children were rescued from their families, letters tended to go nowhere. One story that runs through the report is about letters that were never sent on, from children to parents and from parents to children. A witness who had spent her childhood in the Cootamundra Girls' Home tells one such story:

> We were all rostered to do work and one of the girls was doing Matron's office, and there was all these letters that the girls had written back to the parents and family – the answers were all in the garbage bin. And they were wondering why we didn't write. That was one

way they stopped us keeping in contact with our families. Then they had the hide to turn around and say, 'They don't love you. They don't care about you'.

(155)

Another witness, Murray, who was removed to Palm Island, says:

I remember when I learnt to write letters, I wrote to my mother furiously pleading with her to come and take us off that island. I wrote to her for years, I got no reply then I realized that she was never coming for us. That she didn't want us. That's when I began to hate her. Now I doubt if any of my letters ever got off that island or that any letters she wrote me ever stood a chance of me receiving them.

(87)

No lie is told, but the refusal to pass on letters fosters the underlying lie: your parents don't love you, your parents are dead. The systematic deculturation of the children is predicated on their social death, and indeed the death of their Aboriginality is the point of the scheme: they are protected, rescued from their blackness because, for the most part, their skins are light and they stand a chance of passing as white. Hence the comments of J.W. Bleakley, Queensland's Chief Protector and Director of Native Affairs from 1913 to 1942, on the dual value of the segregation of Aborigines on reserves run as missions: 'Not only do they [the missions] protect the child races from the unscrupulous white, but they help to preserve the purity of the white race from the grave social dangers that always threaten where there is a degraded race living in loose conditions at its back door' (73). But protection, as the Canadian Royal Commission on Aboriginal Peoples notes, is 'the leading edge of domination' (*People to People*, 1996: I, 7). Bleakley's confused metaphor which places the 'child races' at once in the innocence of an earlier evolutionary stage where they are threatened by rapacious white people, and in the degradation of a late stage which has fallen from innocence and thus threatens the sexual purity of those same white people, unconsciously reflects the double-bind structure on which the Black Gulag[2] is operated.

V

For if these children are protected by means of a social death, there is nevertheless for most of them no social rebirth. Their place is mapped out by the games that define its impossibility. The structure of the double bind runs right across the system, from the initial game in which local councils drive Aboriginal people from their camps near the towns and then declare the children thus made homeless to be neglected and subject to removal, to the prison-farm structure of the reserves in which they are to be protected, to the systematic punishment and physical and sexual abuse handed out in their best interests by the churches and 'the Welfare'. But above all it is their race that is defined in self-cancelling terms by a double negative. Millicent D. again: 'They tried to make us act like white kids but at the same time we had to give up our seat for a whitefella because an Aboriginal never sits down when a white person is present' (116). Sarah:

> We were constantly told that we didn't have families and that we were white children. It wasn't until we went across the road to school that we were called the names of 'darkies' and 'niggers' and those sorts of names. So when we were at school we were niggers and when we were at home we were white kids.
>
> (173)

And Tony says in his poignant testimony: 'I'd ask her [his adoptive mother] why I was dark. She would tell me it was because I kept playing with aboriginal kids at school' (426).

Racial identity is thus simultaneously a kind of Original Sin and a state of shame for which the children themselves are held responsible. In truth, these kids are driven crazy, and part of their craziness consists in the theft of the very language that would allow them to clarify and to state the wrong done to them. Their knowledge of their own languages is systematically eradicated, and so are the social relations that are bound up with them. Fiona, reunited with her birth mother after thirty-two years, has to speak through an interpreter, but also finds that she has lost interest in all the questions that she once thought she wanted to ask her (130). The Jawoyn Association's submission discusses the problems caused by the removed children's loss of the 'knowledge about the law; knowledge about country; knowledge about "the system"; and a social connectedness to all things Jawoyn' that allow a person to 'speak for country' (219). Again, this was the point: the children 'were to be *prevented from acquiring the habits and customs of the Aborigines* (South Australia's Protector of Aborigines in 1909); *the young people will merge into the present civilization and become worthy citizens* (NSW Colonial Secretary in 1915)' (202). What the young people become, of course, is something else. For one witness, 'what you see in a lot of us is the shell' (177); another tells the Inquiry that it is as though 'you've just come out of nowhere' (13). And Carol, who tried to document her stay at a mission reserve and was told there was no record she was ever there, says: 'I haven't got anything to say I've been to Beagle Bay. It's only memories and people that I was there with. I don't exist in this world. I haven't got anything, nothing to say who I am' (404).

VI

What the Report documents is both a remembering and an absence of memory. It is itself in one sense a ceremonial act of remembrance, reading – as its Terms of Reference require it to – the 'traces' of past 'laws, practices and policies', 'tracing' the histories of the stolen generations of children, reading their 'scars' (3). All of these metaphors, with their indexical link between past and present, suggest a reference to a pre-existing reality; but if de Certeau's argument is correct, this reference is *also* a way of instituting a reality in and for the present, and this reality is, as Krygier (1997: 96) puts it, a moment in 'a contemporary conversation'. The Report involves itself in this conversation by means of a constructive activity of writing which, rather than simply referring to a past located within its own pastness, brings together a heterogeneous assemblage of times: first, the time of enunciation, with its two-fold division between the time of speaking and the time spoken of; second, a series of chronological times from nineteenth-century Australia through to the more heavily documented postwar period and the historical present within which the address and the reception of the Report take place; and third, the diverse set of social temporalities that I referred to at the beginning of this chapter: the dominant temporality of a modernization process that, in subjecting them to global market forces, has unevenly transformed both urban and rural Australia; that of national citizenship, where changing patterns of migration have generated a multicultural doxa in which indigenous Australians have only an imperfectly defined place; that of indigenous peoples themselves, seeking both to recover a cultural identity which for many has been obliterated, and to assert a place in the national polity other than the purely negative one of dispossession and dependency; and that of a right-wing backlash, rooted in the country and the urban fringes and finding its political voice in Pauline Hanson's populist *ressentiment*, a combination of overt racism and opposition to the modernization and globalization of the Australian economy. These temporalities run at different speeds and according to different social imaginaries, and their intersections, which don't happen simultaneously, are in many respects random and contingent. In other respects they are not, and the Report documents, in particular, the clash between an assimilationist project which assumes the inevitable absorption or extinction of the indigenous population,[3] and the resistant survival of a dispossessed and disoriented people living on stolen time.

These pasts continue to exist in the present in that 'contemporary conversation' in which they are invoked to authorize a move within a serious game: the game of working out, or not working out, a just settlement between an invaded and displaced people whose dispossession has never been formally codified, and the people that now possesses the land and which has a diversity of interests including a moral interest in a just settlement ('just' within certain definite limits). As it happens, the sides in this game have been rather clearly defined around the question of history itself.

For Michel de Certeau (1997: 2), the writing of history 'aims at calming the dead who still haunt the present, and at offering them scriptural tombs'. The Commonwealth's submission to the Inquiry and Prime Minister Howard's address in May 1997 to the Australian Reconciliation Convention enunciate a number of principles governing the relation between the living and the unplaced dead. The first is a stricture concerning the relativity of systems of value: the Government submission cautions against anachronism by suggesting that in evaluating the laws, practices and policies prevailing at earlier times in Australian history 'it is appropriate to have regard to the standards and values prevailing at the time of their enactment and implementation, rather than to the standards and values prevailing today' (*Commonwealth Submission*, 1996: 30); one epoch may not judge another. The second principle concerns the absence of moral responsibility on the part of governments or peoples for actions committed at an earlier time. 'Australians of this generation', says Howard (1997), 'should not be required to accept guilt and blame for past actions and policies over which they had no control', and the discrediting of laws and policies is not a basis for the recognition of liability on the part of subsequent governments (*Commonwealth Submission*, 1996: 32). The third principle has to do with a distinction between symbolic and pragmatic fields of action. John Howard contrasts 'symbolic gestures and overblown promises' with the need to address 'the practical needs of Aboriginal and Torres Strait Islander people in areas like health, housing, education and employment'. It is these areas, rather than 'past injustices' or indeed an entire history of physical and cultural dispossession, which are 'the root causes of current and future disadvantage among our indigenous people'. Past wrongs should be acknowledged – or, to put it more precisely and in a way that does not 'apportion blame and guilt', *both* sides should 'acknowledge realistically the interaction of our histories';[4] but extrapolation from the past to the present is useless in addressing those 'practical needs'. If there is 'sorrow' for the 'blemishes' on our 'past history', it is 'personal', not collective, and is thus without effect. Implicit in these formulations is a kind of brutal imperative to stoicism: stop whining about your life and get on with changing it.[5] More fundamentally, the Government's position depends upon a historical relativism which seals past and present in their separate and internally homogeneous temporalities; in the words of the columnist Frank Devine, the 'eternal truth' in play here is that 'that was then and this is now' (*Australian*, 15 January 1998). This stark division between the present and a series of self-contained pasts repeats a classically historicist position for which each period is, in Ranke's words, 'immediate to God'. Its purpose is to render the past at once quite strange and quite inconsequential: 'These ghosts find access through *writing* on the condition that they remain *forever silent*' (de Certeau 1988: 2).

The critique of historicism that Gadamer develops in *Truth and Method* emphasizes its theological character: it is only from the perspective of God that each historical time can be seen in its completeness and its separateness from every other time. But our encounter with the past, he argues, is not a relation between two isolated points; it is an encounter with an open-ended process *of which it is itself a part*.[6] One of the images through which the Report imagines this intrication of historical times is that of *resonance*, the passage of a succession of overlapping sound waves outwards in echoing repetition from a point of departure. 'The actions of the past', it proposes, 'resonate in the present and will continue to do so in the future.' And it goes on, in what is clearly a response to a criticism of the uselessness of historical knowledge, to say:

> In no sense has the Inquiry been 'raking over the past' for its own sake. The truth is that the past is very much with us today, in the continuing devastation of the lives of Indigenous Australians. That devastation cannot be addressed unless the whole community listens with an open heart and mind to the stories of what has happened in the past and, having listened and understood, commits itself to reconciliation.
>
> (3)

The argument has to do with establishing that continuity of will and responsibility that defines the self-identity of the nation state. The authority invoked by the Report in its support is the Governor-General, Sir William Deane, who – against the Commonwealth's attempt to turn away from 'symbolic gestures' in order to address the 'practical' issues of the present (a turn which, as Gaita (1997a: 48) notes, 'treats as irrelevant the fact that the Aborigines are landless because they were dispossessed rather than because of a natural catastrophe') – calls for a form of historical settlement which would be the precondition for policy measures to take effect, and which would involve the rendering of discursive justice. He thus stresses the need for 'appropriate redress for present disadvantage flowing from past injustice and oppression' (4), one element of which – the Report concludes – should be a formal apology made to indigenous peoples by all Australian parliaments.

VII

But the act of apology is ambiguous. Etymologically the word refers to a speech given in one's own or another's defence, and only secondarily does it have its modern meaning of acknowledgement of a wrong. The definition that comes closest to this modern meaning is a subsection of the *OED*'s: 'a frank acknow-ledgement of the offence with expression of regret for it, by way of reparation'. There are three parts to this: something happened; it should not have happened; by saying so, I make it up to you. The apology is thus a complex performative which seeks to de-institute a past reality 'which should not have happened' and to transform the conditions of a present relationship. But given the semantic ambivalence that connects self-protection with concession of error, there can be no guarantee that this is what will occur: an apology may be a way of acting symbolically which feigns weakness in order to defend or even to strengthen power. 'Forgive me', it says, 'I was wrong': the apologist gains honour, and nothing changes.

Like all gifts, the apology thus has the potential to work coercively (Frow, 1997: 102–32). Tavuchis (1991: 34–5) stresses its non-reciprocality by calling attention to 'the morally asymmetrical positions of the protagonists, the essentially symbolic character of the transaction, and the unpredictability of the outcome'. The crucial question then – since any speech act may be inefficacious[7] – becomes that of the conditions under which a historical apology will work, and of what that might mean. This is above all a question of the relation between different and incommensurable kinds of costs and values: between rhetorical penance and actual humbling, and between discursive and material costs. Norma Field's careful argument about Japanese apology for the war and particularly for the enforced prostitution of Korean 'comfort women' points to an inverse relation between apology and material compensation: a good apology must include effective reparation, but material cost must be subsumed within a real symbolic cost to pride, rather than being either a payment for services rendered or a non-committal expression of sympathy – both of which effectively undo the apology. In the case of a bad apology, it is 'as if the words themselves were simulating money' (Field, 1997: 13). Material compensation must thus be secondary to symbolic reparation, but it is only when the apology gives rise to a chain of consequences that it can be said to be successfully accomplished. Past realities are not changed in and for the past, but they can be changed in and for the present; 'apologies are made *to* the victims of past wrongdoing but *for* the shared present of victims and apologizers, and most of all, for the sake of a common future' (Field, 1997: 37).

Apology is one of the forms taken by discursive justice, but it has two more elementary forms. The first is the reception and recording of testimony, which the Report recommends should be continued by the establishment of an archive similar to the Shoah Foundation's project of recording the victims of the Holocaust. The second is that 'listening' which the 'whole community' must undertake and for which the process of the Inquiry as it heard from witnesses in each state was a metonym. Listening is a form of ethical responsiveness which recognizes a duty to the story of the other (Terdiman, 1992). The model of storytelling that holds in the Report is that of a narrative catharsis triggered by the release of memories. If the experience of forcible removal is repeatedly referred to in terms of 'scarring' and 'trauma', the process set in train by the recording of and attention to the testimonies of the victims is one of healing: 'The experience of the Shoah Foundation and of this Inquiry is that giving testimony, while extraordinarily painful for most, is often the beginning of the healing process' (22), which extends from those directly affected to the larger trauma of the body politic.

The almost unspeakable word here is 'genocide'. While seeking to remain strictly within the legal framework of the time and to avoid a retrospective moralism, the Report nevertheless concludes that a principal aim of the child removal policies was the elimination of indigenous cultures, and that in the sense given the word by the relevant international convention this aim constitutes genocide.[8] The Government, by contrast, has consistently refused the applicability of the term (Stekete, 1997). Gaita (1997b: 45) acutely observes the puzzling absence of any call, from either the left or the right, for those guilty of genocide to be brought to trial; trials are literally unthinkable, and that they are so, he concludes, 'is the most persuasive evidence that the significance of the crimes against the Aborigines has not been fully appreciated'. But the point is more than a legal one: the assumption made by the Report is that collective acknowledgement of a nation's past criminality – and this means, in the first place, naming it – is essential to something like the honour or the moral integrity of nationhood. The making of reparation to the victims, both materially and discursively, is equally a repair of the wounded body politic. There is also a sense that, in the opening of memory and the restoration of a voice to the dispossessed, a kind of redemption can take place, a cathartic release from the pain of damaged lives.

But perhaps there can be no redemption. Speaking of the 'first-order narratives' of the survivors of the Holocaust, Jay (1992: 104) writes that, unlike the 'second-order' discourse of the historian who seeks to make sense of them, they 'must approach a kind of incoherence because of the fundamental unintelligibility of what happened to them'. And Langer (1991: 83) coins the term 'humiliated memory' to describe 'an especially intense form of uncompensating recall' amongst Holocaust survivors, a form of remembering which, far from restoring a sense of power or control over the past, torments the survivor, 'reanimating the governing impotence of the worst moments in a distinctly non-therapeutic way', and refusing to lend itself to the ennobling uses of history.

VIII

The writing of history, says de Certeau (1986: 199), takes place midway between two poles: one of them, which he calls dogmatism, 'is authorized by a reality that it claims to represent and in the name of this reality, it imposes laws'; the other, which he calls ethics, 'is articulated through effective operations, and it defines a distance between what is and what ought to be. This distance designates a space where we have something to do.' It is the interstitial placing of historiography that is important here: its ethical function, which has to do with the time of writing and the contest of forces within and for that time, will have little purchase if it is detached from a sense of the irreducibility of the past to this writing, and if it thus projects too easy a redemption of stolen time. Elsewhere de Certeau (1986: 217) speaks of the task of history as being that 'of articulating time as the ambivalence that affects the place from which it speaks and, thus, of reflecting upon the ambiguity of place as the work of time within the space of knowledge itself'. This, I take it, is – with all necessary qualifications – not dissimilar to Gadamer's (1972: 283) argument that genuinely historical thought must include its own historicity as a component of that history it seeks to understand.[9]

This ambiguity, and this inclusion of one historicity within the objectification of another, are what has here been called 'resonance', and de Certeau has quite explicitly drawn out its implications for the use of orally relayed stories. Instead of a transcription and exorcism of those voices 'whose disappearance was formerly the condition of historiography', the historian may learn to listen to them, and so to discover 'interlocutors, who, even if they are not specialists, are themselves subject-producers of histories and partners in a shared discourse. ... A hierarchy of knowledges is replaced by a mutual differentiation of subjects' (de Certeau 1986: 217). This, again, is perhaps too easy. There are no solutions to the theft of time; nothing gives it back, nothing redeems the lie that systematically falsified the world of the stolen children, and language is never equally shared. But a history which is rigorously committed to ambiguity may open up, as the 'Stolen Generations' Report does, a space of listening which will define that ethical distance that gives us 'something to do', and it will do so by unsettling its own enunciative relation to the disparate voices and the heterogeneous pasts that structure it. 'Time is precisely the impossibility of an identity fixed by a place. Thus begins a reflection on time' (de Certeau 1986: 218).

NOTES

1   All further citations from this Report are given as bracketed page numbers in the text.
2   A term I use advisedly. For documentation on the control and confinement of indigenous people, cf. Austin (1997), Haebich (1992), and Kidd (1997).

3   There is detailed documentation of this project, and of the languages of primitivism and of eugenics in which it was expressed, in McGregor (1997).

4   Cf. Luke (1998); and Gaita (1997b: 45): 'How extraordinary ... that this government and much of the right treat reconciliation as though it were substantially a two-way affair, as though we had something for which to forgive the Aborigines.'

5   The argument is made explicit in Brunton (1997).

6   'Alle Begegnung mit der Sprache der Kunst [ist] Begegnung mit einem unab-geschlossenen Geschehen und selbst ein Teil dieses Geschehens' (Gadamer, 1972: 94). Gadamer's formulation refers to the historicity of the work of art, but it can be extended more generally to the historical event, and indeed its implicit referent is surely the Holocaust.

7   Cf. the argument made by Butler (1997: 16–19), that inefficacity is a necessary condition of performatives.

8   The 1948 Convention on the Prevention and Punishment of the Crime of Genocide was ratified by the Australian Government in 1949.

9   'Ein wirklich historisches Denken muß die eigene Geschichtlichkeit mitdenken.'

4

FROM TIME IMMEMORIAL

Narratives of nationhood and the making of national space

Nuala C. Johnson

In the opening chapters of C.S. Lewis' *The Lion, the Witch and the Wardrobe* Lucy's great difficulty in convincing her siblings that she entered the land of Narnia through the wardrobe partly revolves around the issue of time. Bemused and worried by her claims, her brother and sister consult the professor, presenting the logical inconsistencies in Lucy's claim to have spent hours in the company of a faun in the country of Narnia, while less than a minute of time had elapsed in the house the children were exploring. The canny professor interprets the seeming inconsistency as a mark of authenticity rather than deceit or madness 'if, I say, she had got into another world, I should not be at all surprised to find that the other world had a separate time of its own; so that however long you stayed there it would never take up any of our time' (Lewis, 1997 [1950]: 48). The prospect that other worlds operating under different regimes of time could exist startles, intrigues and captivates the children of Lewis' imagination and the reader of this tale. But it is not just in the literary text that a suspension of disbelief around the axioms of time has been activated.

To think about time for many of us is paradoxically to think both in our own time and yet to attempt to step outside of it. Historians, museum curators, film-makers, biographers, to name but a few, are acutely aware of the dilemmas attendant on narrating and imagining times other than the one in which they are located. Indeed our descriptive lexicon for dealing with time, as geographers are well aware, is intimately bound up with a vocabulary of space, one which is powerfully captured in the title of David Lowenthal's book *The Past is a Foreign Country*. For nationalist movements times past may not be so much foreign countries as resources from which cultural imaginations can be excavated. These excavations, however, take many forms and in this essay I seek to underscore the fluid ways in which interpretations of time's passage become the idiom and the accent of a nation's trajectory.

NUALA C. JOHNSON

Putting shape to the past

In a recent book *The Shape of the Past* the philosopher Gordon Graham presents a defence of the conceptual validity of what he refers to as philosophical history, perhaps more commonly known as universal history. Graham takes seriously the objections raised by scientific-academic historians, particularly to the *apriorism* that appears to be embedded in any attempt to provide a grand shape to times past, that is to say the presupposition that the past has a shape. He presents a compelling case for how historical narrative (apart from the type of accounts which exclusively relate the past as first this, and then that and persistently confuse history with mere chronology) invariably confronts questions of a hermeneutic variety which are not solely answerable to brute facts of history. He speaks of the meaning of history beyond its atomistic particulars. Questions such as the *emergence* of democracy, the *triumph* of liberalism, the rise of modern science or indeed the *genesis and spread* of nationalism all necessitate some engagement with questions of interpretation and judgement as much as they do with hard evidence. The need to be hermeneutic, to be an interpreting creature, arises out of what Gadamer (1975) has referred to as 'the alienation of historical consciousness'. For Graham even the present concern with postmodernism invokes a grand historical classification where postmodernity is predicated on the existence of a period of modernity. He claims: 'it is quite implausible to understand by "modernity" and "postmodernity" simple cultural or political events; they are, rather, eras or epochs which are to be identified and character-ised by the ideas which dominated them' (Graham, 1997: 17). Whilst one does not have to accept the thrust of Graham's thesis in defence of a universal history, the dominant shapes of the past which he identifies as progress, decline, recurrence and providence have had a profound impact on our conception of time's role and the theoretical schema underpinning many of our accounts of the past. In particular in nationalist discourse these four shapes have had a prominent influence in structuring explanatory frameworks. In this essay I shall elucidate the ways in which progress, decline, recurrence and providence have anchored many discussions of nationalism and the sources of national identity. While these four shapes of the past regularly overlap and intermix in studies of nationalist discourse, for the purposes of clarity in this essay I will treat them individually. In other words, though these four discrete categories of time are being used heuristically to elucidate how they surround ideas of nationhood, for many particular nationalist movements more than one shape of the past can be adopted to define national space and national identity. The typology, to put it another way, is suggestive rather than exhaustive.

For each of the four shapes I will briefly elucidate their relevance through a specific case-study. First, in the instance of progress I will examine Turner's thesis regarding the frontier and the evolution of a unique American identity. Second, in the case of decline, the example of linguistic nationalism and the mapping of decline of language will be analysed. Third, the ideology of the retrieval of the Volk in Nazi and conservation thinking in Germany will be discussed. Finally, the invocation of providential time and sacred space will be dealt with in relation to Israel. Before moving on to this discussion however, I wish to pause for a moment to briefly overview the emergence of clock and calendar times as these are the principal vehicles for measuring time and they particularly emphasise time's relationship with space. Their structure also connects with the four shapes developed by Graham.

Clocking time

While the earth's rotation within the solar system may be the largest timepiece we possess (space calibrating time), regulating the length of day and night and seasonal transformation, the scheduling of time through mechanical devices is closely linked to the desire to order life according to the precepts of a Christian God. As early as the sixth century Benedictine monasteries began to divide the day into hours to enable monks to devote themselves to prayer as much as possible. These early schedules dividing the day into abstract units of time gradually emerged as significant tools in the regulation of everyday life (Zerubavel, 1981; Mumford, 1934). While early regimens had to be varied to correspond with the changing length of daylight, the introduction of mechanical clocks superseded this schedule and they became the first devices to mark abstract units of time accurately. Their first appearance in the monasteries and cathedrals of Europe in the fourteenth century marks a clear link between metaphysical considerations of time, Christian theology and religious practice.

While mechanical clocks could measure time independent of the movement of the earth within the celestial sphere, they encouraged the view that any regular motion could serve as a measure of time. The facility of the mechanical clock to offer precise determinations of time and the length of time that had elapsed over a given period, encouraged the use of spatial metaphors to describe lengths of time. This ability to mark time with some accuracy provided the basis for labour disputes in the textile industry of the fourteenth century, where employers substituted piece pay for time-based pay which bred conflicts over the accuracy of timekeeping devices. The economic conflicts about time, however, were interlinked with the theological. St Augustine's view of time as the divine creation of the soul, where salvation could be secured through the maximum contemplation of God's eternity by prayer and meditation, represented the direct intersection of the economic and spiritual worlds. The question of usury that occupied the minds of theologians in the thirteenth and fourteenth centuries directly confronted this issue; to collect interest on money just because time had elapsed was regarded as theft because the usurer steals time which belongs to God. Thus, while the clock offered an abstract measurement of time, even with the chronometer the meaning of time was socially mediated through debates over the questions of profit, labour, investment and divine contemplation. The issue of Sunday opening for commercial transactions today continues to revolve around some of these types of issues.

In the end, though, the clock can mark time's passage but 'timepieces don't really keep time. They just keep up with it, if they're able' (Sobel, 1998: 35). Clock time is circular time, constantly rotating around a set of numbers in a cycle to which it constantly returns. By contrast calendar time is square time, comprising a series of boxes that represent individual days which can never be recovered. In this sense 'Calendar time has a past, present and future, ultimately ending in death when the little boxes run out' (Duncan, 1998: xvi). The linearity of one sort of time measurement and the circularity of the other parallels Graham's taxonomy of history's shapes; progress and decline mark time on a linear axis while recurrence marks time in a cyclical pattern and providential approaches can incorporate all three shapes of the past. In discourses of nationhood linear and recurrent time formulations have co-existed and at times competed in particular contexts. Dominant explanations of the rise of nationalism and its spatial manifestation through the nation state have consistently employed grand historical frameworks which have been underwritten by a tension between ideas of temporality and the partitioning of space into discrete territorial units.

The relationship between time and space is perhaps most tellingly revealed in the efforts to determine longitude. While nature's design provides an adequate vehicle to identify latitudinal position through an examination of the position of celestial bodies, the longitude meridians are governed by time and the difficulty of measuring these meridians, in the absence of an accurate clock, left many a sailor literally at sea until the late eighteenth century. The heart of the problem resided in the necessity to know both the precise time on board a ship at sea and also the time at another place of known longitude at the same moment. The navigator would then be enabled to convert the time difference into geographical separation, with distance and time translating into degrees of longitude. Time and space were umbilically connected. Clockmakers, astronomers and a variety of other specialists spent at least the best part of a century trying to resolve this conundrum. Unlike latitude, where the position of the sun in relation to places on the earth determines the zero-degree parallel 'the zero-degree meridian of longitude shifts like the sands of time' (Sobel, 1998: 4). That it currently resides at Greenwich Mean Time, passing through the heart of London may reveal the historical geography of European power struggles for control over the prime meridian, but the location itself has no particular advantages where time itself is concerned. The longitude question, then, crystallised debates about time, space, technology and politics. The question of national space similarly engaged with these types of issues. In the remainder of this chapter I will examine time and the metastories surrounding nationalist discourse through the four shapes of the past identified above. I shall rehearse them through specific examples of nation-forming episodes. The first of these, the notion that time measures the progress of the nation, is the matter to which I now wish to turn.

Time, progress and the American frontier

> For a moment, at the frontier, the bonds of custom are broken and un-restraint is triumphant. There is not tabula rasa. The stubborn American environment is there with its imperious summons to things to accept its conditions; the inherited ways of doing things are there also; and yet, in spite of environment, and in spite of custom, each frontier did indeed furnish a new field of opportunity, a gate of escape from the bondage of the past.
>
> (Turner, 1894)

The conception of the past as an exercise in human progress has had a diverse range of adherents over the centuries. The belief in progress, however, won powerful intellectual and popular respectability in the nineteenth century. Thinkers such as Hegel and Marx formulated progressivist universal theories, early anthropological research operated under a progressivist agenda where indices of tradition and modernity could be charted across space, imperialist political and economic projects were regularly underwritten by a resounding commitment to progress, and the driving force behind much reform in domestic educational and social policy arose from the desire to advance a more equitable and better society. While twentieth-century historians have distanced themselves from the 'Whiggishness' characterising some of these approaches to the past (Butterfield, 1979 [1933]), the popularity of progressivist perspectives continues to inform some forms of economic and social analysis (see for example Fukuyama, 1992). Progressivist accounts conventionally employ some idea of temporal sequence and a principle of evaluation where change from worse to better can be determined. Graham points out however that 'Progressivism does not have to invoke anything in the way of a teleology, or grand design' (Graham, 1997: 50). It can, at its simplest, act as a descriptive account of the past where there is 'perceptible, recountable, steady progress from worse to better' (Graham, 1997: 50). Progressivism can be divided into three distinct kinds: first, there are analyses which employ a straightforward linear progress that is continuous and whose direction is constant; second, there are evolutionary theories of progress where overall progress occurs but one which is tempered by intermittent periods of decline or reversal of fortune; finally, there are revolutionary theories of progress where over long periods of time there may be no steady pattern of progress but the direction of the past is marked by intense periods of revolutionary change where sharp advances are achieved.

One of the finest formulations of an evolutionary theory of change and in this case the pattern for nation-building and the constitution of a national identity in America is found in Frederick Jackson Turner's frontier thesis published in 1894 as *The Significance of the Frontier in American History*. While historians have exhaustively examined his theory to assess its robustness as an explanatory paradigm for the development of a national character (for a comprehensive overview of this debate see Gressley, 1958), geographers have been especially interested in the geographical or spatial component of his hypothesis both as a legitimation for their discipline and as a key organising principle for geographical research, particularly in the United States (see Zelinsky, 1973; Eigenheer, 1973–4; Gulley, 1959; Block, 1980). My purpose, however, is not to adjudicate on the merits of the thesis *per se* but to examine the links between time and space developed in this theory and to comment on the universalistic and particularist threads that interweave in this progressivist analysis of nation-building. While Turner may have conceived of his project as a modest exercise in scientific history, outlining the dynamic of settlement along the American frontier, the structure of his account represented the frontier 'as process [which] stood universal and omnipotent' and it answered the key question in Turner's thinking '"How does civilisation march?"' (Coleman, 1966: 23). While seeking to propose an explanation for the particularity and historic specificity of the development of American national identity, the theory that he employed domesticated the intellectual and universalising principles of evolutionary science and human geography. It achieved this by proposing that the unique characteristics of the unsettled West, the frontier, provided the vehicle through which a peculiarly American form of identity – characterised by individualism, democracy and nationalism – flourished over time. According to Turner 'The frontier is the line of most rapid and effective Americanization' (1962 [1894]: 3–4).

The concept of a frontier dynamically engendering cultural change, however, was activated by a specific set of ideas centred around the potential of humans to experience evolutionary progress through the biological metaphor of the social organism. Drawing upon the evolutionary biology popular in his day (and in particular a Lamarckian interpretation of evolution), Turner employed the notion of evolutionary social stages in his thesis. The temporally sequential occupation of spaces along a frontier – moving across the meridians so to speak from east to west – each contributed to the overall constitution of an American national character. Coleman observes that (1966: 30) 'Hence American frontier history, perhaps indeed all historical development, advanced along the scale of evolutionary stages, moving from hunting, to settlement, and finally to urban manufacture.' While each encounter with a new frontier environment may necessitate a temporary and temporal reversal of progress the human and social organism would eventually adapt to these conditions and produce a uniquely American culture. Time then was calibrated by changes in adaptation to new environments. Turner claimed: 'The history of our political institutions, our democracy, is not a history of imitation, of simple borrowing; it is a history of the evolution and adaptation of organs in response to changed environment, a history of the origin of new political species' (Turner, 1962 [1896]: 205–6). Unlike European nation-building projects, where the relationship between ideas of culture and political organisation in the nineteenth century frequently revolved around questions of a common ethnicity, for Turner American national identity could not invoke a common cultural inheritance but it could appeal to a common experience of successive adaptation to environmental conditions by settlers along the frontier. From the seventeenth through to the late nineteenth century, American nationhood was continuously following a trajectory of evolutionary progress along the American West, which Turner regularly projected through the myth of the great forest. The simultaneous deployment of a scientific and romantic discourse to account for the development of America's political institutions and national character rendered his thesis intensely seductive to contemporaries, by appealing both to well-rehearsed commonsense arguments about the relationship between human societies and their environment in parallel with a highly romantic rendition of the virtues of possessive individualism.

While Turner's thesis ultimately promoted a theory of change which focused on American exceptionalism, the intellectual antecedents predating his frontier hypothesis therefore relied much more heavily on some of the universalistic principles of natural science and evolutionary biology in particular. From the viewpoint of time and space, Turner's conception of the development of American nationalism inextricably conjugated time around space. Spatial categories – the West, the frontier, the wilderness – all became synecdoches of temporal progress. Change occurred fundamentally through human encounters with the spaces of nature and nature's timepiece in evolutionary terms regulated the pace and type of change. For the biologist this might translate into evolutionary stages of development through the progressively adaptive qualities of organisms and through scientists' progressively adaptive tools for monitoring and making sense of evolution. For the historian like Turner, armed with the tools of the map, the census report, and eye-witness account of the frontier, the adaptive evolution of the human species as a social organism was being observed as the frontiersman (*sic*) gradually evolved to develop a unique set of political institutions, cultural traits and national characteristics which overcame the superficial (or skin deep?) differences of ethnicity and responded to the precepts and guiding hands of nature. Time across space precipitated this evolutionary change.

That the environment did not impress itself so hopefully on the native peoples of North America was not a part of Turner's calculus. Perhaps his ultimate aim to account for a particular form of American nationalism ironically precluded consideration of the logical consequences of his thesis *vis-à-vis* America's indigenous population. Why native peoples along the American frontier did not display the same evolutionary impulse as white European settlers is not directly accounted for by Turner. Maybe one can only speculate that a racial hierarchy of the human species within the evolutionary chain was implicit to his thinking. No less than progress, however, decline and in some cases complete collapse, has also provided the guiding principle underlying some discussions of nationalism.

Linguistic nationalism and the trope of decline

> Has nationality anything dearer than the speech of its fathers? In its speech resides its whole thought domain, its tradition, history, religion, and basis of life, all its heart and soul.... With language is created the heart of a people.
>
> (Herder, 1783)

The vocabulary of decline has frequently attended debates surrounding the relationship between language and national identity particularly in a nineteenth-century European context. A discourse of decline has animated both popular and intellectual views of the past, often supported by the belief in a Golden Age found in the trajectory of times past. The Book of Genesis, for instance, suggests that there once was a Garden of Eden followed by 'the Fall' from Grace; similarly Rousseau's *Social Contract* begins with the contention that humans were born free, although they now find themselves enslaved by chains. While we frequently conceive of decline in negative terms, as a form of degeneration and loss associated with the passage of time, there is no necessary link between decline and regret. Decline in the belief in magic, for instance, may be welcomed as might a decline in the incidence of infant mortality (Graham, 1997).

In nationalist discourse the role of language in the stimulation and maintenance of a national community is well documented. Inspired by the Romantic movement and especially by the writings of Johann Gottfried Herder (1744–1803), language came to be seen as an expression of both individual and collective identity, the external badge which would differentiate one 'nation' from another. Indeed Anderson (1983) locates the rise of nationalist thinking with the emergence of print capitalism and the proliferation of print cultures in the vernacular once the market for a 'universal' language, Latin, had been exhausted. Hastings (1997) rather tellingly suggests however that capitalism alone provides an inadequate explanation for the rise of printed vernaculars and claims that a strong case can be made to support the contention that 'the single most effective factor was the desire of many Christians, clerical and lay, to translate the Bible or produce other works conducive to popular piety' (Hastings, 1997: 22). Hastings' argument is particularly relevant for the role of the translation of the Bible in effecting the awakening of a national consciousness in Protestant Europe, where vernacular religious literature was used in a regular way in the service of a state church as it has more recently in the consolidation of African nationalisms.

For Catholic Europe, where Latin remained the language of religious worship, the impact of this argument is considerably diminished. But a vernacular language could serve an important function in the cultivation of an idea of nationhood in other important ways. Firstly it could create a temporal link between language communities of the present and those of the past. Time then could be conceived as a manifestation of authenticity. In the days preceding the printing press manuscript sources written in the vernacular provided empirical confirmation of the existence of specific language over time, albeit in a modified form from contemporary usage (e.g. Old English, Medieval English, Modern English). Secondly the emphasis on vernacular languages could create a link across space between people living in different places but sharing a common language. It could offer a territorial basis to national identity which simultaneously linked time and space, connecting orality with the written word (Fishman, 1972). The introduction of standardised grammars, received pronunciation and common spelling at least in principle, if not in practice, established clearer connections between oral and written cultures. As Fishman (1972: 77) observes:

> With the passage of time, and with control over media and institutions of society, it [language planning] converts the new into the old, the regional into the national, the rural into the urban, the foreign into the indigenous, the peripheral into the central and merely efficient into the authentic.

The preservation and cultivation of a specific vernacular thus could be regarded as imperative in the nation-building practices prevalent in nineteenth-century Europe.

If a vernacular language can be imagined as a legitimate basis for cultural identity, the decline of a language aroused widespread concern about the potential decline and ultimate death of a nation. The trope of decline has aroused nationalist movements from the Basque country and Catalonia in southern Europe to the Celtic regions of northwestern Europe. For these movements time became an index of the gradual annihilation of particular linguistic communities across space. Census enumeration data became the measure of the temporal and spatial pattern of decline and this decline could be mapped.

In mid-nineteenth century Ireland, for instance, the intellectual and cultural nationalist Thomas Davis claimed that 'a people without a language is only half a nation' (quoted in Fishman, 1972: 49). His remark was influenced by the Romantic tradition of which he was a part but was equally prompted by the visible decline in usage of the Irish language. While language decline had been gradually occurring since the seventeenth century, the nineteenth witnessed a substantial substitution of Irish for English brought about by the impact of the Great Famine, the institution of state-funded primary education in 1831 and the accelerating integration of Ireland into the British space economy. The regional pattern of decline corresponded broadly with an economic east/west polarisation of the island, with the west representing the cradle of Gaelic speakers. Consequently time and space became inextricably interlinked as linguistic decline came to be measured in spatial terms and nostalgia for a period before the impact of Anglicisation began to be mapped on to the western seaboard of the island. For a mid-century intellectual elite the west came to represent the 'unspoilt beauty' of virgin territory, where the influences of modernity were at their weakest (Brown, 1981). Landscape artists began to gravitate to these regions to capture their quintessential features lost to modernity in other parts of the island and '[to] search for unspoilt scenery and rural life beloved of Realist painters' (Sheedy, 1981).

Antiquarians, anthropologists and creative writers all cultivated an image of the western seaboard as a space located back in time. This trend led one commentator to suggest that the 'western island came to represent Ireland's mythic unity before the chaos of conquest: there at once were the vestige and the symbolic entirety of an undivided nation' (Foster, 1977: 65). Continental scholars working within a framework which suggested that the western seaboard of Ireland represented one of the last 'outposts' of the continent came in search of the final remnants of an archaic Indo-European culture. The apparent physical isolation of the region and the continued use of the Irish language led them to think that the most ancient features of a Gaelic culture, now in decline elsewhere, might still be visible in this place. Anthropologists, in particular, devoted much energy towards studying the 'real peasant' culture, a popular theme in functional anthropology where the dichotomies of tradition and modernity seemed to be most clearly in evidence.

Coupled with this theme of temporal decline, however, was the equally influential trope of revival and renewal. In the late nineteenth century revivalist movements took hold in Europe especially in areas where decline was seen to be most marked. Hutchinson (1987: 117–18) declares that 'these cultural projects were led by secular intellectuals in contact with the international community of Celtic scholars – Rhys in Britain, de Jubainville in Paris and Zimmer in Germany'. The alliance of antiquarians, secular scholars and nationalists gave an intellectual coherence to the image of declining cultures and provided the impetus for revivalist movements. In Ireland the Gaelic League, established in 1893, sought to reverse the trend of Anglicisation and restore the language to the centre of cultural and political life across the island. Its first president, Douglas Hyde, captured the spirit of fatalism regularly associated with decline when he claimed that:

> The moment Ireland broke with her Gaelic past, she fell away hopelessly from all intellectual and artistic effort. She lost her musical instruments, she lost her music, she lost her games, she lost her language and popular literature and with her language she lost her intellectuality.
>
> (Quoted in Ó hÁilín, 1972: 96)

While the League set about its task of reversing the process of language shift in oral culture that had taken place over the previous three hundred years, its efforts were matched by a literary revival, which sought inspiration from Celtic antiquities. Together these movements transposed shifts over time and expressed them as shifts over space. They played a significant role in articulating an Irish 'imagined community' and allowed the west of Ireland to act as the synecdoche of that imagination. We must acknowledge, however, that much of the stimulus for this process emanated ironically from an urban-based environment. Thompson (1967: 66) rightly asserts that:

> The Gaelic League was the strongest and most popular [society] for the simple sociological reason that its celebration of folklore appealed to the intellectuals while its moralistic rejection of civilisation, decadence and empire appealed to the Catholic lower middle-class.

In the context of universalistic theories of decline it is towards the question of morality that some of the most doom-laden prophesies of decline have been articulated. Graham (1997: 86) suggests that 'as well as the belief that specific moralities and sets of values have declined, social theorists from Nietzsche to MacIntyre have postulated the inner collapse of morality itself'. For nationalists too the outer manifestation of decline expressed through linguistic contraction often mirrored an inner decline in the authenticity of the nation itself. Time marked this descent from cultural and moral effervescence to cultural and moral morbidity, occasionally prompting efforts at cultural retrieval and renewal. This connects well with our third shape of the past – recurrence – one which has had a significant influence on both the practice of the rituals of national identity and in the articulation of nationalist ideology.

Tradition and the cycle of recurrence

The concept that the past comprises a series of historical cycles has antecedents in ancient Greek thought, although the Italian theorist Giambattista Vico (1668– 1744) is most closely associated with the full exposition of the idea of history as cyclical. Dismayed by the hegemony of Cartesian science and mathematics, Vico sought to offer a view of cultural history which emphasised a perpetually recurring pattern. He claimed that 'Our Science therefore comes to describe at the same time an ideal eternal history traversed in time by the history of every nation in its rise, progress, maturity, decline and fall' (quoted in Graham, 1997: 146). While strictly cyclical versions of this theory necessitate a return to the same point (e.g. economists' views of cycles of boom and bust), recurrence does not necessarily imply the return of an individual set of historical conditions but the recurrence of the general pattern under a variety of different circumstances: 'What is recurrent is not the civilisation itself, but the pattern which the histories of otherwise disparate civilisations all exhibit' (Graham, 1997: 147).

In terms of conceptions of time, cyclical views of the past can be seen to be analogous with clock-time where there is a return to the same point in the clock's dial or to the same point in the rotation of the earth, whereas recurrent theories employ a seasonal analogy. The past corresponds with nature's seasons – spring, summer, autumn, and winter – each season returns but each year is not the same year. Seasonal time then has both elements of cyclicity and linearity where there is a one-digit annual change to the calendar. Organismic analogies of time marking the course of civilisation have enjoyed particular popularity in the social sciences and history where the past is likened to the life course of birth, maturity and death. In political geography stadial organic theories of the state enjoyed considerable popularity in the earlier part of this century. It is in the arena of nationalist political thinking and practice, however, that the imaginary of recurrence has had the most profound influence. The cultivation of tradition, the re-enactment of ritual, the exploitation of symbol and the anchoring of myth have all contributed to the cult of politics associated with nationalism and especially its more extreme forms in the twentieth century.

It is in the area of Volkish retrieval in Nazi ideology that one encounters a clear manifestation of the idea of time marking the regeneration of the national soul through the trope of return. It is with respect to nature conservation and the environmental movement that the intersections between National Socialism and the recultivation of a rural identity were clearly articulated. Both the Nazis and nature conservationists shared a desire for national regeneration and throughout the First World War conservation publications repeatedly stressed that it was in the unspoiled homeland of Germany's natural environment that the fighting spirit of the people resided. On Germany's defeat in the war the naturalist Konrad Guenther claimed that the country's collapse could be accounted for by the break in the close relationship between soil and people. He alleged that 'the ground upon which a people stands can only be the soil of the homeland. The more firmly it is rooted, the harder it is to uproot. ... But what does one love in the homeland? ... One loves Nature!' (quoted in Dominick, 1992: 87). Drawing from nineteenth-century parallels, Guenther warned that if the breach between the German people, nature, the customs and traditions of rural life, were not restored Germany would quickly become the cultural fodder for other European nations. This articulation of the relationship between 'blood and soil' found strong adherents within National Socialism and provided part of the rationale for Nazi propaganda on nationhood. In particular, the rural peasantry provided the cultural backbone of Germanness because the national character developed in close interaction with the land. The physical and ideological tie with the soil differentiated this group of people from both the degenerating effects of city life and the nomadic 'character' of the Jews.

From this anthropocentric view Nature provided the vigour for the retrieval of the German soul and conservation would be paramount to this process. For Guenther the element of Nature most closely connected to this soul was the German woodland. Otto Großjahn, director of the Alliance of German Natureschutz, claimed that 'A beautiful and rich Nature is the original source of all the people's powers, and the protection of Nature is the touchstone of the heart of the Volk' (quoted in Dominick, 1992: 87–8). In promoting a case for conservation then the environment in which the Volk would be best retrieved was the 'primordial homeland' (87). The blood and soil rhetoric served to harness a diverse set of complaints about modern society. Rejecting at one level a progressivist view of history and society, the Nazis 'extolled the virtues of a rooted peasantry while it condemned the erosion of tradition, the collapse of the social order ... the destruction of the natural environment' (93). This viewpoint was held by Richard Walther Darré, Minister for Agriculture under the Nazis and Alfred Rosenberg, Director of the Fighting League for German Culture. The latter, deeply concerned about the degenerative effects of metropolitan life and modernity, defended against the importation of ideas, architecture, music or other cultural practices from outside Germany. He claimed that despite the move towards internationalism 'the greatest event of our time is the emergence of the new consciousness of the primordial deutsch-German [sic] values of the soul, the new discovery of the deep-rootedness of every culture' (Dominick, 1992: 93). Similarly Darré promoted the idea of racial retrieval of the peasantry, not through transplanting urbanised Germans to the countryside but by propagating existing rural populations (Bramwell, 1985).

If a return to Nature, the rural landscape, represented part of the world view of National Socialism, the links between conservationist thinking and fascism also found expression in a policy of retrieving the 'natural' geographic boundaries of the state. Many advocates of nature conservation saw that the territories outside the official map of Germany (from Switzerland to Estonia), but rooted with ethnic Germans, also represented part of the German nation. Before the First World War the Nature Park Society had united Germans from the Second Empire and Austria and the parks it founded were in both countries. The greater Germany which underpinned much of Hitler's foreign policy was shared by this group. The restoration and return of the Volk necessitated the retrieval of Germany's 'natural' lands and this had profound geopolitical consequences for European society between 1939 and 1945.

Although there are numerous inconsistencies and incongruities within fascist and conservationist thought, the emphasis on a retrieval of the volkish soul rotated around the spatial setting of rural, agricultural life and the woodlands. Whilst conservationists parted ways with the Nazis on a variety of issues, they shared a nationalistic solipsism which claimed that 'the natural environment existed to create the German Volk' (Dominick, 1992: 114). From the temporal perspective an organismic view of history emerged where return, retrieval, regeneration and the cycle of time became the dominant leitmotif. This recycling of a set of historical conditions found spatial expression in the life and lands of the rural peasantry and in the preservation of the country's woodlands. Time as recurrent cycle for the retrieval of the Volk provided one of the organising frameworks for a German sense of nationhood and for some of the policies used to promote this under National Socialism.

The final category, providential time, links directly with religious and sacred renderings of the past and it too has found expression in nationalist ideas and practice.

Providential time and sacred space

Although the term 'sacred history' was coined by St Augustine and elaborated in his monumental work *City of God* within the Christian tradition, the notion that divine providence plays a central part in shaping the past has a long history across a variety of religious traditions. Leaving aside the question of whether God actually exists, sacred histories and the theologies underpinning them have also had to contend with the problem of how a God whose eternity exists beyond time and space, can bring to bear effects in time and space. This metaphysical issue remains unresolved and continues to vex the minds of theologians and philosophers. Providential time, unlike the other three categories, can view time as progression, decline or cycle. In some Eastern religious traditions time is conceived as cyclical (e.g. reincarnation) and in the Judaeo-Christian tradition, elements of linearity and cyclicity regularly co-exist.

In the context of myth, Eliade (1975: 5) states that 'Myth narrates a sacred history; it relates an event that took place in primordial Time, the fabled time of the "beginnings" … myth tells how, through the deeds of Supernatural Beings, a reality came into existence.' For sacred histories the myth of origins is fundamental to the constitution of the narrative. Strictly chronological time is sometimes replaced by primordial or sacred time, the time when an event first took place, when something of significance was made manifest. For instance the story of Adam and Eve sits outside strictly chronological time, whereas the birth of Christ, at least according to some, is situated within a Western linear time frame. Coupled with this in the Judaeo-Christian tradition is the importance of both individual and cosmic eschatology which combines the circularity of liturgical time practised through religious ritual (for instance the liturgical repetition of the birth, life, death and resurrection of the Lord in the Mass) with an acceptance of the linear time of history where the world is created only once and will experience one end, with one Last Judgment.

The books of the Old and New Testaments recount examples of sacred histories where the relationship between the divine and the worldly are explored. The significance of these accounts resides in a narrative structure where the human and the divine both work as agents yet with different roles and where key episodes or events in the story reveal both divine intention and the attachment of immense human significance to these events. In addition the Bible provides a rich narrative of the sanctity of place (Japhet, 1998). In particular sacred places are either a place where God dwells (e.g. temple or church) or a place where the divine is revealed to humanity (e.g. site of revelation).

The notion of divine providence and teleology has been a powerful trope in a variety of nationalisms. For instance, in Zionism the centrality of the notion of the 'return' of Jews to their ancestral homeland – to the sacred land of Judaism – is an assertion of temporal origins and simultaneously spreads time out into space. Jerusalem represents the site of the sanctuary and the chosen city. While Biblical scholars may query some of these interpretations of the Bible text, for nationalists such reservations are unnecessary in defining national space. David Ben-Gurion, for instance, sought to 'design a Hebrew nation in the homeland that will be an example and model for new and old nations' (quoted in Azaryahu, 1998: 135). The desire to combine the providential conception of time with the historical revivification of Jewish national territory is exemplified in the drive to revive Hebrew as the idiom of cultural rootedness and to adopt a resettlement policy in chronological time. As one commentator has observed 'The transformation of the

"Land of Israel" (Jewish Palestine) from an imagined Jewish homeland that had existed for Jews only in the Jewish liturgy and sacred texts, into an actual one, necessitated both relocation and vast settlement activity. ... As the language of the Old Testament, Hebrew presented an authentic historical-cultural Jewish option that also provided the greatest measure of national unity' (Azaryahu, 1998: 136–7).The religious text provided the rationale for demarcating national territory but the actual occupation of that territory necessitated the settlement of Jews in new towns and villages across this land. The transformation of biblical time into secular time was mirrored in the articulation of a Hebrew calendar which combined Jewish and Zionist elements into a coherent national history including the celebrations of Jewish festivals such as Hanukkah and Passover into festivals of national liberation.

Judaism, of course, is not alone in its embracing of sacred and profane conceptions of time into its discourse. Christianity, no less, has employed the syntax of providential time in its narratives of national identity. Paradoxically the universalising message underpinning a Christian conception of history has regularly been nationalised in the rituals, practices and rhetoric of particular nation states. Alasdair MacIntyre has perhaps hit at the central appeal of sacred conceptions of history when he claims that: 'To share in the rationality of a craft requires sharing in the contingencies of its history, understanding its story as one's own, and finding a place for oneself as a character in the enacted dramatic narrative which is that story so far' (MacIntyre, 1990: 65). Perhaps nationalism, no less than religious interpretations of the past, seeks to provide answers to the existential questions at the heart of individual and collective identities. The perceived need to belong and to attach significance to one's existence can be partly answered through the imagined community of nationhood and the anchoring of that idea of nationhood to a shared conception of time, to a providential deliverance and to a cultural authenticity made manifest through the time-space of the nation state.

Conclusion

While this essay has considered nationalist renditions of time through the narratives of progress, decline, recurrence and providence, theoreticians of nationalism have mirrored some of these tropes in their predictions for nationalism's future. For those with a progressivist frame of reference nationalism is but a stage in the modernisation of societies and will be superseded by other forms of political organisation which may serve the needs of international capital or indeed social justice. For others nationalism is an anachronistic ideology adhered to and practised by the backward and the ignorant. Decline, in a positive sense of the word, is its inevitable fate and is to be welcomed as multicultural and multiethnic forms of political and cultural life take hold. Indeed the legal boundaries defining states may become blurred out of extinction. Writing from Belfast at the turn of the millennium perhaps renders one more circumspect about the value of prediction where nationalism is concerned. If understanding times past can be likened to entering a foreign country, predicting times future is no less an alien universe and perhaps best explored through the literary rather than the social scientific imagination.

Acknowledgements

Many thanks to David Livingstone, Jon May and the staff and students of Cambridge University and Queen's University of Belfast for their really helpful comments on an earlier incarnation of this essay.

5

REFLECTIONS ON TIME, TIME-SPACE COMPRESSION AND TECHNOLOGY IN THE NINETEENTH CENTURY

Jeremy Stein

Introduction

Social scientists have long been aware of the significance of developments in transport and communication for the reorientation of temporal and spatial relationships between places (Janelle, 1968; Falk and Abler, 1980). The concepts 'time-space convergence' and 'time-space compression' were developed by geographers to describe the cumulative effects of historical improvements in the speed of movement of goods, services and information. The former concept refers to the increased velocity of circulation of goods, people and information, and the consequent reduction in relative distances between places. The latter concept describes the sense of shock and disorientation such experiences produce (Harvey, 1989, 1990). As Harvey describes it, time-space compression refers to:

> processes that so revolutionise the objective qualities of space and time that we are forced to alter, sometimes in quite radical ways, how we represent the world to ourselves. I use the word 'compression' because a strong case can be made that the history of capitalism has been characterised by speed-up in the pace of life, while so overcoming spatial barriers that the world sometimes seems to collapse inwards upon us.
>
> (Harvey, 1989: 240)

Harvey's discussion of time-space compression relates mainly to contemporary capitalism. He views new ways of experiencing time and space as coinciding with intense periods of technological change, a consequence of capitalism's need to speed up the circulation of capital and information. Other writers emphasise how new communications technologies shape contemporary economic, social and political processes. These writers point to the increasing speed, volume and significance of capital and information flows in recent times (Castells, 1989; Leyshon, 1995). Enabled by instantaneous communication between places, organisations are increasingly able to integrate and co-ordinate their activities on a global basis, captured by the term 'globalisation'. At their most extreme, writers suggest that an era of ceaseless global flows of money, capital and information heralds the end of geography (O'Brien, 1991). However, as Thrift and others argue, flows of capital and information rarely lack a geography of their own, and because communities of experts are required to interpret their significance, this often reinforces traditional national, institutional and urban geographies (Thrift, 1995, 1996; Michie, 1997).

The main aim of this chapter is to show that contemporary notions and experiences of time-space compression were prefigured in the nineteenth century. In the next section, I will briefly summarise these experiences and their significance. I suggest that these accounts of changing experiences of time and space may need revision in at least two ways. First, interpretations of time-space compression typically rely on accounts of privileged social observers, and are thereby elitist. Is it right to assume that the experiences of time and space documented by privileged travellers were equally felt by the general population, or that the rapidity of change was the same for everyone? Second, changing experiences of time and space are often assumed to result from the advent of new technologies. This is a form of technological determinism. The danger here is to assume and to exaggerate the consequences of technology. Are new technologies always faster and more efficient than earlier ones? Moreover, do interpretations of time-space compression that see it resulting from improvements in technology risk ignoring the social processes in which technology is embedded? After reviewing historical accounts of time-space compression, I will draw on my own research on a nineteenth-century Canadian textile town to develop a more nuanced interpretation of the historical experiences of time and space.

A shrinking world

As Harvey and other writers have suggested, experiences of time-space compression are not limited to the contemporary period. Giddens' interpretation of modernity, for example, emphasises the role of technology and institutions associated with it (state agencies, businesses, news media) in extending economic, social and political influence over time and space during the nineteenth and twentieth centuries (Giddens, 1990). Berman stresses the unifying role of technology, defining nineteenth-century modernity as the shared experience of unity and disunity (Berman, 1991). For several writers the period between 1880 and the end of the First World War witnessed an intense phase of time-space compression. Marvin describes how the introduction of electrification at the end of the nineteenth century led to intense cultural anxiety concerning the effects of these technologies on Victorian and Edwardian societies, governed by strict systems of hierarchy and social etiquette (Marvin, 1988). Kern demonstrates how during this period a set of new technologies, including the telephone and the wireless telegraph, generated intense feelings of simultaneity coincident with novel ways of thinking about time and space in areas of social thought as diverse as physics and psychology, contemporary art and music (Kern, 1983).

The rapid introduction of a set of new technologies at the end of the nineteenth century no doubt contributed to intense feelings of simultaneity during this period. However, this does not mean that a revolution occurred in the experience of time and space. It is more appropriate to see these changes in the context of ongoing processes of change occurring from at least the late eighteenth century. In the next few paragraphs I will summarise the most salient aspects of these changes and their consequences. Before doing so it is worth noting that authors such as Kern have been criticised for other reasons: for their overreliance on elite sources, such as the electrical trade literature that not surprisingly glorifies technological advances; and for their technological determinism. Kern, for example, argues that, because of its immediacy and democratic qualities, technologies such as the telephone invaded privacy, resulting in nervous dispositions and undermining social and spatial barriers. Critics question whether the consequences of technology are inevitable or predictable, and whether the characteristics of a technology can automatically be assumed to be impressed on to its users (Fischer, 1985, 1992; Marx, 1997a, 1997b).

Without doubt, the cumulative effects of technological change, especially over the past century and a half in the field of transport and communication, have been impressive. This was particularly the case in Europe and North America during the nineteenth century when the introduction of railway, telegraph and steamship services radically reoriented geographic and temporal relationships. 'The annihilation of space and time' was a common mid-nineteenth century phrase used to describe the experience and significance of these changes. Karl Marx, writing in the 1850s, used similar terminology to describe the significance of improved transport and communications for the circulation and reproduction of capital. In the *Grundrisse* he wrote that 'while capital must on one side strive to tear down every spatial barrier to intercourse ... it strives on the other side to annihilate this space with time, i.e. to reduce to a minimum the time spent in motion from one place to another' (Marx, 1973: 539).

Developments in transport and communications in Britain during the eighteenth and nineteenth centuries had profound consequences for the economy, society and culture. The railway's impact, for example, was noticeable by the

1830s and 1840s, most significantly in the increased speed of the trains with consequent dramatic reductions in travel times between places (Perkin, 1970; Thrift, 1990a: 463). The effect was to shrink national space. The 'annihilation of space and time' was a common term used to characterise the experience of railway travel in the mid-nineteenth century. The metaphor of 'annihilation' evoked the sudden impact and violence of the railway as it overturned existing notions of time and distance. On the one hand, the railway opened up new spaces and made them much more accessible. On the other, the railway seemingly destroyed space and diminished the uniqueness of individual places (Schivelbusch, 1977: 41). The uniqueness of place was further eroded by the introduction of uniform railroad time. Growth of the railway network, the consequent complexity of railway scheduling and the wish to avoid accidents, gradually made the existence of multiple local times untenable, and led railway companies to introduce 'standard time' along their routes (Bartky, 1989; Kern, 1983; Stephens, 1989; Thrift, 1981).

The railway was intimately associated with developments in communications technology, particularly the electric telegraph and the telephone. These technologies inaugurated simultaneous communication and a phase of 'time-space convergence' (Janelle, 1968; Falk and Abler, 1980; Harvey, 1989, 1990). Railway companies required first class communications and from an early phase in railway development the telegraph was used as a safety device and as a means of traffic control. The Great Western Railway used the telegraph experimentally in 1839. It was used throughout on the Norfolk Railway in 1844 and during the same period on the South Eastern Railway. The telegraph, and later the telephone, was an important means of internal communication allowing railway companies to better manage their staff and complex operations. Telephone systems allowed the central office of a railway company to be in constant and immediate communication with every signal box, station, yard, office or other point in its organisation. The railway companies were also quick to take advantage of teleprinters and exchange switching technology, giving them greater flexibility and capacity to handle information (Ellis, 1959; Parris, 1965; Kieve, 1973).

Elsewhere, authors have speculated on the cultural and ideological consequences of new communications technologies. Carey, for example, argues that the introduction of the telegraph in America had profound ideological significance. The telegraph in America was dominated by Western Union, the first great industrial monopoly, which established the principles upon which modern businesses were managed. Carey suggests that the telegraph had a set of ideological consequences. Culturally, it brought changes in language, knowledge and awareness, leading, for example, to 'scientific' newspaper reporting, with news stripped of its local and regional context. By separating communication from transportation it changed the ways in which communication was thought about, for example, contributing to organic and systematic modes of thinking. Politically, it made the idea of 'empire' practically possible by allowing distant colonies to be controlled from the centre. Economically, it evened out commodity markets, diminished the significance of local conditions of supply and demand, made geography seemingly irrelevant, and shifted speculation from space to time, making possible the emergence of a futures market (Carey, 1983). Cronon usefully extends this argument and the metaphor of 'annihilating space' to the Chicago grain and meat-packing industries. He demonstrates how, in conjunction with the railway and the introduction of grain elevators and new technology in the abattoirs, the telegraph speeded up the delivery of goods to urban markets. The result was regional integration of the wheat and meat markets and the emergence of standard abstract product qualities, for example, September wheat, that could be easily traded on Chicago's stock market (Cronon, 1992).

As this last example suggests, it was a combination of technologies that gave nineteenth-century commentators the sense of a shrinking world. Hobsbawm describes, for example, how in the middle decades of the nineteenth century the world was unified by a combination of land and subterranean telegraphs, the construction of trans-continental railroads, steamship services and the completion of major construction projects such as the Suez canal (opened in 1869) (Hobsbawm, 1975: 49–67). It was these tremendous mid-century improvements in transport and communications that enabled Jules Verne to envisage his fictional character, Phileas Fogg, circumnavigating the globe in eighty days (a feat that would have taken eleven months in 1848) and for the results of the 1871 Derby to be transmitted from London to Calcutta in five minutes (Hobsbawm, 1975: 52–3). Improved communications in the nineteenth century were accompanied by greater global interdependency. The upside to this was the creation of a genuinely world market and significant reductions in the speed of movement of goods, people and information. The downside was the increased uncertainty of events and the threat of global slumps. The acceleration in the speed of communications intensified the contrast between those areas of the globe having access to improved technology and those that did not: 'in widening the gap between the places accessible to the new technology and the rest, it intensified the relative backwardness of those parts of the world where horse, ox, mule, human bearer or boat still set the speed of transport' (Hobsbawm, 1975: 60). Greater global interdependency, a consequence of improved communications, increased the frequency of contact and the apparent contrast between technologically advanced and technologically backward regions of the globe, contributing to emerging nineteenth century ideologies of Western dominance (Headrick, 1981; Adas, 1989).

To summarise, time-space compression was apparent in Britain from at least the late eighteenth century. By the mid-nineteenth century, with the coming of the railway to Europe and North America, it was commonly experienced, and its effects were increasingly global. However the changes described above took place over several decades so that it would be wrong to view experiences of time-space compression in any one period as being revolutionary. It is also questionable how widely felt these experiences were. Writers, such as Schivelbusch, typically dwell on the dramatic episodes of technological change such as the coming of the railway (Schivelbusch, 1977). We learn that contemporaries' reflections on these events were captured by the metaphor of 'annihilation': symbolising, on the one hand, technological and societal progress, and on the other, the downside of rapid social and technological transformation. However Schivelbusch does not indicate how widespread these reflections were, whether they varied along gender and class lines, and whether the ideology of progress was espoused by specific social groups.

Advances in transport and communication were not limited to communications between places. This was equally a feature within cities. Beginning in the early nineteenth century European and North American cities constructed an infrastructure of pipes and wires for the systematic centralised distribution of fuel, light, power and information (Tarr and Dupuy, 1988). These early 'networked' or 'wired' cities may be regarded as forerunners of today's electronic or informational cities (Castells, 1989; Stein, 1996, 1999). Improved intra-urban communication is typically not discussed in academic treatments of time-space compression, and is generally under-researched and poorly understood. Moreover, detailed studies of the diffusion of new communications technologies often challenge exaggerated accounts of the effects of these technologies. During the nineteenth century, for example, the majority of telegraphic and telephonic communication remained local (Stein, 1999: 47–51). Similarly, Fischer's study of the diffusion of the motor car and the telephone in early-twentieth-century America shows that far from undermining localism, these technologies sustained it, by extending the volume of local and extra-local contacts (Fischer, 1992: 193–221).

Time, space and time-space compression in nineteenth-century Cornwall, Ontario

In the last section of this chapter I want to present a more nuanced account of changing human experiences of time and space, by drawing on elements of my research on a nineteenth-century Canadian textile town (Stein, 1992, 1995). The aim of this research was to discover how Canadian urban experiences of time and space changed during the nineteenth century. A longitudinal study of a single urban community undergoing urbanisation and industrialisation was designed to provide insights into the historical significance of temporal and spatial restructuring. The context for my discussion is Cornwall, Ontario, situated on the St Lawrence river, seventy miles upstream from Montreal. From the late 1860s Cornwall became a thriving manufacturing town. This was a direct result of investment by a group of Montreal businessmen who were looking for suitable sites for manufacturing. Cornwall was chosen as a site for textile manufacturing because of its proximity to Montreal and to Canada's industrial heartland, the main concentration of population and consumer demand, and because of its available water power and cheap labour. Contemporary newspapers, travel accounts and Cornwall's Town Council records provide evidence of how residents experienced time and space differently during the nineteenth century. These documents reveal how Cornwall residents grappled with a set of new technologies, ranging from improved forms of transport and communications to urban street lighting and waterworks systems. In their experience of new technologies Cornwall residents reveal how technology was valued. Because technology was typically associated with improved capacity to overcome the physical barriers of time and space, these discussions equally indicate how time and space were socially valued. I will discuss these themes in two contexts; firstly, in relation to changing experiences of travel and communications; and secondly in terms of the experience of work in Cornwall's textile factories.

Probably the most conspicuous change that Cornwall residents witnessed during the first few decades of the nineteenth century was improved travel and communication. During the mid-nineteenth century a spirit of improvement, originating in Great Britain and the United States, swept over the province of Canada. By 1848 there existed a fully operational and enlarged set of canals along the St Lawrence (Ross, 1991). The introduction of steamship, telegraph and railroad services during the 1840s and 1850s further integrated the fledgling province, and tied Cornwall and settlements like it closer together and to the outside world. Although not a necessary precondition for it, improved communications no doubt contributed to Cornwall's subsequent industrial expansion.

Cornwall residents would have felt these changes in several ways. Travellers in the early decades of the nineteenth century experienced journeys along the St Lawrence as a set of journeys. Treacherous rapids prevented the continuous passage of steamboats, forcing travellers to endure alternate passages by steamboat and stagecoach. In 1842 Charles Dickens observed that 'the number and length of these portages over which the roads are bad, and the travelling slow, render the way between the towns of Montreal and Kingston, somewhat tedious' (Dickens, 1989: 249–51). Improvements in riverboat technology and the building of canals and railways gradually dispensed with long stagecoach rides between places. Improved design and more powerful steamboats made journeys more comfortable. Although changes were sometimes sudden and dramatic – for example, the inauguration of new steamboat services or the opening of a stretch of canal – time-space compression was experienced as an incremental process, the effect of cumulative changes observable over a period of years.

The most noticeable cumulative change was the steady reduction in journey times. This would have been apparent to the regular traveller, who made routine recourse to travel guides. *Disturnell's Railway and Steamship Books*, published annually from the late 1840s, provided travellers with comprehensive logs of routes, fares and timetables. These guides made shorter journey times popular knowledge. The *Disturnell Railroad, Steamboat and Telegraph Book* for 1850 quoted twenty-six hours as the usual steamboat passage from Montreal to Kingston; three years later twenty-four hours was sufficient for the journey (Disturnell, 1850: 67; Disturnell, 1853: 79). Shorter journey times implied greater access to a wider range of economic goods from outside the Cornwall area and greater ease of movement for the transportation of local produce to distant markets. Thus the benefits of time-space compression were not restricted to wealthy travellers but were indirectly, if not directly, felt by the majority of Cornwall residents.

Improvements in communications were equally apparent. Whereas in the eighteenth century St Lawrence settlements received mail perhaps once in two months in winter, and once a month in summer, by 1812 mail was despatched weekly. This increased to three times a week by 1830, and with the completion of the Grand Trunk Railway in 1856, this increased again to twice daily. There was a corresponding increase in the number of post offices in the district, rising from six in 1827 to twelve in 1838, with a total of 113 by 1888 (Osborne and Pike, 1984; Pringle, 1972: 165–70). Cornwall's connection to the telegraph in 1847 heralded an era of instantaneous communication. One of the first instances of telegraph use in the Cornwall vicinity was recorded by the English travel writer, Isabella Bird, on a visit to North America in 1854. Having lost her watch on a lake steamer she noted that:

> As an instance of the way in which utilitarian essentials of a high state of civilization are diffused throughout Canada, I may mention that when we arrived at Cornwall I was able to telegraph to Kingston for my lost watch, and received a satisfactory answer in half an hour.
>
> (Bird, 1966: 246)

Bird's comment nicely illustrates mid-nineteenth century associations between technology, time and speed with ideologies of European dominance (Adas, 1989; Headrick, 1981). How widely held such views were is uncertain although these opinions were likely shared by social elites, wealthy enough to afford leisure travel and to take advantage of new communications technologies. Access to the latter was initially largely restricted by cost to businesses and wealthy individuals. For example, the first institutions in Cornwall to install telephone services (introduced in 1878) were the town's railway station, three textile factories and a local hotel. Cornwall's first telephone directory, published in 1887, listed forty-three subscribers, mainly businesses and a few homes, and it was only in the 1890s that the hundredth telephone line was installed (Senior, 1984: 270; Stein, 1992: 76–7). Thus in the case of new communications technologies there was a significant time lag between the introduction of an innovation and its widespread social diffusion. This should at least make us suspicious of writers such as Kern whose interpretations of the telephone and similar technologies tend to exaggerate their social and economic effects.

Work in Cornwall's textile factories is the second setting in which I would like to examine changing experiences of time and space. It is first necessary to compare Cornwall's factories with its other industrial establishments. At mid-century Cornwall was described as 'not a place of any great business' (Smith, 1851: 387). Its industrial structure was typical of a small Ontario town, comprising a collection of small establishments processing agricultural goods for a local market. By 1871 the situation was little different. Most firms were small employers, modestly capitalised, running for between ten and twelve months a year and producing for the local market. The only exception was Andrew Hodge's flour mill, capitalised at $41,000, employing five people, with its annual production valued at $59,000. This contrasted with the Cornwall Manufacturing Company which was capitalised at $390,000, employed 145 men, women and children, and with an annual production worth $200,000 (Canada, 1871). The factory's size, machinery and organisational structure differentiated it from Cornwall's other industrial establishments. By 1882 Cornwall had three textile factories, all owned, financed and administered from Montreal. These factories were modelled on the Waltham system, a branch of the New England textile industry, typified by the construction of large integrated plants, professional corporate management, power-driven technology and factory villages (Prude, 1983; Scranton, 1983: 3–51; Tucker, 1984; Ware, 1931). Cornwall's textile factories were smaller in scale than the major New England factory towns but exhibited all their principal features, most notably their systems of social discipline. Besides physical size, economic scale, mechanised production, type of ownership and management structure it was the heightened level of social discipline in evidence in Cornwall's textile factories that most differentiated them from Cornwall's other industrial establishments.

Cornwall's factories featured strict systems of time and work-discipline (Thompson, 1967). Clocks and bells symbolised the work routine. In the late 1880s this comprised a twelve-hour day and a 65-hour work week which was fairly standard in the Canadian textile industry at the time. Work commenced at 6.30 each morning and ceased at 6.30 in the evening, with an hour allowed for lunch at midday. On Saturdays employees worked until noon, being granted the afternoon as a 'half-holiday'. By 1888 Cornwall's textile factories employed 1,400 people so that by this time this was a common work routine for 20 per cent of the town's labour force (Royal Commission, 1889: 1,058, 1,062–3, 1,068). This affected the temporal ordering of other urban activities because of a series of 'coupling constraints' that prevented textile workers from taking advantage of local shops and services (Pred, 1981a: 11). For example, in March 1883 Cornwall's post office started to open during the early evening to allow factory workers to collect mail on their journey home (*Cornwall Freeholder*, 16 February and 2 March 1883). A second example was the inability of Cornwall's merchants during the 1870s and 1880s to institute early closing arrangements, and to reduce shop employees' typical fourteen-hour day (*Cornwall Reporter*, 9 February 1876, *Cornwall Freeholder*, 4, 14 and 18 May 1883, *Cornwall Standard*, 12 and 19 August 1886). This would have made it difficult for textile workers, prevented from shopping during the day, to shop in the evening. The capacity of Cornwall's factories to dominate urban work routines was a manifestation of their considerable local economic power. It was this that was significant rather than the novelty of their work schedules. Evidence, in fact, casts doubt on the latter. Since 1830 Cornwall's Presbyterian church bell was rung three times daily, at 6 a.m. 12 noon and at 9 in the evening to sound the curfew (Pringle, 1972: 215). This suggests that Cornwall's textile factories adapted a pre-existing urban routine to suit a more purely economic function.

Although they set the basic urban routine, the work schedules of Cornwall's textile factories therefore co-existed with other routines. There was in reality no single uniform urban time but multiple times and multiple routines. Factory work intersected with religious, family and domestic routines (Hareven, 1982; Parr, 1990). In December 1876, for example, the new superintendent of the Canada Cotton Mill instituted the arrangement of ringing the factory bells on Sundays, at times that coincided with the start of church services (*Cornwall Reporter*, 16 December 1876). Store clerks, as we have seen, worked a regular fourteen-hour day. Labourers employed by the Town Council in the early 1870s worked a ten-hour day. Despite the existence of multiple times and routines it is possible to recognise the greater use of more precise and increasingly standard measures of time. We see this in the way Cornwall's Town Council made payments to its labourers for municipal public works projects. In 1868 labourers were paid by the day with the minimum unit of time paid for the 'quarter day'. By the early 1870s the Town Council Minute Book increasingly records payments in hours and half hours rather than as fractions of the day (Corporation of Cornwall Town Council Minutes, 25 October 1868, 14 August 1871 and 14 September 1873).

A further example of the transition to standard measures of time was the introduction of uniform public time. Arguably this was one of the most significant temporal readjustments of the modern era, described by one commentator as 'the most momentous development in the history of uniform public time since the invention of the mechanical clock in the fourteenth century' (Kern, 1983:11). Systems of standard time were initially introduced on the North American railroads to reduce accidents in the mid-nineteenth century and the confusion of having multiple local times (Bartky, 1989; Kern, 1983; Stephens, 1989). The continental North American railway system adopted uniform public time on 18 November 1883. The switch was clearly significant, being the basis of our current time zones. The extent of time-space compression by the end of the nineteenth century was demonstrated by this capacity to co-ordinate time nationally and internationally, enabled by telegraphic time signals. While it is known that opposition to standard time erupted on the margins in the United States, where the discrepancy between local and standard time was most conspicuous, little is known about how the change was experienced in urban Canada (Bartky, 1989: 50–2).

The Grand Trunk was the last of the four lines serving Montreal to adopt the new system of measuring time. The railway adopted standard time at noon on 18 November 1883 (*Montreal Star*, 19 November 1883). To avoid the confusion of different local and railroad times many towns and cities across North America declared their civil time would match that of the railroads in their vicinity (Bartky, 1989: 49–50). This was the case in Montreal and in neighbouring centres. Points east of the Eastern Standard Time zone had to retard their clocks, points west, to advance. Thus Montreal switched its clocks back six minutes when, at eight o'clock in the morning on 19 November 1883, the city adopted Eastern Standard Time (*Montreal Star*, 19 November 1883). Quebec City adopted the standard for civic purposes on the eighteenth and put its clock back fifteen minutes (*Ottawa Daily Citizen*, 19 November 1883). In Kingston the city's caretaker advanced the public clock by five minutes and fifty-five seconds at 12.30 p.m. on the nineteenth (*Kingston British Whig*, 17 November 1883). In Cornwall one of the city's newspaper editors reminded readers that if they wished 'to be in time for the trains, and down to time in other respects [they] should put [their] watch or ... clock back five minutes and forty seconds' (*Cornwall Freeholder*, 16 November 1883). Except for these news items, the adoption of standard time received limited comment in the press, and there was no identifiable public opposition to the change.

But this does not imply that the change was insignificant, as Sir Sanford Fleming, the chief propagandist for standard time, realised in 1889 when he described its adoption as 'a noiseless revolution ... effected throughout the United States and Canada' (Fleming, 1889: 357). The change was symbolic of the increasing integration and interdependence of distant places and of the ascendancy of clock time. That it went largely unnoticed does not detract from this argument, for major social and economic transformations, especially in their early stages, do not always generate significant social comment. The adoption of standard time in urban central Canada was not experienced as a dramatic or sudden event. This was partly because the new system of time-keeping was steadily introduced on the North American continent over several decades, and because it was compatible with existing local work routines.

Inside Cornwall's factories workers faced other aspects of time and work-discipline. A typically inflexible form of social discipline characterised the workplace. This was more formal and the degree of supervision greater than that prevailing in Cornwall's other industrial and mercantile establishments. A list of rules and regulations was posted up in every room of the factories and printed on the back of pay envelopes. Fines were imposed for a variety of offences, including destroying company property, spitting, neglect of work, and absence without notice. The ultimate sanction was dismissal. Evidence shows that overseers dismissed workers for incompetence, stealing and for inappropriate behaviour. One man lost his job when a bobbin he threw in the direction of a young girl caused severe cuts to her head. Employees were equally subjected to forms of spatial discipline. They were prevented from leaving the factory compound during work hours. While at work they endured the supervision of overseers, who enforced work-discipline and work standards. In addition, the integrated system of textile production favoured an internal spatial division of labour that assigned groups of workers to different floors of the factory and to specific tasks and machines. These forms of discipline were never total. Supervision could never be constant. Work routines were disrupted by a range of stoppages, including labour disputes, lack of orders, raw material shortages, and winter freezes that prevented the generation of water-driven power. This demonstrates the continuing difficulties late-nineteenth-century capitalists experienced in being able to control the conditions of work, and questions the effectiveness of systems of social discipline in instilling workers with capitalist conceptions of time and space (Stein, 1995: 286–94).

A further aspect of work in Cornwall's factories was the experience of new technology. This was a doubled-edged sword. On the one hand, technology symbolised progress and modernity, and often led to greater economic efficiency. On the other hand, technology created new problems. Cornwall's textile factories employed up-to-date technology. The integrated system of textile production of the 1860s and 1870s employed water and subsequently steam power to drive spinning and weaving machines by belts and shafting. This meant that workers had limited control over the pace of work. This intensified in the 1880s when overproduction in the Canadian textile industry led to consolidation and to the introduction of faster machines, such as the high-speed gravity spindle and the automatic loom (Ferland, 1989: 29–33). This was designed to increase labour productivity. One consequence was the greater incidence of industrial accidents where workers caught fingers or limbs on rapidly moving machinery (Stein, 1995: 291–3). These were routinely reported by the local press and documented by the Ontario Factory Inspectors. In one extreme case, while altering a piece of machinery in the weaving room of the Canada Cotton Company James Smith became entangled in its shafting. Despite attempts to release him his left arm was severed below the elbow (*Cornwall Freeholder*, 9 March 1883). This kind of accident diminished after the passing of the 1884 Ontario Factories Act but in a very real sense they demonstrate that 'time-space compression' in the workplace could have negative and positive consequences.

Other examples of new technology in the factory illustrate this theme of the double-edged nature of time-space compression. As has already been noted, Cornwall's factories were among the first institutions in the town to install telephone services. This was presumably to improve business efficiency and management by connecting the factories to such places as the railway station and to their headquarters in Montreal. But the telephone was also a potential means of social control. One of the first uses of the telephone reported in the local press was to send for the police to arrest a factory employee caught stealing (*Cornwall Freeholder*, 23 April 1886). So too with the introduction of gas and electric lighting systems. Designed to facilitate improved conditions for night-work, these technologies allowed factory managers to further exploit their workers by extending the working day. The Stormont and Cornwall Manufacturing Companies installed gas lighting systems in 1883 (*Cornwall Freeholder*, 19 January and 19 October 1883). The Canada Cotton Company opted for electricity and extensive sky-lighting in its new weaving department (*Cornwall Freeholder*, 2 February 1883). This was the first reported adoption of industrial electric lighting in Canada. It was a public and civic event, an opportunity to display Cornwall's progressiveness as a community in matters of technology. Over 400 people from Ottawa and Montreal were invited for the occasion. Thomas Edison was among those present (Senior, 1984: 231; *Cornwall Freeholder*, 30 March and 6 April 1883). Full coverage of the day's proceedings was reported by the Ottawa and Montreal press. These newspaper reports are examples of the 'technological sublime', the authors marvelling at the speed, light and power of electricity, as 800 looms were simultaneously switched on and above them 400 electric lights (Nye, 1994). The *Ottawa Daily Citizen* described the scene as rivalling 'the best lighted streets of the Dominion' (*Ottawa Daily Citizen*, 5 April 1883). The *Montreal Gazette* described the noise as 'indescribable' and the light 'brilliant' (*Montreal Gazette*, 4 April 1883). The only dissenting voice was the *Cornwall Freeholder* which for party political reasons chose to emphasise electricity's disadvantages, its 'too intense and flickering brilliancy', that it produced 'all the most dangerous and afflicting forms of opthalmia among mill operatives', and that 'electric light had had its day' (*Cornwall Freeholder*, 2 March 1883).

Conclusions

Tremendous advances in transport and communications since at least the late eighteenth century have restructured human experiences of time and space. The effects are global. The processes are on-going. The nineteenth century was significant in this regard, creating for the first time in human history a truly global capitalist market linked together by telegraph wires, steamships and railroads. The 'annihilation of space and time' was a phrase employed by nineteenth-century observers to describe the changes they witnessed. The metaphor of 'annihilation' symbolised both technological and social advancement and the disruption that inevitably accompanies sudden change. One consequence of improved global communications was the greater contrast between technologi-cally advanced regions and those that were relatively technologically backward.

However, traditional views of time-space compression in the nineteenth century also need revision. Firstly, it is inaccurate to assume that any one period of the nineteenth century witnessed a revolution in the experience of time and space. Evidence from Cornwall, Ontario suggests that time-space compression was evolutionary not revolutionary. Technological improvements and reductions in travel times between places were typically cumulative and gradual. Secondly, although the effects of these changes were widely felt wealthy social observers were most able to take advantage of them. Their reflections on new technology reveal assumptions about social progress that were not necessarily shared by other groups in the population.

Evidence from Cornwall, Ontario, shows that urban communities experienced a restructuring of time and space during the nineteenth century. The adoption of uniform standard time was the most conspicuous example of this. Industrialisa-tion and the advent of large-scale factory production witnessed greater efforts to instil time and work-discipline. These systems were not revolutionary or entirely new but adapted to pre-industrial urban work and religious routines. Although more precise and standard measures of time were in evidence by the early 1870s it is questionable how effective these methods were in instilling capitalist notions of time and space. Despite the dominance of Cornwall's factory work schedules to reorder urban activities they failed to entirely displace other urban routines so that in reality multiple times and multiple routines continued to co-exist.

6

'WINNING TERRITORY'

Changing place to change pace

Jenny Shaw

The foreign exchange dealer reported as having chosen to live on the Isle of Wight and commute to the City of London because he thought his family would be 'safer on the island (where) the pace of life is slower' may be unusual, but deciding where to holiday, bring up children or retire because of the pace of a place is not (Griffith, 1995). In the 1960s a Location of Offices Bureau advertised on the London Underground with pictures of fathers playing football with their sons in a 'new town' setting – a suggestion that in such places there would be enough time for family life. Since then pace has become a selling point for a number of industries, particularly those like tourism and construction concerned with place, and as I write I have brochures in front of me for holidays in Ireland, Greece and the Isle of Wight respectively which coo: 'This Autumn we have just the solution for anyone feeling the need for breathable air, miles and miles of unspoiled beaches, stunning landscapes and a pace of life always in tune with your own', 'We are the last people to hurry you. After all, we know better than anyone that the real Greece is about relaxing and adapting to a slower pace of life' and, for an activity holiday company, 'The pace of life is slow but the outdoor action intense.'

The idea that places have a particular pace which is part of their character is widely held and, as Kenneth Lynch (1972) notes, implied whenever market towns are described as 'sleepy'. But, until recently there has been little which directly explores this relationship. Indirectly, it is implied in much of the work on modernity, postmodernity and globalisation, most of which challenges the idea of time and space as fixed (Harvey, 1989). But, because of the abstraction of much of this work the middle-range institutions which link pace or temporality to place rarely or barely feature. To help plug this gap this chapter concentrates upon changing place as part of the search for a better quality of life and the growing tendency to equate the good life with the slow life. It argues that the association of pace with place is partly a matter of projection and partly a way of managing the life course, of 'doing' family life and retirement as people change places to change pace whether for a little while (when going on holiday, for example) or on a more permanent basis (for example, when retiring).

Using material from Mass Observation (MO) it suggests that, for those who can afford it, internal migration is becoming a popular coping strategy or method of time management resorted to because of increasing time pressure on the one hand and increasing pressure to take more control over one's passage through the life course and 'manage it' well (for example, by taking out private pensions) on the other. Passage through the life course is not automatic. It has to be worked at; identities and relationships have to be made and maintained under varying circumstances and success is always contingent upon other people playing their parts and upon the possession of those 'props' which the dominant culture deems appropriate (Goffman, 1967a). The 'doing' of family life or retirement does not just depend upon individuals assuming appropriate social identities but also on it being done in the right place. At certain stages in the life cycle, it is the pace of a place which would appear to render particular places more appropriate for such identity maintenance than others as place itself becomes one of the 'props' through which identities are made and maintained.

The pace of places and Mass Observation

Although this chapter starts with examples of what is essentially a discourse about pace and place, the common perception of parts of Britain or the world as faster or slower than others is widely held. This relationship between pace and place is by no means wholly imaginary or constructed only at the level of discourse. At the same time, the imaginary is clearly important and not least – as the writings of the MO demonstrate – where people would seem to find 'place' easier to envisage or think about than 'time', so people often describe their experiences of the changing pace of life through the figure of place.[1]

As a research method the attraction of MO is that it combines some of the benefits of both qualitative and quantitative methods. The panel of voluntary writers is, at around 550 and with a response rate of between 60 and 70 per cent, large enough to see any pattern in the distribution of pace by age, sex or class.[2] But its main value is the extremely rich qualitative data it generates. Contributors write about themselves and their lives exclusively around themes suggested by the MO Archive. The writing is largely unstructured, very detailed and often very moving. The panel has existed for a long time and the writers feel a long-term commitment to it and confidence in a system which ensures confidentiality. The accounts are not, of course, wholly unproblematic or unmediated by the conditions under which the writing is produced, but because of the trust built up between writers and the Archive contributors write freely and are at liberty to stray from topics if it seems relevant and necessary to them. Thus MO not only gets at meaning, in this case those attributed to pace, but also at the contexts in which those meanings are constructed.

Though it is hard to measure pace directly, a number of studies have found the pace of life to vary geographically. Thus, for example, the average time taken to complete routine tasks varies both between and within countries (Morelli, 1997). So too, walking speeds differ between urban and rural areas (Bornstein, 1979), whilst Robert Levine (1989) has found that where American cities differ in their time urgency (and rates of coronary disease) certain 'faster' cities (such as Boston) also attract more Type A (aggressive, impatient) than Type B personalities. Referring to its role in maintaining social integration, Levine (1989) calls the tempo of a place its 'heartbeat' (see also the work of Gerald and Valerie Mars, discussed in Douglas, 1997).

Thus, in countries where a relaxed approach to time is dominant (like Brazil) meetings will neither start nor end on time whilst it can be acceptable to arrive for dinner two hours late without society grinding to a halt. In contrast, the same behaviour in say Germany – where there is a national pride in good time-keeping and punctuality – would render hosts speechless and social relations would be permanently ruptured (see Hofstede, 1991). Values about time-keeping are, in other words, part of national and local cultures and although it is tempting to think that it is simply richer countries which are the faster not all countries at the same level of 'development' have the same pace. In Norway, for example, pedestrians have priority and walk freely and slowly across the road making the cars stop for them and the country as a whole has a distinctly slower pace of life than equally rich Japan. Culture is clearly one of the institutions which links pace to place and this was much commented upon by the MO writers. For example, one woman wrote that where she lived – mid-Wales – there was a concept of 'Radnor time' which excused locals for being late for everything.

The slow life as the good life

Not everyone wants a slower life. The young especially, for example, often move to places perceived as faster just as much as the old move to places perceived as slower. But there is more commonly discussion of life as too fast and of a need to slow down rather than as a desire to speed things up. Evidence of that desire is to be found as it becomes clear that the villages and small market towns of East Anglia or the West Country are chosen by the elderly not only for their milder climes but because they are thought to offer a better quality of life as they are perceived as places where one may live at a slower pace. Like downshifters, the newly retired do not head for where property is cheap (as in parts of the inner city) so much as for more remote or beautiful spots on the grounds that the slower pace of such places is better. For example, and fairly typically, a man described moving north from the south of England as part of his search for a more relaxed way of life and a teacher, who had retired to a market town in Wiltshire from Epsom (Surrey), wrote of the slower, more peaceful form of life she now enjoyed and how it had enabled her to 'find herself'. Another woman emphasised: 'By the by, one of the things that my husband and I value about living in Norfolk is that most "natives" don't rush and won't be rushed ... I would not want to move back south again.' Yet another woman, recently settled in the Shetland Isles, describes valuing the long summer evenings and how often visitors to the island commented on the slower pace of life there. A parson's wife wrote of 'needing holidays in Scotland or at a friend's cottage in Eire' because 'Nothing happens in Ireland – no one bothers about anything – or that is how it seems to the visitor.' And a middle-aged clinical psychologist who had moved to Cornwall from London noted with some affection the motto 'Cornish men do it dreckly – which means in months rather than weeks', though he also added that 'while Cornwall is relatively slower by tradition, the emphasis in the services and companies that I have contact with is to do everything fast'. For many writers places like the Isle of Wight were fused with periods or epochs like the 1950s, symbolising a certain sort of stability and calm. Though it is impossible to return to the 1950s, people therefore often choose to holiday in places which seem to them like stepping back in time and to a slower pace of life.

Abroad, differences in pace and place are even more striking. One man noticed that pedestrians in Madrid were allowed more time to cross the road than in Britain and that being fifteen minutes late for an appointment in that city did not matter. While another, who had spent four years in the West Indies in the 1950s, noted:

> Life there was slower, we were slower and pressures were lower. So physical state as well as mental state is a factor. I could do almost any job quicker than the locals (and all our staff were paid very much higher wages or salaries than generally accepted). Most of our staff had their own car, owned a home, etc. – i.e. equivalent to English C2 standard. But when I got home – what a shock – everybody moved and worked about three times as fast as me. My 65 year old father walked so fast I had nearly to run to keep up.

In much the same way a woman who had gone with her family to the Sudan was initially very annoyed by the slower, more relaxed tempo but came around to valuing it more positively:

> When we first arrived there it irritated us considerably. Waiting for service in a bar or restaurant took the patience of Job. Meetings were

arranged but there was no sense of time and people would turn up two hours late. Even at school, which was meant to start at 7.30, children would be brought late by their parents and tell us they'd stopped off to do this or that. A Sudanese friend was much amused by our obsession with clock time and in the end made appointments with us by British or Sudanese time. If we were to meet by British time that meant on the dot. If it was Sudanese time, then it could be any time at all. ... I doubt very much whether, even with their immense problems, any of (the Sudanese) die of the effects of stress. We used to notice this particularly when the electricity supply went off at regular intervals. The Sudanese just took it in their stride while the ex-patriots screamed in frustration. And then, in a couple of months, we realised that we had adjusted without even noticing. Once we had adjusted, it altered our values. We had time to see what was important and what wasn't. Most of our lives in the West are filled with trivia and we are bowed down under the weight of worries about everything from our health to whether the bus will be on time. I can't believe this does us any good. Since we have returned we've had to re-adjust to the Western pace because people here just won't tolerate you ambling in to work when you feel like it or failing to mark books, do reports, pay bills or organise meetings.

Because time-keeping is profoundly embedded in everyday life, habits and values, accommodating to a different tempo challenges what is expected and can produce intense feelings of dislocation in those forced to march at an unfamiliar pace. In the MO material the angrier descriptions of slowness nearly always came from people who had been obliged to work in different places, such as a woman who – though she had said she learned to expect varying time zones – wrote of how she was infuriated by how: 'the further East I travel, the sloppier the perception of time becomes. It irritates me in Poland and drives me gibbering in the USSR.' The pace of market and socialist or 'planned economy' societies were clearly very different such that, to John Bornemann (1993) prior to the fall of the Berlin wall the sense of time seems 'petrified' in the East and 'quickened' in the West, accelerating in both sectors after reunification as East Germans strove to make up for 'lost time'. Anecdotally, I was told by a Leipzig economist that the most dramatic change since 'die Wende' had been an increase in the pace of life, which he attributed to an increase in traffic and the emergence of cafés on the broad streets and squares which gave the city a buzz it had not had before. Clearly, culture – which includes a sense of time – changed along with the political and economic systems.

Although it can seem a fussy or petty concern, time-keeping is a crucial aspect of social structure. It is a matter of culture, the 'way things are done' and discussion of it quickly leads to a deeper critique of how we live our lives and what we take for granted. An ex-college lecturer who had taken early retirement in order to write, describes both some of the stages of self-re-evaluation which a different pace of life can trigger and the two sorts of time which crop up in so many of the descriptions:

> A few years ago we rented a house for the summer in the Ariege, the least developed part of the Midi. It was looked after for its owner by her sister, an unmarried 75 year old who befriended us. She got news of the wider world via her TV, but her own routine must have been unchanged for years. She took her cow to pasture each morning and brought it back in the evening. During the day we could see her scything the grass or doing other regular tasks. She had little time for the late twentieth century, had never travelled further than Toulouse in her life. She spoke, critically, of the 'age of speed, the age of going to the moon'. When the Tour de France came through on the way to Luz d'Ardiden all it meant to her was the inconvenience of having the road to the pasture blocked by the cyclists and their media entourage. But that kind of interruption was the exception, I think. As a rule there was no kind of dislocation for her between chronological time, seasonal natural time and the accustomed rhythms of her experience. She was a charming, friendly person who brought us flowers, was active and contented: but to me life in a depopulated village, from which the young had moved away to the towns and broader more varied perspectives, would be an imprisoning stasis. … Our chatelaine, though a lovely person, was poorly informed and her preferred way of life narrow. Yet it's not the 'pace of life' that matters to me, more the degree of control. We all live in both clock time and subjective time, and my aim is to always win territory from the clock and – by either speeding it up or decelerating it – put it to my own use.

The battle for time

This image of life as a battle between the two sorts of time which has to be won for self-respect to be gained echoes what has come to be seen as a broad cultural conflict or historical shift from pre-industrial and task-based ways of working to the clock-based time or industrial ways of the factory. Described in a seminal essay by E.P. Thompson (1967) the non-standard and flexible rhythms of agriculture, driven by the seasons and the needs of crops and animals, have been overtaken and succeeded by capitalist and industrialist practices based on piece rates and the labour theory of value. Such changes are argued to have led to a profound shift in the social construction of time with time becoming altogether less elastic and forgiving and such that – whatever the long term trends, and many have argued that the two types of time co-existed long before the industrial revolution (see Zerubavel, 1981) – it is clear that the two now exist in an uneasy relationship with each other.

This unease is most apparent on the home/work front, because almost singularly it is the family which is still understood as embodying the idea of a generous, flexible or 'pre-industrial' notion of time (May, 1994). This connection between a more 'natural' 'pre-industrial' time and the family is a clue as to how the experience of family life mediates the relationship between pace and place. Like pre-industrial time, family time is widely believed to be qualitatively different to work time. It is often understood as vivid and suspended and, like a dream, to carry with it a degree of resistance to change and progression (Forrester, 1992). In unconscious terms, family time is essentially anti-linear and opposed to work time, which is linear and progressive. The ensuing opposition or tension between family and work time appears in many forms, but it is mapped most clearly on to place. Working late, at home, at the weekend or on holiday – though increasingly common – almost always leads to bad feelings because, done in the 'wrong' place, it represents a basic incompatibility between work feelings (which are about moving on) and family feelings (which are about staying put).

With family time ever more squeezed or threatened by 'industrial' or work time (that is, by longer, more intense and/or more flexible hours) the quality of time expected from and attributed to the family is substantially different to that associated with the workplace and becoming more so. With a quarter of all fathers with young children in Britain working over 50 hours a week (Smith and Ferri, 1996) the stakes are higher for what time is left after work. But as important as the lengthening working day is that, whilst family life was to some extent traditionally *underpinned* by the spatial division between home and work, it is currently *undermined* by the difficulties of maintaining that boundary as more and more people take work home or on holiday. As a result, the public/private divide is increasingly experienced both internally and in temporal terms (the longing for more 'own time') leading to increased attempts to identify suitable coping strategies. The strong sense that family time needs to be different to and separate from work time is most evident in the idea that if parents spend 'quality time' with their children then this will compensate for there not being very much of it.

As 'finding time' gets harder in daily life, holiday time becomes more important and more frequent, and certain places rather than particular spaces (for example, the 'domestic') come to stand for time. Part of the slowness which is sought on holiday, and from places where they are taken, stems from the definition of holidays as family times: often we want holidays to go slow, last a long time and be remembered because we feel guilty that in our everyday lives we do not give enough time to children or to making and repairing family relation-ships. The MO writers were quite clear about this, as one woman wrote:

> The time I most enjoy is when we're away on holiday. There is so much more time just to be with one another without having to do anything. We went to Tenerife this year – fortunately our resort was not a very commercialised one – and there was very little to do. Dave (her partner) didn't like it very much because the place was so small and, I admit, quite uninteresting. But I liked it because of the sheer nothingness of the place. There was time just to be.
>
> I think it's important – time to be, time to be alone. And it's some-thing I miss these days. Even if I sit in the house when Sarah (her daugh-ter) is asleep I am always aware that she is there. When Dave and I were on our own it was something we almost achieved together – if that's possible – just sitting reading. I'm sure he misses the 'being' bit too. For all these things are important in their own right. Time to be together as a family, time for Dave and I and time for each of us as individuals. Our problem is we have no time for these kind of times. That I suppose is one of the biggest problems in today's world, where women work, jobs are more demanding, people more pushy and time just slips away.

Such sentiments are repeated many times over, not just in the MO material but also in responses to the British Household Panel Survey which shows that what most people see as the main purpose of holidays is, simply, to preserve family unity (Buck *et al.*, 1994). The idea of family time as slow time easily gets wrapped up with feelings about places and holidays because, as the above musing indicates, 'family time' leans heavily on the gendered opposition of 'being' and 'doing' (compare, for example, Dave's desire to 'do' with the respondent's own desire simply to 'be'). The times and places most associated with women carry both a sense of timelessness and 'oneness' because of their shared dependence upon primitive emotions stemming from infancy (see Kristeva, 1981). Whilst families, holidays and places all lend themselves to being viewed from a psychoanalytical perspective, family holidays especially lend themselves to that perspective because they usually involve travel and – as a number of writers have pointed out – all travel starts with a separation and is at some level carried along by the desire to return and make good a loss (see de Certeau, 1984; Game, 1991). Extending the psychoanalytical interpretation, Barry Curtis and Claire Pajaczkowska (1994) show that much of what goes to make up a holiday (shopping, eating and sightseeing) is about incorporation, and that many holiday brochures appeal directly to primal states when claiming that everything a tourist could want (beach, old quarters, shops, etc.) are only 'minutes away': an unconscious appeal to the infantile inability to wait. Hotels cater for the same sort of primitive feelings and longing for a return to mother with their offer that almost every physical need will be met by someone else in a deeply maternal way; the room service, the clean sheets and the abundant food.

Similarly, much leisure activity involves giving in to the desire for cessation, stasis and unity as well as to a compulsion to repeat – common in behaviour still largely driven by primitive or early feelings. Once such feelings have become attached to place they provide a rich vein which can be, and is, mined by the tourist and construction industries. In Ireland, for example, postcards bearing captions like 'Rush Hour in Ireland' or 'Life in the Fast Lane' which picture a farmer strolling along a country lane behind a straggle of cows or a pony and trap on an open, empty road are sold to reinforce the idea that a holiday in Ireland buys a quality of life based on unhurriedness, time to talk, look and listen. Heavily marketed products in particular build upon a logic which exploits the compulsion to repeat (Mellencamp, 1992) and the annual holiday is nothing if not compulsive and repetitive and based upon a desire for a return to a state of calm oneness. Thus, even with a history of fraught holidays and, at some level, knowledge that they are not an escape but a continuation of everyday life with all its travails, many people and women in particular approach holidays in the belief and hope that they will be restorative (see Deem, 1996). Psychologically this greater investment by women in finding the 'right place' for holidays, and their creative fantasising and planning of them, makes sense: it is part of their gender identity.

For the family as a whole rather than just the individuals within them, places can become important and be returned to repeatedly or compulsively as when going back to the same place for a holiday year after year because, like a family story, they help create or re-create a common feeling of unity. And family stories, as Jerome Bruner (1990) explains, are part of the 'micro-culture' of the family which helps keep it together. If stories aid the act of remembering and so perform a function of maintenance, so too do 'family places'. The terrible sense of loss that is felt when the cottage rented for twenty years is sold, for example, comes about because such accommodation stands in for a broader sense of place that means security (Bertaux-Wiame and Thompson, 1997).

There is clearly an element of displacement at work here; places which are remembered as where life was leisurely tend to be associated with the idea of the family having a good time and it is no accident that people often retire to the places where they remember happy family holidays. This is nostalgic, of course, and part of a broader longing which is central to modern consciousness and which has boosted the heritage industries and a search for the primitive or pre-industrial idylls in a variety of ways (May, 1996b). But the point is that people seek unity in numerous ways and anything which threatens that unity will be resisted. All forms of separation and moving on are threats to the family and if places represent stasis they tend to become valued by families for their role in playing down that threat. Like family myths, in which they often play a major part, certain places are a way of preserving the past and stopping the clock (Byng-Hall, 1990).

'Doing' the life cycle

The routine association of tempo and place sketched above suggests both projection and introjection of feelings on to places and places in to feelings which allow them to be used in programmes of both family maintenance and personal change. Physical moves ease people in to new identities and stages of life, both practically and imaginatively. Traditionally young adults were 'farmed' out as labourers, apprentices or young knights to other households in order to learn how to grow up and take on adult responsibilities. Now students follow a similar sequence as they move from home to college and, if the move is successful, often remain in the places where they took their degrees. Though it is women who most often talk of biological clocks ticking by, there is therefore a strong and continuing social pressure on both sexes to observe the 'right time' to get a 'proper' job, marry, have children or stop work and so on, and for each stage or shift to be marked by some visible change – like a house or place move. For those in the middle years of parenting, the desire to see their children grow up safely and not to go 'off the rails' can lead to more decisions to move, away from danger.

Whether it is the slowing down of old age or the speeding up of youth, places are an important prop and, as Erving Goffman (1967a) argues in his essay 'Where the action is', not only is the world divided into 'safe and silent' places and those which increase 'fatefulness and consequentiality', but behaviour itself (and the meanings subsequently ascribed to that behaviour) depends upon the type of place in which it is enacted. If going somewhere risky where the 'chances that (one) will be obliged to take chances' is a way of finding a 'true self' and increases the chances of 'enhancing moral character', so is going somewhere to where the action is not. Going to the country in order to realise an ambition to be an artist or to give grandchildren a taste of a certain rural idyll are just as much strategies for gaining the true self as the gamblers studied by Goffman, and the process just as dependent upon place. Retirement requires 'doing' just as much as does family life, and it requires other actors to 'do' their bit too and to relate differently to the 'new' individual. Signals have to be given out and received if the change of identity which is necessary for progress through the life course is to be accomplished (McAdams, 1990). Retirement must be planned and different daily routines established. If it is intended to be a period of slowing down and, perhaps, of keeping death at bay by living more sensibly (i.e. with lower stress) this also has to be worked at and a change of place can be helpful – first imaginatively and then practically, by permitting escape from existing commitments and routines. A move at the end of life in order to find a slower pace of life is not only a move away from the places and pain associated with work (a need to put a distance between one's self and one's past) but should be understood as an expression of the need or desire to become a different person. A retired engineer and managing director, who would probably be very impatient with employees who were not good time-keepers, delighted in describing how:

> Brockenhurst, in the heart of the New Forest and an important centre for residents of outlying hamlets (is) brought to a halt for an hour or so because two cows, nominally out to graze on common land, chose to rest in the middle of the main street chewing the cud. It would have caused irritation in most parts. It was tolerated with amusement. There is (also) a family of donkeys which has as its range the village of Beaulieu and its environs. When the grass on the verges to the lane to Lymington is sweet one must be prepared for delays which are not announced on the morning radio.

In addition to enjoying the rural image, this respondent hints at the pleasure in seeing busy lifestyles subverted. He had chosen to put himself in the way of being enchanted by the donkeys. By contrast, a deputy head teacher living in a village outside of Huddersfield was struck by what a social and time-consuming event visiting the local shop was – lots of chit chat, first names and preferences noted and catered for. Reflecting on why she did not shop there more often, she concluded that it was due to pressure of time which meant she could not live according to village tempo. To choose the pace of one's life, especially in a world where time competition is increasing and those at work are increasingly losing control over their time, is a privilege and for the retired one of the privileges of getting older.

The safe and silent place as the 'positional good'

In his analyses of tourism, John Urry (1990) refers to the zoning of places according to the quality of their time, arguing that as new scarcities emerge (scarcities that now include both time and space) new social divisions will be redrawn around them. The high house prices of commuter villages and pretty towns in Devon or Cumbria are examples of the way in which pace has become an 'added value' and commodified, a way of keeping up prices just as the fast pace of an 'urban lifestyle' helps maintain the prices of the gentrified city district. Simple distance may no longer 'buy' time, and time and space may no longer serve as limits to one another in a straightforward sense, but different places *are* bought because of their different subjective and temporal qualities (Urry, 1995). This commodification of place and time in a blend which allows, or is thought to allow, life to be lived at a more desirable pace (whether faster or slower) is an illustration of Fred Hirsch's concept of the 'positional good' (Hirsch, 1977). Using the example of the empty beach which loses its charm once it is accessible to the masses, Hirsch explores the 'paradox of affluence' which means that even the rich have to run to stay still because, as most wants are successfully satisfied by economic growth, competitive pressure switches towards those positional goods or experiences which lose their exclusivity (and thus their function as a marker or status) once expanded to meet the demand of a mass market. A continual process, the search for new positional goods lends itself to an ever-increasing pace of life.

The process of time and place zoning which follows this competition for positional goods also exacerbates other social divisions, of course. Here, however, I have focused mainly upon psychological processes because pace has both objective and subjective components and, as a construct, emerges from the relationship between an individual and their environment. Like time, pace does not exist outside of or independent from the individual and, whilst remarks about the ever increasing pace of life have become a short hand for the difficulties many now experience in gaining control over their lives in a rapidly changing world, so too that discourse (or 'culture of complaint') in turn helps create the subjective experience of a world moving too fast.

Conclusion

Exactly how much the pace attributed to place is an inner quality projected on to the outside world and how much a quality of the environment introjected and used to produce a change in the individual remains unclear. What is clear from the references to pace and place in MO is that physical moves are felt to help people reform themselves psychologically and to adapt to life-cycle related change. The process of projection enables places to serve a range of symbolic and imaginative functions – evoking or containing feelings, triggering memories or promising fresh starts, offering solace or inducing recuperation. A change of place enables people to change themselves or their relationships (how many of us have gone away to start or to save a relationship?) and the more people are expected to take responsibility for themselves, the more they will look for solutions which include going to a place for a purpose or to change the pace of life. Moving through the life course involves a series of transitions which require personal change and in this chapter I have argued that changing place to change pace can be as much a part of 'doing' family life as it is of 'doing' retirement or downshifting. Faced with the fact that we cannot change when we live, cannot turn the clock ack, we look for alternatives. As place is easier to imagine, remember and 'manage' than time, it is worked upon in the process of self-reinvention. Not everyone can follow Prince Charles to the nostalgic re-imagination of a better, slower life that is Poundbury, but increasingly people are choosing where they live in order to better control the pace at which they live – if only for a time.

NOTES

1 The data drawn upon in this chapter was collected during the project 'Age, Pace of Life and Social Change' funded by the ESRC grant no LC-14–25–0027. This enabled a directive on the 'pace of life' and what that meant to the contributors to MO to be commissioned. Overwhelmingly, the panel felt that the pace of life was increasing both for themselves and in general.
2 Notably, whilst the current project revealed few such differences by age or class, more women than men complained about the pace of life being too fast.

7

RESPONSIBILITY AND DAILY LIFE

Reflections over timespace

Karen Davies

Everyday life is, in all its simplicity and complexity, structured by timespace. As a sociologist, I have found the study of everyday life fascinating. Rather than beginning in an abstract, conceptual world, it has meant starting in individuals' own accounts and reflections and in their understanding of the world in which they live. As a feminist researcher, it has meant inhabiting the world of women. It has involved taking as my starting point the 'real bodies' of individuals that are enmeshed in timespace.

The concrete reality of women's daily lives, and the importance they place on responsibility or a rationality of caring, has led me to try and extend our theoretical understanding of timespace from a feminist perspective. The first part of this chapter will attempt to show why gender needs to be incorporated into discussions of spacetime and takes as its starting point a critique of the standard time-geography approach. With an emphasis upon an atomistic and resource-based notion of time, the approach fails to fully account for a relational construction of time that lies at the heart of (women's) everyday life as shaped by a rationality of caring. The temporal and spatial implications that arise from women taking care of others in both the public and private spheres results in difficulties in carving a space for a 'pause' – the focus of the second part of the chapter. A major characteristic of present-day society is argued by postmodern or late modern theorists to be reflexivity. Yet reflexivity in the late modern age requires time – or a 'pause' – for self-reflection and for women this may well be difficult due to the nature of women's work and by the ways in which timespace is gendered.

The standard time-geography approach

The time-geography approach, originally developed in Sweden by Torsten Hägerstrand, has played a fundamental role in helping us understand space-time. The basic tenets of time-geography are surprisingly simple but fundamental to social theory. All individuals' actions take place in time and space. Since an individual is indivisible, she/he cannot be in two places at the same time. Time and space therefore impose basic constraints for freedom of action and movement. Despite the possibility today of transcending classical one-dimensional time-space frames with the help of cyberspace and new technology, the above premises still hold true for most social actors navigating their daily lives. The time-geography approach has been seminal since it gives prominence to the significance of *both* space and time. It allows us to understand how women's and men's daily lives are actually shaped in practice, while also focusing on the constraints or enabling factors that influence choice and action.

Hägerstrand broke with an earlier geographical tradition by inserting and emphasising the importance of a *social* framework. Time in itself, he argued, was not sufficient for a wider understanding:

> The landscape, as we see it, or the region, which we only get to know more indirectly, is the zone of action as it appears at a particular moment. It is not enough to add time to the spatial dimension. Not only the visible participates, but everything that is *present*. To a great extent human actors, their knowledge and intentions, belong here naturally, which not even natural geographers can ignore any longer.
>
> (Hägerstrand, 1985b: 6)

While the above quotation suggests the potential of dealing with individuals' activities in space and time so that complex interconnections become visible – and this indeed was Hägerstrand's vision – the model continues to build on a somewhat limited and traditional understanding of time and this subsequently mars its possibility of making sense of the gendered nature of individuals' lives in a satisfactory manner.

In the time-geography approach, time is identified as a *quantity* – in terms of the calendar or clock – against which events in space are related. Its concept of time is *linear*, drawn as the y-axis in its diagrams, which may be divided into mathematically exact units. While Hägerstrand emphasised that individuals' actions are limited by their inborn and learned personal qualities, accessible techniques, resources in the social environment, other individuals' choices, as well as by rules and norms developed within governing institutions and organisations, individuals in time-geography studies tend to be regarded as social atoms rather than acting subjects. In summarising many of the criticisms directed at the time-geography school, Tora Friberg (1993: 78) writes: 'Time-geography does not take social processes into account and has an undeveloped view of contradictions and power. This means that social transformation remains unexplained. Its view of time is oversimplified.'

A relational construction of timespace

The emphasis on time as a resource – equally available to all – is also inherent in the time-geography model. This understanding of time emerged in part with the modernist project which emphasised the individual and his/her identity as well as his/her individual life trajectory. The assumption that time is an equally distributed resource where each individual receives his/her allotted share is however, as Ewa Gunnarsson and Ulla Ressner (1985: 109–10) argue, somewhat problematic:

> Even in working life we start from the premise of individual time. But if we take a look at how women use their time, it's obvious that it is rather a question of 'collective' time which others, for example, their families, have a right to lay a claim to.
>
> (my translation)

In other words, how we in fact use our time and locate ourselves spatially (the two being inseparably related to each other) is dependent upon the social relations in which we are embedded.

The spatial relations between home and work as well as the temporal demands in women's lives have been the subject of a wealth of studies in recent years. These studies illustrate the ways in which differential access to space and time are inextricably gendered; that control over time and space, or at least the ability to be able to freely choose how to use one's time and space, are for women substantially influenced by the interlocking of the public and private spheres and by women's structural position in these spheres. Who takes the car to work, how often the buses run, how far away the workplace is, who drops the children off at day-care, who does the shopping and when the shops are open, are mundane, everyday considerations where time and space are pivotal and where gender leaves its imprint in a tangible fashion.

Hanson and Pratt (1995), for example, show how women's shorter commuting distances may lead to the creation of a number of separate labour markets *within* a single metropolitan area as well as how women may become 'trapped' in place as their domestic responsibilities curtail the time available for commuting to work. Others have shown that the need to work close to home may mean that women choose jobs beneath their qualifications (Davies, 1990; Freeman, 1982). Women may decide to take jobs in the public sector as opposed to the private because full-time work in the former entails less hours and times of starting and finishing work are more estimable (Elchardus and Glorieux, 1994). Spatial and temporal considerations may even influence the choice of home-working, temporary work or work sharing and of course of part-time work. Cynthia Negrey

(1993) found in her study of work sharing that for men this form of employment could mean increased personal autonomy as well as a feeling of being on vacation, even the possibility for doing specific household projects arose – while for women this was rarely the case. Finally, as Aitken and Carroll (1996) found in their review of several research projects concerning the motives for teleworking, whilst women often emphasised the wish to combine paid work and family/home in a more flexible fashion, men tended to emphasise the possibility of escaping the control of the workplace and avoiding 'face-to-face-authority'.

One of the more useful concepts to emerge within discussions about more flexible working practices is the notion of time-sovereignty. For Garhammer (1995) current changes in the organisation of working life raise the possibility of recovering one's own time from all foreign-determined orders of time, though that possibility is by no means equally enjoyed by all (Garhammer, 1995). Thus, whilst time-sovereignty is enjoyed by tele-homeworkers and time pioneers (the latter being: the self-employed, freelance workers and executives, all of whom it is argued can plan their work time relatively autonomously) at the other end of the scale we find 'work on demand', where the unpredictability of working hours is extreme, as well as rigid shift schedules or traditional weekend work where external determination is at the uppermost.

For Garhammer (1995) one of the most significant problems facing those who would reorder the world of work, providing for greater flexibility and greater time-sovereignty, is the continuing separation of the world of paid work from the world of the family. Yet time-sovereignty cannot only be discussed in relation to the demands of paid work. From a feminist perspective, *work* does not only refer to paid labour carried out in the public (or private) sphere but to unpaid work carried out in the home, informal work carried out in the community, as well as to what feminists have called emotional work – which is carried out more or less everywhere and in all contexts (see, for example, Davies, 1990, 1996a; Hochschild, 1983; James, 1989; di Leonardo, 1987; Leira, 1983; Lynch, 1989; Oakley, 1974).

It follows then that women's use of time and their temporal consciousness are strongly related to the time demands of *significant others* and this has, of course, important implications for women's possibility to control their own time. In a study looking at flexible working hours on an intensive care unit, for example, nurses who were single were extremely positive about a newly implemented model within which one could not always be sure when one would be working, whilst those who had responsibility for others (children, partners, sick parents) were more apprehensive (Davies, forthcoming).[1]

Nor is it only time that is of issue here. As with their use of time, women's control over space is similarly moulded by their responsibility and caring for others – both on the job and at home – and by their structural position more generally. Both time and space are, in other words, not abstract concepts in women's lives. Rather they are importantly shaped by the presence and needs/demands of an other, be it a child, a sick parent, a patient, a boss, etc. Space and time are thereby not individual resources as such, rather they must be understood in a relational manner. Put in simple terms, *where* women find themselves and *when* they find themselves where they are, are importantly determined by the needs of others.

The feminist contribution to an understanding of time

An individual notion of time thus builds upon a particular temporal discourse where time is seen as linear, finite and quantitative and consisting of discrete units which can be divided between work, leisure and personal time. A number of researchers, including myself, have tried to show that a linear conception of time is grounded in gendered power relations and in a discourse of masculinity (Davies, 1990; Forman and Sowton, 1989; Knights and Odih, 1995; Kristeva, 1981; Nowotny, 1982). Indeed, earlier I have argued (Davies, 1990) that time-that-we-take-for-granted – the present-day dominant linear and quantitative temporal consciousness – grew out of various religious, scientific and economic interests, all of which were male dominated. Or put another way, the social construction of current dominant temporal patterns has its roots in the central concerns of various male-dominated hierarchies which were interested in solidifying and retaining their positions of *power* in society.

By arguing that time cannot only be seen in individual terms, following the linear yardstick, feminists have shown that women's time features simultaneous activities and overlapping temporalities (Adam, 1995; Davies, 1990, 1994, 1996b; Forman and Sowton, 1989; Leccardi and Rampazi, 1993; Leccardi, 1996). It is characterised by plurality and interdependence. Women weave complicated temporal tapestries consisting of, as Carmen Leccardi and Marita Rampazi (1993) point out, temporal threads that are biological, social and chronological, processual, cyclical and linear:

> It is a time 'contaminated' by cosmic dimensions, linked to the rhythms of nature, synchronised with the times of economic activity and its ra-tionality and also with the logic of caring.
>
> (Leccardi and Rampazi, 1993: 354)

When looking at everyday life, we therefore need to analyse more carefully how we negotiate and switch between *different* temporal orders, how we weave together different temporal patterns, how temporal meaning is constituted through social interaction and how gender as well as discourses of femininity and masculinity are part and parcel of all this. Returning to time-geography, it becomes obvious that the approach has difficulties in capturing the multi-dimensionality of women's time since it builds upon a discrete temporal framework: one that functions independently of what it actually means to live, as Friberg (1993) puts it. It is also questionable whether it captures the reality of men's lives. To expand upon these arguments and, more specifically, to illustrate this gendering of timespace, I turn now to the concept of care and the rationality of caring.

Responsibility and a rationality of caring

> The nurse also sometimes fires furnaces and mends the plumbing, i.e. she does tasks of people below her or outside the role hierarchy of medicine. It hurts her, but she does it. Her place in the division of labour is essentially that of doing in a responsible way whatever necessary things are in danger of not being done at all. The nurse would not like this definition, but she ordinarily in practice rises to it.
>
> (Hughes, 1984: 308)

While much has changed in the nursing profession since Everett Hughes' account of the profession was written in the 1950s, he none the less captures what is still the essence of servicing work – be it as a nurse, in other care professions or in the home. Women's work is importantly premised on the well-being of *others*. While Hughes' words were written long before the emergence of latter-day feminist writing, the passage therefore links to a central area of thought and research that was to arise within feminist research, namely the issue of caring.

Women's thinking and actions are often fired by a rationality of caring or responsibility as it has been called in the Scandinavian countries (see, for example, Sørensen, 1982, Wærness, 1984) or an ethic of care as it has been named in Anglo-Saxon countries (see, for example, Gilligan, 1982; Tong, 1993). A concrete example of this type of rationality is illustrated in the following words of a nurse:

> If I sit on the edge of a bed and see that the patient is depressed, well then I've got to spend time with him and talk to him. Perhaps I'll sit there for half an hour after my hours are over. But it wouldn't enter my head to claim overtime. It isn't a work duty.
>
> (Liljeström and Jarup, 1983: 80; my translation)

Liljeström and Jarup argued that in many women's jobs, the similarity between what women do for free and what they do for a wage makes it difficult for them to distinguish between what belongs to their private life and what belongs to the job. Regardless of whether the actions of the nurse as described above make sense or not from a labour union or economic rational point of view (and certainly such actions reinforce women's under-paid and lowly valued positions in the labour market) they are understandable in terms of a rationality of caring. This is an ideological issue: for women there exists the moral imperative to care for those who cannot care for themselves (cf. Cunnison, 1986).

While a rationality of care has been highly fruitful in examining care occupations and in understanding many of the invisible components in these types of jobs, it has also been useful when appraising all sorts of service jobs – where men too may be found in large numbers, such as janitor, taxi driver, or community police officer – as well as in women's experiences of (male dominated) skilled industrial work (Gunnarsson, 1994; Motevasel, 1996). A rationality of caring means that one is very much 'other oriented'.[2] It involves being sensitive and showing empathy and being able to put one's self in the other's shoes. It means being able to meet the other's needs in a spatial-temporal here-and-now. Thought and action grow out of social praxis. This 'other orientation' and women's understanding of its importance, means that women will often waive rules and routines which are grounded in linear and commodified time. A municipal home-help, for example, receives instructions from her superiors about which jobs she may and may not do for the care-receiver and how much time she may spend on these activities. But the home-help's understanding of the actual needs of the pensioner means that these instructions are often put aside. She designs her own schedule. For example, cleaning the windows may be one of the duties that the home-help may *not* carry out. However, the pensioner's birthday is coming up and relatives and friends will be visiting for a small celebration. For the pensioner in question, it is immensely important that the windows are polished. The home-help understands the importance and is prepared to carry out the duty.

This complex relationship to time is closely related to specific types of knowledge, i.e. 'knowing in action' and 'reflection in action', to use Donald Schön's (1983) concepts. It means that women display a flexible relationship to their work and to relationships within that work. It allows a woman to know how she should best act in a certain situation, based upon her own judgement and her understanding of the needs of the other – rather than upon a bureaucratic and economic rationality. Here I am drawn to the concept of a rationality of caring as it links rationality with feelings – a connection central to a feminist epistemology, as others have noted:

> the claim that emotion is vital to knowledge both challenges positivism's construction of knowledge with its split between feeling and knowledge, and is part of the move to overcome the historical separation of

139

the faculties: of reason and emotion, thought and action, evaluation and perception. The faculties, which have been constructed as separate and arranged in hierarchical dualities, need bringing together in a way which is both non-hierarchical and anti-foundationalist.

(Rose, 1994: 50)

A growing literature both from women of colour and from postmodern feminists has taught us to be wary of universal accounts and of side-stepping difference. Yet, whilst the forms a rationality of caring takes obviously vary according to context and position, it appears to be an important component in most women's lives[3] regardless of whether they are middle class or working class (see Davies, 1990; Friberg, 1993; Franssén, 1997). As Beverley Skeggs (1997: 164) argues in relation to the British working-class women she studied, for example:

> Another challenge to traditional theories of the subject is made through the women's exercise of responsibility which was opposite to the traditional form of individual responsibility outlined by Mauss (1985) in his discussion of the development of the concept of the self. Rather than being the free, autonomous, independent selves which he suggests accompany the modern self (and on which modernist theories of justice and morality are built), their selves were full of duty and obligation generated through their relationships to others rather than legally enforced.

In the second part of this chapter I will be illustrating some of the features of women's timespace with the help of empirical material from studies that I have carried out in Sweden. The examples are taken almost exclusively from the practices of women working in the care professions where the rationality of caring comes to the fore and where the problem of finding time for a 'pause' are highlighted. But the patterns that emerge are more general, illustrative of the patterns of women's lives more broadly and set by the wider gendering of the public and private spheres.

Pauses

Pauses are moments at work or at home when the usual flow of work is halted for a period. Snow and Brissett (1986: 12) delineate the importance of pauses in the following fashion:

> the most far-reaching consequence of pauses is that they are essential in establishing a rhythm in one's personal and social existence. The fact that rhythm is ubiquitous in all life forms may belie its importance. At the

very least we feel that pausing provides the contrast, emphasis, and energy that aid in developing and sustaining meaning in any arena.

Thus, Snow and Brisset (1986) emphasise that pauses should not simply be conceptualised as breaks in action or periods of *inactivity*. Rather pauses are important for 'recharging the batteries', for rest and relaxation. They also fill another important function, namely the space they provide for *reflection* – a process essential for the development of one's professional self as well as for self-growth more generally.

I will attempt to show that while pauses are essential for an individual's well-being, they may in fact be difficult for women to achieve because of the nature of women's work where caring for others is often central. Having illustrated the difficulties women find in carving a space for a pause, the chapter ends by considering the implications of such difficulties for those who would propose a 'reflexive turn' to the (post)modern subject. Specifically, the remainder of the chapter juxtaposes the reality of the lives of the women interviewed below with the allegorical constructions of Giddens (1991) and Bauman (1995) and their (essentialist) discussion of the reflexivity of the postmodern self.

There's never a spare moment

'There's never a spare moment'[4] succinctly summarises the reality of care-work and of process time (see Davies, 1994, 1996a). Tasks are not easily divided off from each other and the nature of care-work presupposes that as a carer you are always *available*. Timespace is not neatly packaged, so to say, nor easily negotiable.

In child day-care, staff find that they are always being interrupted, both when involved with the children and when talking to other staff. The nature of the work allows no time for reflection or 'time for digestion', as one person put it.

> The few occasions when we get a chance to talk to each other, it's in passing ... but you never get a chance to finish a conversation. You never get a chance to finish a thought.
>
> (my translation)

At one of the child day-care centres I studied, each unit washed up after every meal. This entailed taking the dirty crockery and cutlery into the kitchen, loading the dish-washer, waiting for the machine to be done, drying certain items by hand, taking everything back to the unit and putting it away in a cupboard. By definition, washing-up is a fairly repetitive and boring activity. It also struck me that this was an additional task that the staff were forced to do and that this was possibly disadvantageous in terms of 'snatching' time from being with the

141

children. Yet with time, I saw that it was a duty that also provided an unexpected gift – in that it provided time for reflection, allowing the member of staff both space and time to slip away momentarily from the never-ceasing demands and care of the children.

Thought takes time, and space for this work – both in terms of a separate time period and a separate place – is often limited in women's jobs. Time allotted for meetings in day-care centres, for example, is minimal with just two hours a month set aside for staff conferences and a quarter of an hour a week for parents. While thinking obviously takes place on these occasions, there is for the most part 'a voluminous agenda' as one interviewee put it and much of the discussion is devoted to the fixing of *practical things*. In other words, talk – in a more open and creative sense – is usually strictly circumscribed.

In daily care-work there is no space devoted purely to thinking as a particular and separate activity. Not surprisingly, women will often therefore explain how they use travelling time to plan their work. As one of the principals of a day-care centre put it, for example: 'I start organising in my mind on the bus in the mornings.' Or as another said: 'You often plan while you cycle to the nursery.' The quotation below explains more clearly why time for reflection/discussion is difficult to incorporate into care-work. Whilst the quotation is specifically about the principal's work, the underlying tension and problem are relevant for *all* careworkers in day nurseries and other caring professions:

> Principal: You really get a kick when you've been on those courses and you say to yourself, 'When I get back to work, I'm going to sit down, reflect – give these things some real thought.' For example, how can we best work with our budget and how can we put our goals into practice, economically. And I think to myself, 'I'm really going to get started on this ...' And then I get here on Monday morning – four written off sick, no one in the kitchen ... 'So, how do we solve this?' I'll take over the kitchen on Monday and Tuesday, and then someone else can take over ...
>
> It doesn't matter how many wonderful words I hear at all these courses about our leadership role and how we should define this role and our function clearly to ourselves – because when I get here, *it's the actual day-to-day running of the day-care which comes first*. I have to solve these questions first if there's going to be any day-care at all and the rest has to be put on the shelf in the meantime. There's no way you can say that the food can wait until next week because I want to go into my office and think for a while. It just doesn't work.
>
> So these things you'd really like to put into practice, they fall flat because there's never any time over. Your first priority is to solve the immediate day to day running of things.

142

But do they get put on the shelf for ever?

Principal: No – yes … the ideas are still there. But the problem is the whole time to find space to develop them. And then if you finally get round to it later, you're maybe not as motivated as when you first got the idea. … So you write one of these memos to yourself that this is something I need to work on when there's time, and then it feels more like a burden in some way – 'ugh, I've got this to think about as well' and maybe several months have passed since it was a salient issue.

(nursery/pre-school principal; my translation)

The quotation highlights the issue that lies at the very heart of things. Care and caring have to be given regardless and cannot be temporally or spatially confined. Care-work can be planned, but we cannot expect our plans to hold. Activities cannot be neatly scheduled; the unexpected repeatedly rears its head and demands flexibility and a process relation to time. Time for reflection for oneself – and with others – by necessity gets put aside. Spatially it is difficult to leave those who cannot care for themselves and concentrate on one's own thought and discussion. And if there is no time for reflection, it is not surprising that women's jobs often tend to display a certain conservatism.[5]

A similar pattern can be found within hospital work where it proves impossible either to set specific time (and space) aside for discussion and reflection or to respond to the need for such discussion as and when the situation demands as the relation to the patient (and his/her relatives) is always given priority. Yet pauses are essential, and not least for an airing of ethical issues. Bischofberger *et al.* (1991) argue that medical ethics have a tendency to concentrate their focus on the beginning and end of life. But, everything that lies between may be equally important. According to Björkqvist and Åhlman (1996), nurses should be trained to reflect over ethical problems even on relatively undramatic and everyday occasions, in order to learn from shared experience and in order to change their everyday way of working. Everyday situations could, for example, entail opinions about forced feeding of senile patients, withholding information to patients, or reactions to a battered woman who seeks treatment to name but a few.

Pauses for the discussion of ethical issues and the philosophy of care more generally should obviously be incorporated into the daily routine. Yet their implementation seems difficult. In part this can be related to a tradition of medical care built around a strictly gendered hierarchy where it was expected that nurses carry out the doctors' orders and not think too much for themselves. But it may also be related to the difficulty the care worker faces in leaving those who cannot care for themselves and creating the time or space for reflection. The situation is exacerbated by present-day cut-backs and rationalisations and by a trend prevalent in all walks of working life where individuals are responsible for a greater array of tasks, increasing the general work-load and eliciting feelings of being split. The letter that was previously typed and posted by the secretary is now the responsibility of each individual with the aid of their personal computer. The demands for documentation, quality assurance, etc. increase the requirements placed, for example, upon the nurse. Time becomes increasingly scarce.

> I said to my head nurse: 'There's loads of things I need to tell you'. 'Yes', she said, 'I do as well, but we don't have time at the moment!' 'OK. We'll talk about it tomorrow.' But there's probably not time tomorrow either. And that's the way it works the whole time ... you just cast a few words at each other.
>
> (nurse, my translation)

While the nature of caring and of their responsibility for others affects these women's access to the time and space necessary for reflection, the differential ability to access such timespace cannot be considered outside of the unequal relations of power that structure the workplace and which, in nursing as in a range of other fields, are powerfully gendered. Thus, there are literally no spaces where nurses may 'hide', their offices usually characterised by glass windows. Refusing to be continually available is, it might be argued, a strategy of resistance as well as an indication of processes of professionalisation in women's jobs and in Sweden doctors have been heard complaining that nurses are just not in their offices any more – and thereby available – as they used to be (Närvänen, 1994).

Friberg (1993: 80) has remarked: 'Those who have power can transfer what needs to be remembered to someone else. Women seldom have that much power.' While here I have focused upon the caring professions, the problematics I have recounted can therefore be recognised in other areas too. For example, and not surprisingly perhaps, secretaries and others in servicing positions also often feel the demands of being constantly available, of having to keep tabs on everything, and of how their time and space are regulated by others as access to one's own time and space becomes an ever more critical problem in the modern workplace, structured by an (often gendered) hierarchy of power.

Time out

Whilst I have so far concentrated solely upon a discussion of working life, the problems encountered by women in taking what might be called 'time out' (time away not just from others but time when one does not have to take the needs of others into account) are clearly evident in other arenas too (see Davies, 1990).[6] For women, there are few 'holes' in the intricate temporal structures of everyday life. And when time out does occur, it will probably take place spatially close to those one is responsible for. Given such difficulties it is perhaps not surprising that needlework, embroidery and knitting have traditionally been the mainstay of women's leisure activities – not time out in the true sense, but rather as Verdier (1981) remarks with regard to embroidery 'an art of waiting'.

In contrast, men's time out (such as frequenting football matches or the local pub) tends to take place away from the home or from those one has a caring responsibility for. Men's own time (in the form of sporting activities or social drinking, for example) may also be a way of constructing masculinity and of facilitating homosociability, which in turn can ensure a collective control of particular resources as David Morgan (1981: 104) has recognised:

> Just as Fleet Street male journalists exchange much valuable gossip, tips, tricks of the trade and so on in the course of their social drinking so too do male academics gain access to valuable career resources in the course of their informal meetings over a pint.

Whilst, with the emergence of extensive women's networking, it might be argued that this situation is changing (and Morgan's account is taken from the early 1980s) it is difficult not to compare men's 'time out' with one of my female interviewees whose access to 'own time' was powerfully prescribed by her feelings of a duty of care to her family (Davies, 1990). A housewife,[7] one of her chief interests was making pottery, which she displayed at exhibitions. To make time for her pottery, the respondent had to get up early in the morning way ahead of the rest of the family as once they awoke 'her time' as she put it had to give way to 'their time'. If she did pottery during the daytime, she reported feelings of guilt (and even though the potter's wheel was itself, and not coincidentally perhaps, strategically placed beside the washing machine). The relational context of women's time affects their ability to enjoy longer periods of 'time out' too. Rosemary Deem (1996) for example, has shown that with the preparation beforehand (for example, packing, buying aspirins and suntan oil, preparing the house for a prolonged absence, cancelling of milk and newspapers, etc.) and the work afterwards (unpacking and washing) having a holiday can also be extremely stressful for women and not least as their time 'away' is itself liable to be spent adjusting their activities to the needs and wants of children and husbands.

The self as 'reflexive project'

> I interpret the reflexive subject as not necessarily being a reflecting sub-
> ject. Confronting oneself and the consequences of one's actions, without

having power to self-reflection at the same time – this I see as constituting the angst evident in the late modern individual's life situation.

(Lindström and Nilsson, 1998: 149; my translation)

Theorists of late or postmodernity have argued for the reflexive nature of
present-day society, its institutions and members. Stripped of traditional ties and
norms where choice was limited and identity clearly staked out, the individual –
in Western society at least – is now left to design, stage-direct and juggle his/her
own life, to use Ulrich Beck's (1994) terms. Or in Anthony Giddens (1991: 5)
words: 'The reflexive project of the self, which consists in the sustaining of
coherent, yet continuously revised, biographical narratives, takes place in the
context of multiple choice as filtered through abstract systems.'

For Giddens the reflexive nature of late modernity should be clearly distinguished from the reflexive monitoring which is part and parcel of all human
activity. Modern reflexivity involves a *constant* revision of and rethinking over
actions and decisions in the light of new information or knowledge. Even whilst
the aim is to influence one's future, a prerequisite for such reflection is therefore
time within the *present* in which one might undertake the task of reflection. If self-
reflection is thus linked to self-actualisation, as Giddens (1991: 77) stresses both
are dependent upon the control of time and, more specifically 'the establishment
of zones of personal time which have only remote connections with external
temporal orders (the routinised world of time-space governed by the clock and
by universalised standards of measurement)'.

But, as became evident in the accounts above, it is very difficult for women to
find these zones of time only remotely connected to external temporal orders as
the needs of others constantly circumscribe their access to their own time and
space.[8] Likewise, for those in the position of carer it is perhaps difficult to trace
that other feature of late-modern life so famously celebrated by Giddens, notably
the 'pure relationship' – central to which is a commitment to others built upon
reciprocity, rather than dependency. Denied the reflexivity essential to self-
actualisation, the relationships of dependency that characterise the care
relationship may also deny the possibility of the more equal relationships that
would in turn help create that space.

The player and responsibility?

To end I would like to suggest some other ways in which the realities of women's
day-to-day lives may complicate readings of a new project of the self (set within a
new framework of space and time) argued by some (mainly male) theorists to
characterise the shift from the modern to a postmodern world. For Zygmunt
Bauman (1995), the essential character of modern and postmodern life is

intimately connected to our understandings of time and space. Modern life is seen as a pilgrimage. Whilst Truth is elsewhere (some distance and some time away) one knows where one is heading, the destination is fixed and the unbending arrow of time supports the pilgrim on his journey. His identity is staked out:

> Time one can use to measure distances must be of the sort that school-boy rulers are – straight, in one piece, with equidistant markings, made of tough and solid material. And such was, indeed, the time of modern living-towards-projects. It was like life itself – directional, continuous, and unbendable. Time that 'marches on' and 'passes'. Both life and time were made to the measure of pilgrimage.
>
> (Bauman, 1995: 87)

In a note Bauman points out that the pilgrim is spoken of as masculine. Self-creation, an important component of the pilgrim's project, was not open to women. '[Women] were consigned to the background, to the landscape *through which* the itinerary of the pilgrim is plotted, were cast in perpetual "here and now"; in a space without distance and time without future' (Bauman, 1995: 87). By contrast, identity building is cast differently under postmodernity. The task now, for Bauman at least, is to avoid being fixed. Here time is fleeting, there is no specific goal. Time is only now, fragmented into episodes and space could be anywhere: 'Time is no more a river, but a collection of ponds and pools' (Bauman, 1995: 91). To capture this new sense of time and space, Bauman replaces the modern figure of the pilgrim with four different, though overlap-ping, roles: that of the stroller, the vagabond, the tourist and the player – roles that, and significantly, may now be taken by women as often as by men.

Yet it is perhaps at least questionable whether women should be included at all in the (post)modern world thus described. Each of Bauman's figures is, after all, based upon a notion of individuality where responsibility for others does not exist. How can one be a vagabond and look after children or a frail mother? Likewise, whilst strolling aimlessly around the mall hardly makes sense for the harried mother and employee who whisks round the supermarket before picking up her children from day-care the world of the player described by Bauman – a world where there are 'just the moves – more or less clever, shrewd or tricky, insightful or misguided' – bears little resemblance to the world of the carer (Bauman, 1995: 98).

Bauman is concerned, of course, with the articulation of a new ethics, a readmission of the other as neighbour; re-creating a 'being for' in place of the 'being with' or 'being aside' that is characteristic of the relationships of the stroller, vagabond, tourist or player. Yet if 'being for' means a true meeting of selves where distance is shattered and time (for others) extended, in place of these grand narratives of epochal change Bauman and others who search for that new ethics may do better to re-examine the everyday practices of modern life. It is precisely a 'being for' that characterises the world of the carer and women everywhere who live (for better or worse) with the rationality of care. Far from framed by a new sense of time and space, the spatio-temporal parameters of the carer's world are familiar ones, carrying with them both opportunities and constraints.

Acknowledgements:

My thanks to Jon May for his helpful comments and criticisms of an earlier version of this paper.

NOTES

1 This applied equally to young fathers who took their caring role (of their children) seriously.
2 The 'other' does not necessarily only refer to another person but possibly to an object (e.g. a machine). See Gunnarsson's (1994) study.
3 And of course in men's lives, although perhaps in different proportions. One of the main contributions of feminist research has *not* been to show male and female differences *per se*, but rather to elicit other ways of conceptualising and understanding – arising from women's lived experiences, which break with traditional modes of understanding.
4 'There's never a spare moment' is translated from the Swedish 'allt går i ett'. The Swedish expression captures both the idea that there is no space for any breaks or 'time out', but also a muddled atmosphere.
5 In my study of Swedish nurseries, I asked staff how they would like to change or develop their work. Their answers, in every case, were that they were quite content with the work the way it was (Davies, 1996a).
6 Women's need for 'own time' is possibly a middle-class concern. The working-class women in Friberg's (1993) and Franssén's (1997) study indicated a wish to spend more time with their family.
7 The category of housewife is actually problematic, which is often the case for women. She had no full-time employment but did earn income sporadically and less sporadically by looking after others' children or by delivering newspapers and advertisements in the afternoons.
8 Obviously women do reflect about their jobs (and lives) as indicated by the reference to Schön's 'reflection in action' earlier. Self-reflection is discussed here more in relation to life-politics and the large and continuous chunks of time and own space needed for this.

8

NEW LANDSCAPES OF URBAN POVERTY MANAGEMENT

Jennifer R.Wolch and Geoffrey DeVerteuil

Introduction

Poverty in the United States is increasingly concentrated in large cities, disproportionately burdening particular people and places. The complex realities that underlie these inequitable processes have attracted considerable academic attention, focusing most prominently on economic restructuring (Kasarda, 1989), underclass theory (Wilson, 1987), and neighbourhood impacts (Anderson, 1990). Encouraged by a new focus on *place* in the social sciences and cultural studies, geographers in particular have begun to interrogate the relationships between urban poverty landscapes, poor people, and larger economic, political and social forces (Wolch and Dear, 1993; Kodras, 1997a). Such an interrogation raises essentially geographical questions about how shifts in processes operating at global, national, institutional and individual scales influence the production of poverty landscapes in large American cities.

This chapter broadly relates how economic, political and institutional arrangements interact over *time* and across *space* to produce current poverty landscapes in large American cities. More specifically, we argue that understanding this production (and negotiation) necessarily involves understanding the interactions between temporal *cycles* and spatial *scales*. Numerous geographers have already pointed to the ways in which poverty landscapes emerge from and are experienced through a variety of spatial scales, ranging from the global to the national (Kodras, 1997a) and local (Anderson, 1990). We move beyond these scale-based explanations, however, by also incorporating a more explicitly temporal framework. In this regard, we argue that current urban poverty landscapes are also the product of a set of divergent time-space *cycles*, ranging from longer Kondratieff waves through the shorter cycles of state policy and institutional practices and finally to Hägerstrands's individual time-space geography. Moreover, these cycles operate at different *speeds*, thereby creating some measure of temporal divergency. Combining this temporal focus with the more traditional focus on geographical scale, we argue that different spatial scales are characterized by distinct time-space cycles operating at different speeds.

Ultimately, current poverty landscapes emerge from the unsynchronized nature of time-space cycles, each operating at four spatial scales: long-wave economic cycles at the global scale; policy cycles at the national scale; new practices at the institutional scale; and the increasingly fragmented individual time-space paths of the poor themselves. Before elaborating upon these ideas we foreground current trends with a discussion of historical antecedents. In the next section, we reveal important theoretical and empirical gaps regarding our understanding of current trends. Our focus here is on the concept of *poverty management*, that is organized responses by elites and/or the state, directed generally at maintaining the social order and more particularly at controlling poor people. Poverty management efforts employ particular *strategies*. Strategies consist of the deliberate attempts by the state, institutions and social elites to exert control over poor people's behaviour, ranging from job training and income maintenance to incarceration and institutionalization. Strategies are based in wider rationales on how best to manage the poor and other marginalized populations. For instance, should certain populations be separated from mainstream society, or should they be integrated? Should they be helped or punished? Not surprisingly, institutional *settings* reflect these larger rationales. A strategy of community reintegration, as an example, would necessarily involve more open or unrestrictive settings, while a strategy of separation would involve settings that spatially restrict client activities and behaviours.

Since the onset of large-scale urbanization in the 1830s, poverty management strategies and settings in the United States have fluctuated considerably, ranging from attempts to isolate so-called 'disruptive' populations into rural total institutions (e.g. asylums, prisons) to integrating these same populations into the community under the aegis of deinstitutionalization. In contrast, current approaches to poverty management are changing, appearing to lack any therapeutic premise and relying instead on circulating poor people across a diverse array of unrelated settings, including standard residential dwelling units, shelters, jails, prisons, hospitals, rehabilitation centers, Single-Room Occupancy (SRO) hotels, and the street.

Theoretically and empirically, however, our understanding of what is behind the emergence of these new poverty management settings remains rather thin. In the following section, we address this lack of systematic knowledge by proposing a conceptual framework for the new poverty management. This conceptual framework expands upon the idea that new poverty landscapes are the result of an unsynchronized series of time-space cycles at four spatial scales (global, national, institutional, individual). More specifically, the framework integrates seemingly disparate theories according to the scale of explanation: long-wave and Regulation theory for the episodic nature of macro-economic restructuring and associated social dislocations (e.g. unemployment, crime, extreme poverty); theories of the capitalist state and policy cycles for the national and institutional scales, to grasp the federal and local responses to these social dislocations; and Hägerstrand's time-geography, for understanding how the everyday time-space negotiations of marginalized and non-marginalized populations are both products of and influences upon these larger processes. Finally, we uncover some of the practices of the new poverty management, relating to the fates of women at an emergency shelter and the indigent mentally disabled.

In the conclusion, we once again return to the theoretical importance of combining temporal and spatial analyses in understanding the production and negotiation of poverty landscapes. We cannot understand processes of spatial restructuring and poverty landscapes without incorporating an explicit temporal dimension that is itself scale-dependent. Empirically, we suggest that our conceptual framework can help guide systematic research that is currently unsystematic, suggesting new and innovative ways to understand the emerging poverty landscapes of large American cities.

The production of poverty landscapes: a historical overview

Landscapes do not simply 'exist' as inert and reified but 'become' in a continuous process of transition (Pred, 1984; Schein, 1997). Landscapes are the concretiza-tion of social relations at a particular moment, representing a palimpsest of contested and restless processes and discourses (Ley, 1988; Knox, 1991; Schein, 1997). Geographers have approached the production of poverty landscapes from many angles, including the role of the state (Kodras and Jones, 1990), geographi-cal unevenness across various poverty landscapes (Kodras, 1997a), and the constraints of poverty landscapes upon marginalized populations (Rowe and Wolch, 1990; Wolch et al., 1993).

In terms of process, the production of (poverty) landscapes reflects an 'his-torically contingent process that emphasizes institutional and individual practices as well as the structural features with which those practices are interwoven' (Pred, 1984: 280). This quote highlights landscape formation as a tripartite process, operating simultaneously and interactively at multiple spatial scales (Thrift, 1983; Pred, 1984; Dear, 1988; Dear and Wolch, 1989, Gregory, 1994a): (1) the structural, relating to longstanding social forces that persist over time; (2) the institutional, referring to phenomenal forms of deep-seated structures, acting as mediators between these and individual agents, both enabling and constraining; and (3) the individual, incorporating both individual responses to the constraints

151

imposed by forces operating at larger scales as well as the creative capacity of knowledgeable actors (agents) to affect these same forces (Dear and Wolch, 1989: 6).

With this schema in mind, individual poverty landscapes emerge as unique repositories of this threefold set of processes. Landscapes, however, are not simply repositories; they also have the capacity to shape societal dynamics, as part of the larger socio-spatial dialectic (Soja, 1980). Dear and Wolch (1989: 9) recognize three aspects to this dialectic: social relations are constituted through space (e.g. opportunities for natural resource extraction influence the location of agricultural activities); social relations are constrained by space (e.g. past sediments of human activity, in the form of the built environment or other similar configurations that may hinder present action); and social relations are mediated by space, whereby propinquity and the friction of distance can help coalesce time-space paths into specific institutions. With these points in mind, (poverty) landscapes emerge as 'complex synthesis of objects, patterns, and processes derived from the simultaneous interaction of *different levels of social process,* operating at *varying geographical scales and chronological stages*' (Dear and Wolch, 1989: 7).

The historical geography of poverty management and poverty landscapes in the United States

The imprints of poverty management upon the landscape are both ideological and physical: a specific strategy (e.g. rehabilitation, punishment, isolation) becomes geographically manifest in terms of its settings, including how these settings appear and how they are arranged in space (e.g. concentrated, dispersed, isolated, centralized). In turn, the nature of poverty landscapes influences both our understanding and ideological constructions surrounding poverty (e.g. late-twentieth-century urban ghettos affect our opinions regarding welfare, crime and race), and strategies/settings of poverty management, acting as 'potent catalyst[s] for social speculation and social action' (Boyer, 1978: vii). In this sense, the presence of extreme poverty can motivate policies as diverse as poorhouses to wholesale urban renewal. Therefore, the appearance and patterns of poverty management settings, such as prisons, asylums or community clinics, have important impacts upon the production of poverty landscapes.

The complex interplays between poverty management and poverty landscapes, however, do tend to stabilize into distinct periods, coalescing around particular institutional patterns. This stabilization corresponds to specific economic, political and societal arrangements at the structural level, subject to particular pressures, and boasts specific strategies of poverty management. In turn, poverty management strategies, and the therapeutic ends they seek, are sustained by particular concepts of place (e.g. countryside as salubrious, cities as diseased), in

Table 8.1　Production of poverty landscapes in the United States

	Pre-asylum era (pre-1830s)	Total institutions (1830s–1960s)	The deinstitutional-ization experiment (1960s–1980s)	Current conditions (1990s): the displacement model
Economic, political and societal arrangements	Agrarian society	Large-scale urbanization and industrialization: advent of anonymous society	Fordist welfare state: federal takeover of local welfare responsi-bilities, advent of psychotropic drugs	Post-Fordist welfare state: federal welfare reform, with devolution, privatization and dismantle-ment
Poverty management strategy	Informal local control	Individual transformation	Community reintegration	Bureaucratic expediency
Concept of place	Local generally best	Urban depravity vs. rural salubrity	Promise of community rehabilitation	Purity of urban public space
Poverty management settings	Diverse and unrelated: 'Ship of Fools', servi-tude, family home	Asylums, prisons, jails, orphanages	Community clinics, group homes, service-dependent ghettos	Unrelated array of treatment settings (jail, shelter, street, etc.)

large part determining the settings of poverty management. Table 8.1 outlines four historical periods in American poverty management history, according to these dimensions.

This periodization of poverty management begins in the late eighteenth and early nineteenth century, when a fundamental shift in the Western world occ-urred in how and where certain so-called problem populations (e.g. the poor, criminal, sick, insane) were to be managed. Previously, low population densities and a weak central state combined with strong local attachments ensured that such populations were treated locally in a non-systematic fashion (Rothman, 1971; Foucault, 1977a). In this *pre-asylum, agrarian era*, social control was informally exerted at the village scale, and generally lacked any systematic institutionalized mechanisms to confine disruptive populations. While poverty management strategies emphasized laissez-faire local control, settings ranged widely, from the family home to indentured servitude and the 'Ship of Fools' (see Rothman, 1971; Foucault, 1973).

In contrast, by mid-century these same populations were now confined to a constellation of spatially isolated institutions, such as prisons, asylums, poorhouses and juvenile halls, representing a shift to an era of *total institutions*. Goffman (1961) defines total institutions as places where 'their encompassing or total character is symbolized by the barrier to social intercourse with the outside and to departure that is often built right into the physical plant, such as locked doors, high walls, barbed wire, cliffs, water, forests or moors' (4). The new principle became one of individual transformation of so-called problem populations through spatial separation from mainstream society and its corruptions (Philo, 1989: 259). For instance, in early-nineteenth-century America, the city was seen as 'dangerously disorganized, threatening and overstimulating' (Dear and Wolch, 1987: 31). By contrast, the countryside was seen as far more salubrious.

As poverty management strategies changed, so did its settings. Given the great fear of cities, the countryside emerged as the favoured environment to locate new institutional settings. Across Jacksonian America rural prisons, asylums, and almshouses sprouted (Rothman, 1971). For example, prisons were built for Auburn NY in 1823, Sing-Sing NY in 1825, Eastern State Penitentiary PA in 1829, Charlestown MA in 1829, and San Quentin CA in 1852. Moreover, by 1860, twenty-eight of thirty-three states had rural, public institutions for the insane. Despite successive attempts at institutional reform throughout the nineteenth century, and particularly during the Progressive era in the early twentieth, the sense that 'disruptive' populations had to be removed from their corrupt (urban) environment would prove surprisingly strong (Rothman, 1980).

The tendency of poverty management strategies to favour isolated spaces of control, however, shifted radically in the 1960s. At that time, the majority of Western nations participated in the deinstitutionalization of former mental patients, criminals, alcoholics and the like (Scull, 1984; Staples, 1990). This *deinstitutionalization experiment* reflected the political imperatives of shifting the financial burden for treatment from states to the federal level while avoiding a potentially costly renovation of an outdated system (Scull, 1984). Deinstitutionalization was also sustained by more individual forces, specifically the introduction of psychotropic drugs and the civil rights movement for clients, along with shifting medical judgments regarding suitable confinement (Lerman, 1982). This experiment would be instituted through a system of decentralized clinics and outreach centres, thus initiating a fundamentally new set of locational dynamics in service provision (Wolpert *et al.*, 1975). As strategies shifted, so did poverty management settings: 'instead of spatial isolation of deviant, dangerous and dependent classes, the modern prescription for client rehabilitation called for spatial integration, engagement in community activities and interaction with the non-dependent population in the course of everyday life' (Dear and Wolch, 1987: 62).

The outcome of this experiment was decidedly mixed, however. Lack of funds and political will, combined with virulent community backlash (Not in My Backyard or NIMBY), led to a drift of services and clients to heterogeneous inner-city locales. The unintended consequences of deinstitutionalization were especially felt in traditional Skid Row areas. Given its 'soft', heterogeneous nature, combined with a large stock of dilapidated housing, traditional Skid Rows served as new service-dependent ghettos, the setting to contain marginalized populations, including parolees, former mental patients, the homeless and the precariously housed (Dear and Wolch, 1987). Although in some ways a coping mechanism, the isolation of Skid Rows from mainstream society led some commentators to call them 'asylums without walls' (Wolpert and Wolpert, 1974; Dear and Wolch, 1987). This lack of financial and rhetorical support for the deinstitutionalization experiment would prove highly damaging and discrediting, leading to calls for reinstitutionalization, and to new, and as of yet somewhat undetermined, poverty management strategies and settings.

A new model of poverty management?

In the past twenty years, the shortcomings of the deinstitutionalization experiment have engendered an increasingly diffuse approach to poverty management. In effect, there has been a dramatic rise in the number of inappropriate, inadvertent and informal settings for marginalized populations, as part of a seemingly new approach to poverty management. This trend is epitomized by the increased transinstitutionalism among the mentally disabled, especially their tendency to move between carceral settings and a makeshift life on the street. This chronic misassignment was recognized by Dear and Wolch in 1987: 'Many groups are being misassigned to inappropriate social settings and reinstitutionalized (for instance, in prisons) because they lack other shelter options' (4). The problem has only worsened in recent years. According to a recent article in the *New York Times* (Butterfield, 1998: 1), the Los Angeles County Jail is 'by default ... the nation's largest mental institution. On an average day, it holds 1,500 to 1,700 inmates who are severely mentally ill, most of them detained on minor charges, essentially for being public nuisances.' Once released, many inmates drift back to the street, only to be arrested again. In a sense, these 'new' poverty management settings represent a full circle, a return to the nascent period of institutionalization when patients suffering from radically different problems were lumped into the same institutional space. Other marginalized groups are also finding themselves in inadvertent, inappropriate and informal settings. For instance, in Los Angeles, there has been a dramatic rise in the number of women with children seeking emergency shelter on Skid Row, traditionally a service hub for service-dependent, unattached males, due to the lack of sufficient family shelters and transitional housing elsewhere (Rivera, 1998).

While increasing attention is being directed towards these anomalous settings, less is known about what is driving these changes. More specifically, how do changes in poverty management settings reflect changes in poverty management strategies? The lack of systematic theorization is addressed in the next section, where we propose a conceptual framework that locates the new poverty management approach as a product of unsynchronized series of time-space cycles at four spatial scales (global, national, institutional, individual).

Understanding the new poverty management

Integrating insights from long-wave and Regulation theory, theories of the capitalist state and policy cycle models, as well as individual time-geography within a single conceptual framework enable us to explain how and why new poverty management strategies are emerging at this particular time and in large American cities, and ultimately how these new strategies are influencing the production of new poverty management settings.

Figure 8.1 outlines the conceptual framework and its main relationships. At the global scale, long-wave economic restructuring creates surges of social dislocation (e.g. unemployment, crime, etc.). These dislocations are then typically addressed at the national level through the welfare state. However,

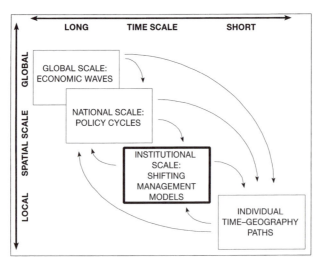

Figure 8.1   Conceptual framework for the new poverty management

because the welfare state tends to operate at a different speed and rhythm than the global economy, national-level policy cycles may (or may not) favour increased social intervention; together, these structural forces determine the scale and scope of marginalized urban populations, such as the homeless, disabled, and other economically peripheral populations. Although marginalized across diverse and multiple dimensions (e.g. disability, race, age, gender, substance abuse), these populations are generally marked by non-participation in or sporadic attachment to the labour market, social isolation, and service dependency. If macro-economic restructuring coincides with shifts in the national welfare state, the result can be a dramatic change in the circumstances of marginalized populations, generating new forms of local poverty management strategies and settings. For instance, as the numbers and presence of marginalized populations have grown, their paths have intersected with the time-space activity patterns of homed residents and business owners, deepening fears for the safety (and purity) of urban space (Stoner, 1995; Zukin, 1995; Ellickson, 1996; Mitchell, 1997). In attempts by urban managers (bureaucrats, police, planners, politicians, real estate agents) to exclude marginalized populations from mainstream urban society, new institutionalized responses have emerged. As a result of the interconnection and lack of synchronicity between larger time-space cycles and individual paths, marginalized populations in cities are presently encountering (and negotiating) a reconfigured poverty landscape. We conclude this section with a preliminary description of the practices and negotiations of the new poverty management.

The global scale: economic change, Kondratieff/Schumpeterian long-waves and Regulation theory

At the largest temporal and spatial scale, certain large-scale and long-term economic phenomena, particularly economic downturns and crises, alternate according to recurring *cycles*, *waves* or *rhythms*. The basic thrust of long-wave theory is that growth in the macro-economy alternates according to persistent and regular cycles. Long-wave models attempt to systematize economic trends and account for economic growth and decline according to a dependable timetable. Such recurring shifts in macro-economic structures are crucial in understanding the timing and spacing of social dislocations and the national, institutional and individual responses they engender. In this regard, Kondratieff proposed a 55-year wave whereby major depressions alternated with major booms, while Schumpeter's theory (1939) of 'creative destruction' focused on how economic periodicity interacts with business and technological innovations.

Long-wave theory, not unlike other time-space theories, tends to simplify many aspects of society, especially the role of the state. In this regard, we complement long-wave theory with Regulation theory, which argues that capitalism, because of its inherent tendency towards episodic crisis, requires social and political intervention (Boyer, 1990; Yaghmaian, 1998). Crises consist of 'ruptures in the continuous reproduction of social relations, and lead to the restructuring of the system and the gradual emergence of a new regime of accumulation' (Yaghmaian, 1998: 244). According to Regulation theory, the twentieth century has been marked by two distinct periods of capitalist accumulation: Fordism, based in large-scale production, consumption and an expansive welfare state, and post-Fordism, based in flexible labour and capital (Lash and Urry, 1987; Storper and Walker, 1989).

This shift to a post-Fordist period has clearly intensified the scale and scope of social dislocation (e.g. unemployment, crime, welfare dependency) (Lee, 1997; Wolch, 1998). In effect, as the economy contracts and/or changes gear, social stressors accumulate and intensify, resulting in the increased prevalence of problems such as mental illness, crime and welfare dependency. For instance, Brenner (1973) claims an inverse relation between the state of the economy and mental illness, that instabilities in the macro-economy are 'the single most important source of fluctuation in mental-hospital admissions or admission rates' (ix), and that this relationship has proved remarkably stable since the 1830s. Increased incidence of mental illness is a byproduct of economic instability, meaning that overall prevalence of mental illness is presumably cyclical in nature. Crime is another, although somewhat muddled, example. Crime tends to increase during economic downturns, although arrest and incarceration rates reflect not only crime rates but penal policy as well (Zimring and Hawkins, 1991). The scale of welfare also tends to fluctuate according to the state of the economy, with caseload surges during downturns and major economic restructuring (Lee, 1997; Wolch, 1998). All of these examples point to the necessarily temporal dimension to social problems.

These social dislocations are traditionally addressed at the national level, and it is here where we encounter a fundamental problem with Regulation theory. In effect, while long-wave theory generally ignores the state, the tendency in the Regulation school is to perceive the state as primarily corresponding to the respective regime of accumulation (i.e. Fordist regime = Fordist welfare state). By contrast, Clark and Dear (1984) understand the state as both embedded in, and partially autonomous from, capitalist relations. At this point, it becomes necessary to explore some of the basic contentions of state-centred theory, focusing in particular on how the state, as both embedded in the capitalist system and relatively autonomous from it, intervenes in the face of large-scale social dislocation.

The national scale – social dislocation, state
theory and policy cycles

Given that social dislocations are typically addressed at the national level, it becomes necessary to expand beyond the narrow, economistic assumptions of long-wave and Regulation theory in favour of more state-centred perspectives, including theories of the capitalist state. Theories of the capitalist state are crucial in addressing the role and timing of the state in managing social dislocation. Two premises inform our analysis: first, that the state must balance the need for private accumulation with legitimizing the social and economic system (Lake, 1997). When downturns occur, this balancing act becomes ever more difficult, stimulating a legitimation crisis of the state (O'Connor, 1973). Despite these constraints, the state always enjoys some autonomy regarding interventions in the face of social dislocations, which brings us to the second premise: that the state is deeply embedded within capitalist relations, but at the same time exercises some relative autonomy (Clark and Dear, 1984). For example, the welfare apparatus intervenes to minimize the periodic surges of inequalities endemic to capitalism, distributing benefits and ensuring the legitimation of the overall system. At the same time, however, the state is constrained in its actions, as it must also ensure the smooth process of capitalist accumulation.

The idea that the state interacts, albeit somewhat autonomously, with larger economic conditions suggests that policy shifts also occur with some regularity. However, state theory does not address the idea of regularity among state actions, the potential for cyclical synchronicity between the larger economic context and the national state. In contrast, a policy cycle perspective explicitly addresses this issue. Berry *et al.* (1995), for example, argue that national-scale social interventions occur at regular intervals according to 55-year Kondratieff waves, contending that macro-economic shifts and political/policy changes are temporally interdependent. Their argument begins with surges of inequality, the result of stagflation crises. As inflation and collapse of old industries take their toll, the political mood becomes more conservative and pro-growth. In turn, this conservative backlash and its emphasis on technological innovation, efficiency and growth is 'replaced by one concerned with social innovations, equity and redistribution' (153). These eras of increased social intervention naturally feature the emergence of new institutional forms to deal with social problems. For example, the fallout from the 1825–6 financial crises and conservatism engendered a significant rise in social intervention, particularly in the rise of total institutions. This sort of significant change in the focus of state action is what Hall (1993) calls paradigmatic shifts in social policy, a disjunctive process whereby formerly stable policy intentions are radically overhauled in favour of new, and generally untested, interventions.

We can now incorporate this explicitly cyclical perspective into capitalist state theory and emerge with a stronger understanding of how the state and larger economy temporally interact to affect the size and circumstances of marginalized populations. As capitalist state theory indicates, the state may exercise substantial control over the timing, scale and scope of these 'paradigmatic' policy shifts. This suggests that although state action is embedded within larger economic conditions, it none the less retains some political autonomy. This insight is particularly useful in explaining the recent demise of the Fordist welfare state, a fundamental context for explaining current poverty management strategies. Especially since the early 1980s, the impetus has been to: (1) devolve welfare responsibilities to lower tier governments; (2) privatize government functions to the commercial or non-profit sector; and (3) dismantle (withdraw, reduce or transform) the federal welfare state (Wolch, 1990; Kodras, 1997b, 80–2). This reduction in the federal welfare state during the Reagan–Bush years, however, came at a time of great economic upheaval, when America deindustrialized and unemployment returned to Depression-era levels. This lack of synchronicity between economic change and social policy would dramatically increase the number of marginalized populations and fundamentally alter their lived circumstances. In the next section, we address the implications of these changes at the individual and urban scales using the concept of individual time-geographies.

The individual level: the time-geography of marginalized populations

According to Wilson and Huff (1994: xv), 'the sources for persistent inequalities in capitalism may have deeply entrenched and far-flung roots, but it is at the scale of local places and settings that these forces are poignantly felt'. In other words, attention must be paid to the localized outcomes (and shapers) of larger dynamics and cycles. The combined force of global economic turbulence and national welfare retrenchment generated, especially after the early 1980s, an expanded pool of marginalized populations at the local level. This increase in numbers was the product of various trends: economic shocks combined with the persistent shrinking of the welfare state at the federal and local levels; the loss of affordable housing in the urban core, due to gentrification and redevelopment; and finally, virulent NIMBYism in suburbs, leading to a growing concentration of human services in dilapidated inner-city locales, most notably traditional Skid Rows (Wolch and Dear, 1993; Wright, 1997). These trends disproportionately affected large cities. For example, in 1995, Los Angeles County had the highest rate of poverty (14.8 per cent) and the largest welfare-dependent population (almost 2.1 million, or 1 in 4) in California (Wolch and Sommer, 1997: 10–11), with both poverty and welfare caseloads sharply up since 1980.

Falling through the cracks, the homeless and the mentally disabled became more apparent in prime downtown areas, their paths intersecting with a variety of homed residents, notably local business people (Stoner, 1995; Ellickson, 1996). At the same time, cities had become more attentive to their image and their attractiveness to global capital in the increasingly crowded urban hierarchy (Harvey, 1989; Kenny, 1995). As a result, a backlash grew against the unfettered use of public space in the city by these newly prominent marginalized populations. A host of new and more punitive urban policies appeared: persistent NIMBYism that now extends to poor inner-city neighbourhoods (Pulido, 1996; Takahashi and Dear, 1997); more aggressive policing against 'quality of life' crimes (Ellickson, 1996); the criminalization of homeless survival tactics (Mitchell, 1997; Stoner, 1995; Takahashi, 1996; Takahashi and Dear, 1997); and the rise of privatized urban caretakers in the form of Business Improvement Districts (BIDs) (Zukin, 1995). As a result, public officials now routinely attempt to seal off urban public space from marginalized populations, one much-publicized example being the physical 'cleansing' of Thompson Square Park in Lower Manhattan in the early 1990s; at this micro-level, the welfare or public city was being replaced by the revanchist one (Smith, 1996).

One way to understand the constrained options of marginalized populations in the revanchist city is Hägerstrand's notion of *time-geography*. Time-geography is both a notation and a heuristic device designed to contextualize collateral processes in time and space. Individuals use time and space as resources to fashion particular *paths*, a trajectory or movement that spans the daily to the yearly and even the lifetime (biographical) scales (Pred, 1996). In turn, these paths interact with larger time-space *projects*, that is the bundling of numerous individual time-paths towards a similar goal (Jackson and Smith, 1984: 50). Projects may include downtown redevelopment or the criminal justice system. By taking into account this intimate relationship between path and project, it is possible to determine an individual's cumulative time-space possibilities or space-time prism (Miller, 1991; Gregory, 1994b).

Viewed this way, the appearance of the homeless, the mentally disabled, and other marginalized populations on the streets of large American cities in the 1980s, as well as the backlash it generated, was the result of the unsynchronized interaction between the global economy and national policy cycles. In effect, as the global economy shifted and the national welfare state downsized, expanding the number of marginalized individuals, their time-paths were more likely to intersect (and clash) with those of the time-paths of homed individuals as well as larger time-space projects, such as downtown redevelopment and the production of consumption and pleasure spaces (Wright, 1997). This dynamic stands in direct contrast to the total institution era, when the time-space paths of marginalized populations were frequently subject to complete time-space control. Based in this time-geography approach, the backlash against marginalized populations in urban areas now appears as a move by urban managers to constrain the disjointed time-paths of marginalized populations by eliminating their access to public space. In several studies of homeless mobility patterns in Los Angeles (Rowe and Wolch, 1990; Wolch *et al.*, 1993; Lee, 1994), the time-space paths of the homeless were subject to important institutional (e.g. law enforcement, service providers) constraints. As part of this larger backlash, local institutions have also begun to impose a more constrained time-space path upon marginalized populations under conditions of global and national restructuring, resulting in shifts to poverty management strategies and settings.

The institutional scale: new strategies and settings of poverty management

We contend that new poverty management strategies, while certainly driven by both long-term, large-scale forces (i.e. the unsynchronicity between global waves and national policy cycles) as well as short-term, small-scale forces (i.e. the disjointed time-space paths of marginalized populations), ultimately take shape at the institutional scale. In effect, we argue that institutional responses and strategies are not exclusively the outcomes of larger structures or individual agents; rather, institutions act with some autonomy and according to their own agendas, and should be viewed as independent agents with considerable influence upon the everyday lives of marginalized populations (Hopper *et al.*, 1997). In other words, they 'do not simply constrain or channel the actions of self-interested individuals, they prescribe actions, construct motives, and assert legitimacy' (Skowronek, 1995: 94). Institutional responses are thus never wholly predetermined, but emerge from professional norms, capacity constraints and client pressures.

In this section, we locate new poverty management strategies as the unintended and unsynchronized results of the intersections of larger time-space processes, individual time-space paths and institutional semi-autonomy. More specifically, new poverty management strategies are based upon three premises: (1) the *localization* of care and (2) *privatized* administration and service delivery, resulting in the (3) increasing *circulation* of clients. Rather than an aberration (e.g. Torrey *et al.*, 1992; Petersilia, 1997), these strategies (and settings) reflect a new and disturbing dynamic within the production of poverty landscapes.

As a first premise, the localization of care is primarily an outcome of federal, and, in many instances, state welfare devolution. Devolution has the implication of downloading responsibilities for poverty management and other 'knotty problems' to states and local jurisdictions so that the central state may avoid the political fallout (Wolch, 1989: 198). In turn, the local state is typically severely constrained in its ability to raise taxes and attract outside capital, and of all levels of government is least able or willing to redistribute wealth (Peterson, 1981; Clark and Dear, 1984). Thus local politicians and urban managers are concerned to minimize their social outlays and avoid attracting outside welfare seekers (the 'welfare magnet' effect), thereby precipitating a 'race to the bottom' in the provision of welfare services (Kodras, 1997b: 84). These trends have only been exacerbated by the passing of the 1996 Personal Responsibility and Work Opportunity Reconciliation Act. The Act codified several secular trends in state restructuring: the localization of welfare provision and administration to the state/local level, individual over collective responsibility for poverty, and self-sufficiency over basic entitlements.

The localization of care for marginalized populations has encouraged further fragmentation of the welfare system, producing an uneven landscape of welfare provision (Wolch and Dear, 1993; Wolch, 1996; Taylor and Bassi, 1998). In an effort to avoid the social costs associated with welfare devolution, many localities are pressuring community institutions to fill the gaps. Moreover, there is increased pressure to transfer services to the non-profit or even private sectors. According to Starr (1988), privatization is more than simply returning services to the private sphere. Privatization holds the promise of replicating the quality of public provision, while at the same time representing an underlying shift of claims in a society and casting into doubt the desirability of public provision. Privatization of human services has generally involved government contracting out responsibilities to the private (or not-for-profit) sector while retaining a primary role in funding, leading to an increased interdependency (Gibelman and Demone, 1998: xi). While decreasing funding for traditional institutions such as the asylums, the federal government has subsidized the spectacular growth in non-traditional institutions, leading to a profusion of privatized, contracted and not-for-profit facilities. The effects of this transfer have been to further diminish the traditional welfare state while abetting the rise of the shadow or contract state, in turn exacerbating inequalities among have and have-not neighbourhoods (Wolch, 1989, 1990; Smith and Lipsky, 1993).

As the pendulum swings in favour of localization and privatization, poverty management has become fragmented across the public, not-for-profit and private spheres. For instance, not-for-profit organizations increased from 309,000 in 1967 to nearly 1 million in 1998 (Weisbrod, 1998: 69). Moreover, the concomitant increase in numbers and reduction in size of service providers, as well as the decline of traditional inner-city service-dependent ghettos (Dear and Wolch, 1987; Wright, 1997), serve to spatially fragment service provision, further disorganizing the time-space paths of service-dependent populations. The localization and privatization of basic service delivery systems, combined with the cutback or dismantlement of local services, have also encouraged the development

163

of a wide, and as of yet poorly studied, world of unlicensed facilities. These facilities, consisting mostly of non-traditional, proprietary institutions (e.g. out-of-home placement facilities and substance abuse rehabilitation centres), work at the margins of the poverty management system (Lerman, 1982; Glover, 1998). Further, the collapse of many state-supported systems of care, including drug rehabilitation, has abetted the rise of more informal and unsupervised settings for personal transformation, including many unlicensed sober-living homes in inner-city areas. For example, recruiting directly from the criminal justice system and street corners, the 'Victory Outreach' organization has 30,000 members, with seventy unlicensed facilities in the Los Angeles area alone (Glover, 1999). Former institutional goals of individualized treatment/recovery and community reintegration are thus displaced by aggregate case management, cost-shifting and caseload shedding, opportunistic fund-seeking and general bureaucratic expediency (Gibelman and Demone, 1998).

While these strategies cut across a broad spectrum of the poverty management landscape, they are couched according to different rhetorics, address different marginalized populations, and, critically, promote different institutional settings. For instance, different marginalized groups are institutionalized according to different rhetorics: emergency aid for shelters, social order for jails, and family values for out-of-home placement facilities. While offering apparently distinct settings, poverty management strategies essentially compartmentalize poverty-related problems, camouflaging their structural origins and confusing policy goals. Moreover, the settings are portrayed as solutions to unrelated problems. As such, client problems may be interpreted as symptoms of individual misfortune, culpability or weakness, thereby encouraging official abandonment, containment, punishment and circulation.

The new poverty management in practice

Using two case studies, we are now in a position to explain why fragmented and unrelated settings of poverty management have emerged at this time in large cities, and why certain segments of marginalized populations are prone to circulating across this reconfigured poverty landscape. To understand how divergent time-space cycles shape the everyday negotiations of the poor, we focus first on evidence of cycling among women applicants to a neighbourhood emergency shelter, and then on transinstitutionalism among the mentally disabled.

Our first example involves the experiences of women upon application to a small emergency shelter located in Central Los Angeles. During 1997 and 1998, the 268 incoming clients had cycled across various unrelated settings (e.g. jail, shelter/transition homes, own apartment, with friends/family, hospital) prior to arriving at the shelter. For many of these women, arriving at the shelter represented another 'station' on a seemingly aimless cycle; for others, the shelter was a strategic move to avoid homelessness and/or having to stay in more restrictive settings. Regardless, the vast majority arrived without financial resources and having suffered a series of personal crises that precipitated a pattern of precarious shelter, institutionalization and homelessness. For instance, a majority of women indicated eviction, loss or decline of income, domestic violence, discharge from prison or hospital, and being asked to leave by family or friends as major factors contributing to their application to the shelter. In effect, by the time they reached the shelter, 20 per cent of the women had hit rock bottom, spending the previous night on the street, at a bus station or in a car. A further 25 per cent had spent the previous night at a shelter or transitional facility, suggesting inter-institutional migration, while barely 5 per cent had spent the previous night in their own apartment or house.

These statistics indicate the general impoverishment of the women, their lack of personal and/or familial resources, and their reliance on institutional settings for survival. For those women who had cycled across at least three or more settings before arriving at the shelter (n = 45, or 17 per cent), the pattern of personal crises emerge even more strongly: 36 per cent had used mental health services at one point in their lives, 44 per cent had suffered episodes of domestic violence, and a full 62 per cent had either been on the street or in other shelters the previous night. Altogether, these patterns suggest a precarious existence, marked by personal crises and downward mobility. In previous periods these women may have found decent employment or been sufficiently supported by the welfare state; however, the decline of both the Fordist economy and the traditional welfare state, combined with the rise of new poverty management strategies, has now greatly limited the women's life chances and induced cycling across unrelated settings. For instance, although most women held at least high school educations (almost 80 per cent), the vast majority of applicants (90 per cent) did not hold jobs and, if they depended on welfare, received insufficient or erratic benefits (i.e. aid of last resort for unattached individuals – General Relief (GR) – set in Los Angeles County at $222 a month plus food stamps). In effect, since the mid-1980s, GR has been targeted by the County for restrictions on eligibility, sanctions and cutbacks in monthly payments (Lee, 1994). In February 1998, fearing a surge of former welfare recipients following federal welfare reform, the County implemented a six-month time limit per year on those so-called 'employable' GR recipients, or nine months if recipients participate in job training.

These limitations on locally funded aid of last resort mirror a nationwide trend to limit or dismantle local welfare benefits. As early as 1981, Pennsylvania had eliminated their aid of last resort (Halter, 1992), while Michigan's General

Assistance was also dismantled in the early 1990s. Moreover, as welfare reform funding now favours job training over income maintenance, those considered 'unemployable' will be increasingly left to their own devices. For many marginalized populations in Los Angeles County, GR was temporary aid of last resort, but for others it represented a fundamental and long-term part of their survival patterns. For those (former) recipients, welfare reform is likely to reconfigure individual survival patterns and how they interact with institutional settings.

The intersections of global economic compressions and federal welfare state retrenchment (including deinstitutionalization and the collapse of the community mental health system) have already radically altered the survival patterns and time-space paths of the indigent mentally disabled. As previously stated, the increasing numbers of mentally disabled on city streets provoked a strong yet uncoordinated backlash, involving greater intervention by the criminal justice system and greater cycling across unrelated institutional settings. Within this context, Hopper *et al.* (1997) undertook one of the few studies to explicitly deal with how the mentally disabled are churned from one institutional setting to another upon release from mental hospitals. Out of a group of 32 mentally disabled patients, 20 had spent 59 per cent of their last five years moving from one institutional setting to another, suggesting a 'durable pattern ... of a life lived on the "institutional circuit" with occasional breaks for temporary housing of their own' (662). In effect, for this subgroup, institutionalization perpetuates, rather than alleviates, individual time-space instability. For the mentally retarded, the collapse of the mental health system also held disastrous consequences: 'many get released from state mental hospitals to community settings with few services and little support. Without services, they flounder and eventually come to the attention of the police and the courts. The result is they end up trading one institutional address for another, and the number of mentally retarded persons in correctional institutions continues to grow' (Petersilia, 1997: 368).

These fragmented time-space paths differ greatly from those of the asylum period. In general, the time-space paths of those deemed mentally disabled were generally quite restrained by the strict schedule of the asylum, in effect 'administer[ing] an ordered routine and hop[ing] to eliminate in a tightly organized and rigid environment the instabilities and tensions causing insanity' (Rothman, 1971: 151). Moreover, many asylums physically isolated patients from each other and the staff. Of course, while many early asylums resembled penitentiaries and were essentially repressive sites, others were far more conducive to mingling between patients and staff, as well as among patients (Philo, 1989). For good or ill, however, the total institution period did impose a *stable* time-space path upon the mentally disabled. As Parr and Philo (1996, 50) state: 'the experience of an asylum is particular and acute, and perhaps to understand this experience is to recognize the status of the asylum as a structured day-to-day reality providing a measure of stability, counteracting the internal confusion endured by many residents ... an external place where individuals find supports to deal with internal distress, an important site of outward reference for the mental patient'.

In contrast, the new poverty management era has imposed an *unstable* time-space path upon the mentally disabled. Once deinstitutionalized, former mental patients who lack outpatient care drift into traditional service-dependent ghettos, and from there cycle among shelters, the street, clinics and jail. In 1991 the *Los Angeles Times* described the wanderings of Michael Brewer, a 33-year old paranoid schizophrenic (Tobar, 1991). He had been to jail fourteen times between 1986 and 1991, staying in special cells for the acutely mentally ill. Upon release, he returned to Skid Row, living on the street until he was once again incarcerated. For him, jail was a chaotic place: 'there's the criminally insane in there. They get picked on ... [The deputies] beat them up. They won't get their treatment [medications]' (A24). On the streets, he is afraid of the police. His fears are not unfounded: in May 1999, a homeless mentally disabled woman in Mid-City Los Angeles was shot and killed by the police, supposedly after 'she threatened officers with a screwdriver when they approached her to ask if the shopping cart she was pushing had been stolen' (Rabin, 1999: A1).

Conclusions

This chapter has focused on the historical and conceptual moorings of the new strategies and settings of poverty management. We have speculated that a new approach to poverty management is being implemented, the result of unsynchronized time-space cycles operating at four distinct scales (global, national, institutional, individual). The new approach to poverty management is driven essentially by bureaucratic expediency and the basic unwillingness and inability of localities to manage extreme poverty. Institutional goals have become caseload shedding, damage control and emergency services, as opposed to care or recovery. As such, this most recent poverty management mode appears dispersed, uncoordinated and unsuited to providing a proper continuum of treatment, especially in contrast to former modes (total institutions, community integration). The advent of federal welfare reform can only amplify these trends.

More critically, however, this chapter has illustrated the importance of combining both spatial and temporal analyses when attempting to understand the emergence of new poverty management strategies and settings, as well as the everyday negotiation of these new strategies and settings by marginalized populations. In effect, new poverty management strategies and settings work through a divergent set of time-space cycles, temporally interdependent yet

167

unsynchronized, from the global and national to the individual and institutional. While it is clear that spatial and temporal perspectives should both be considered in understanding new poverty landscapes, our analysis has also revealed the utility of a more explicitly temporal approach that highlights how different scale-based processes operate at divergent 'speeds', and how their lack of temporal synchronicity can inadvertently work to create new urban poverty landscapes.

Part II

LIVING-THINKING TIMESPACE

9

ANXIOUS PROXIMITIES

The space-time of concepts

Elspeth Probyn

We agreed that perhaps distance in space or time weakened all feelings
and all sorts of guilty conscience.

(Diderot, in Ginzburg, 1994: 110)

Time/space

The enormity of the terms strikes as I wander the arcane space of a university
library. The vastness of the collision of the two terms fills rows upon rows, an
entire section that stands as a frontier bastion of philosophy. These are hard
concepts: 'time' and 'space' sort out the men from the boys, the real philosophers
from the mere cultural theorists. At the same time, the argument that there is
'little sense to be had from making distinctions between space and time – there is
only space-time' (Thrift, 1993: 93) may sort out the geographers from the rest.

Here I want to take up the problematisation of space-time in order to think
through questions about the nearness and farness of conceptual terms: the space-
time of concepts. Through the notion of 'relations of proximity', I will question
how we currently use the concepts that are to us, second nature. Roughly, one
could call these concepts the sites of identity formation and theorisation: gender,
sexuality, class, race. While it has been assumed that this is a set of concepts and
sites that are mutually affective and effective, that is, that it is their interaction that
is important, increasingly it seems that our privileged sites of analysis and inquiry
are becoming static. In this chapter, I am particularly worried that familiar uses of
sexuality as site and concept have lost the power to connect both conceptually and
empirically. Either reified or castigated, can sexuality still generate relations of
proximity? In the face of the explosion of sexual celebration, some scholars are
now returning to a vision of the theoretical landscape in which sexuality, class,
gender and race are separated off from each other and exist in fixed points of
opposition to each other. This then allows for a new theoretical moralism which, I

will argue, blocks the way to envisioning an alternative ethics of existence. Using Foucault's work on space and time, I want to connect questions of proximity to his thinking on attitude and ethics.

Space-time, sex and ethics

Of course, none of these terms is without complication, and such is their appeal. As geographers have argued in regards to certain non-geographical writings, the free and easy ways with which 'space' is bandied about have tended to stunt its reach. In her key text, 'Politics and space/time', Doreen Massey points out that while there is 'a variety of uses and meanings of the term "space", ... particularly strong and widespread ... is the view of space which, in one way or another, defines it as stasis, and as utterly opposed to time' (1994: 251). In particular, Massey is concerned with the implicit ways that Fredric Jameson's and Foucault's definitions favour 'a notion of space as instantaneous connections between things at one moment' (155). Her argument is that 'in the end the notion of space as *only* systems of simultaneous relations, the flashing of a pin-ball machine, is inadequate' (155). These types of definitions, she argues, continue to render space and time as opposed terms, or at least insist on holding apart notions of temporal movement and spatial connection. Objecting to the idea of surface, Massey calls for ways of seeing things in terms of their four-dimensionality. This is a powerfully generative argument and has inspired much important work. It also runs alongside other research that implicitly at least operates within a four-dimensional sensibility.[1] Acknowledging the scope of her argument, I none-the-less want to query the way in which Massey skims over a crucial point in her characterisation of Foucault's position. Her point of reference is his article 'Of other spaces' (Foucault, 1986a). This most cited of texts in terms of cultural questions of space, starts with the well-known and rather portentous statement that 'the anxiety of our era has to do fundamentally with space' (1986a: 23). The line that Massey takes as demonstrating Foucault's extraction of time from the simultaneity of spatial connection is the following: 'We are at a moment, I believe, when our experience of the world is less that of a long life developing through time than that of a network that connects points and intersects with its own skein' (1986a: 22; cited in Massey, 1994: 264). Far from implying the conceptualisation of space in the absence of time (264), it seems to me that this has to be read as further evidence that 'it is not possible to disregard the fatal intersection of time with space' (Foucault, 1986a: 22). This is a line that clarifies for me the connection between issues of space-time, questions of the geo-genealogical, and a conception of corporeal ethics. The folding or doubling of time-space here, the idea of a network that 'intersects with its own skein', replays Foucault's thinking on 'the fold', and the doubling of lines of subjectification that

Deleuze has so clearly exposed (Deleuze, 1986). Rather than discounting space-time interactions, this may extend their theoretical and political positivity. This is especially so when Foucault's musings on space are connected with his argument about the 'limit-attitude' of modernity, and the 'etho-poetics' of an alternative ethics of existence (Rabinow, 1997). In terms of this latter project, his notion of the 'dietetic regimen' (elaborated in *The Use of Pleasure*) can be used to ground more abstract ideas about space, time and ethics.[2]

Now my point is not to be churlish about readings of Foucault; rather it is to follow through on the underlying problem that Foucault identifies, and which in fact travels alongside Massey's own provocative conclusion. Read in the context of his oeuvre, it is clear that neither space nor time *per se* constitute the major site of anxiety for us; rather, it is the changing nature of relations of proximity that has become the central site of concern and intensity. The question of 'how close?' or 'how far?' animates thinking about a range of issues: from personal issues and identifications, geo-politics and economics, the changing nature of the imbrication of the 'private' and the 'public', the 'global-in-the-local', to the reconceptualisation of the work of identity categories. It is the nature of their proximity (to each other and to us) that makes these questions so disquieting. For instance, in Australia the arena of Aboriginal land claims leaves few unconcerned. The issue of land claims is threatening to whites precisely because it scrambles any notion of the past as 'faraway'. It places history in the present, and demands that spatial relations be thought in terms of historical and physical proximity. In elemental terms, the refrain of 'not in my backyard', or in Australia, 'not my backyard'[3] captures a visceral reaction to relations of proximity that have come 'too close'.

In the realm of Foucauldean thematics, the notion of relations of proximity can be seen as the linking movement between the themes of governmentality, discipline, power, and processes of subjectification. In short, it recalls Foucault's insistence on the movement of ethical relations that opposes static figures of morality, and that may even undo them. As an initial clarification of this term, we can understand relations of proximity to be quite simply the calculus of the distance and closeness between and amongst different social sites. It may be helpful to think of this in terms of a force field whereby the sites are themselves zones of classification, which while they order practices are also themselves sites of organised practice. As individuals we are ourselves being reordered just as we find ourselves in new relations of proximity to others, and to ourselves.

Only connect?

Of course this has been recognised in one way or another by many. In Massey's terms, it is clear that 'we need to conceptualize space as constructed out of interrelations, as the simultaneous coexistence of social interrelations and interactions, from the most local level to the most global' 1994: 264). In Nigel Thrift's ambitious project for regional geography (1990b, 1991, 1993), a logic and a feel for proximity, especially amongst ideas, fuels his writing. At a more pedestrian level, there is a widespread argument that we are now in a 'post-traditional' society which is ruled by 'connectedness'. In Anthony Giddens words, 'This extraordinary, and still accelerating, connectedness between everyday decisions and global outcomes, together with its reverse, the influence of global orders over individual lives, forms the key subject-matter of the new agenda' (Giddens, 1994: 58). Others are less sanguine about how connected we really are. For David Smith, the question is 'how far should we care?' (1998). It is a question doomed to the negative, or at least the ambivalent, given his summation of 'the combination of ethical hedonism and resurgent parochial self-interest into which much of the world seems to be sinking' (1998: 35). In Carlo Ginzburg's argument, 'distance and closeness are ambivalent concepts: moreover, they are submitted to temporal and spatial constraints'. 'Distance', he writes, 'if pushed to an extreme, can generate a total lack of compassion for our fellow humans' (1994: 116). To Smith's concerns about the limits of a spatialised ethic of care, Ginzburg adds the question of 'what are the historical limits of an alleged natural passion such as human compassion?' (ibid.). The nature of this 'alleged-ness' is pithily underscored in Richard Rorty's remark that 'our sense of solidarity is strongest when those with whom solidarity is expressed are thought of as "one of us", where "us" means something smaller and more local than the human race' (1989: 191; cited in Smith, 1998: 22).

How far do we hurt?

If Rorty lays down the limits of 'how far we can care', in a recent exchange Nancy Fraser and Judith Butler take issue with each other's position on the subject of recognition and justice. Fraser's argument centres on the 'normative distinction between injustices of distribution and injustices of recognition', which she quickly adds are both 'equally primary and real kinds of harm' (1998: 141). Later she will state that the distinction itself is heuristic, but first she takes up the issue of misrecognition, and explicitly defines it against Butler's position. 'Misrecognition is an institutionalized social relation', Fraser argues, and she adamantly adds, it is 'not a psychological state' (141). Of course, given the psychoanalytic frame of Butler's argument, and the socialist-materialist optic of Fraser's they were bound to clash, especially on the grounds of the materiality of recognition and identification. To briefly characterise Butler's argument in *Bodies that Matter*, recognition, or identification is 'a phantasmatic trajectory' (1993: 99). Drawing on the work of Slavoj Zizek, the '*phantasmatic* promise of identity' is always accompanied by disappointment (188), thus recognition is always constituted in the inevitability of misrecognition. These are central terms that reoccur in Butler's work and, at a very abstract level, they are attempts to capture some of the visceralness of life. The process of iteration and its accompanying failure is also central to her conception of materiality, or the 'notion of matter ... as a process of materialization that stabilizes over time to produce the effect of boundary, fixity, and surface that we call matter' (9).

Even as she incorporates Marx, it is a conception of the material that many 'materialists' wouldn't recognise.[4] It is certainly at odds with Fraser's argument about materiality which seeks to return the terms of the debate to the economic ground of injustice and inequality. Questions of identification and misidentification in Fraser's interpretation are grouped as injuries of status. As such, a status injury 'is analytically distinct from, and conceptually irreducible to, the injustice of maldistribution, although it *may* be accompanied by the latter' (141). For Butler, this redrawing of the boundaries feeds into what she perceives as the ever present, and seemingly everlasting, Marxist 'tendency to relegate new social movements to the sphere of the cultural', indeed to dismiss them as being preoccupied with what is called the 'merely' cultural. She argues that this is to construe cultural politics as 'factionalizing, identitarian, and particularistic' (1998: 33). In particular, she critiques Fraser's book, *Justice Interruptus* (1997) for reproducing 'the division that locates certain oppressions as part of the political-economy, and relegates others to the excessively cultural sphere' (1997: 39). Not surprisingly, gays and lesbians are relegated to the excessive bit, and in Butler's characterisation, Fraser divorces homophobia from the realm of the political-economy, and quite clearly marks out the latter as more pressing.

While the debate may look like the spectre of the undead to anyone with a passing acquaintance with the large bodies of work on the old and not-so-old social movements,[5] Butler and Fraser both have valid points. However, none is strong enough to put a stake in the heart of that old division between the political-economic and the cultural. Butler's leading question is 'Why would a movement concerned to criticize and transform the ways in which sexuality is socially regulated not be understood as central to the functioning of political economy?' (39). For her part, Fraser contends that for Butler 'gay and lesbian struggles against heterosexism threaten the "workability" of the capitalist system' (1998: 146). That said, both of their arguments then run off on unproductive lines. In Butler's case, we go back to Marx and Engels on the family and reproduction. This then renders the category of sexuality impossibly broad. In Fraser's, we have the rather thinly argued case in which the fact that some multinationals court the 'pink dollar' becomes *empirical* proof that 'contemporary capitalism seems not to require heterosexism' (147).[6] Fraser may be quite right that 'the principal opponents of gay and lesbian rights today are not multinational corporations, but religious and cultural conservatives, whose obsession is status, not profits' (p. 146). At a conceptual level, however, the division between status and profit, misrecognition and maldistribution constructs sexuality and class as separate and closed off analytic spaces. To paraphrase Elaine Showalter's remark about 'postfeminism' (1986), the distinction between misrecognition and maldistribution has the potential (for some the benefit?) of becoming code words for 'let's not talk about sex anymore'. It equally closes the door to thinking about class in more productive and diverse ways.

To recall the terms of the distinction, misrecognition is for Fraser 'a status injury, it is analytically distinct from, and conceptually irreducible to, the injustice of maldistribution' (1998: 141). In her book, *Justice Interruptus*, she presents class and 'despised sexuality' as two opposing ideal-typical cases. The former is exemplary of 'socio-economic injustice, rooted in the political-economic structure of society', the latter illustrates the injustice of the 'cultural or symbolic, rooted in social patterns of representation, interpretation, and communication' (1997: 14). The remedies for these forms of injustice are equally distinct: on the one hand, 'political-economic restructuring of some sort', on the other, 'upwardly revaluing disrespected identities and the cultural productions of maligned groups' (1997: 16). The proof of her model is found in the case of class: 'The last thing it [the proletariat] needs is recognition of its difference'; whereas the case of homosexuality demonstrates 'a mode of collectivity ... of a despised sexuality, rooted in the cultural-valuational structure of society' (1997: 18). Set out in these terms, it is hard not to agree with Butler that Fraser's analysis 'presumes that the distinction between material and cultural life is a stable one' (1998: 36). In more hyperbolic terms, Fraser runs the risk of dismissing decades of feminist work on the intersection of class, sexuality, race, gender, etc. Of course, such is not her stated intention. Rather, it is to rectify the 'postsocialist' condition which she characterises as a 'shift in the grammar of political claims-making'. 'Claims for the recognition of group difference have become intensely salient in the recent period, at times eclipsing claims for social equality' (1997: 2).

Many would, I think, concur with Fraser that by and large we are faced with 'the absence of any credible progressive vision of an alternative to the present order' (1997: 1).[7] Equally, it is hard not to agree with her claim that we need to develop 'a *critical* theory of recognition' (1997: 12).[8] However, the terms of her argument seem, on the one hand, calculated to inflame, and on the other, are based in an unapologetically normative framework that renders the categories of class, sexuality, gender and race as static points of opposition. We are squarely back in the zero-sum game, whereby reparation of *your* harm will always be at *my* expense. Thus her *new* approach would identify and defend 'only those versions of the cultural politics of difference that can be coherently combined with the social politics of equality' (1997: 12). While one understands the wish to 'clean up' the messy and sometimes tedious terms of critical theory, setting up an antagonism and a distance between those who *really* hurt, and those who profess to be hurt by 'representation, and interpretation' is unproductive. While of course she is arguing conceptually about categories, not individuals, the spectre of bourgeois gays versus starving innocents haunts her text. For instance, listen to her images of those who 'starkly' hurt in terms 'of caloric intake and exposure to environmental toxicity, and hence life expectancy and rates of morbidity and mortality' (11). At a frivolous level, this recalls the Monty Python sketch about sucking on a dry tea bag ('oh that was a luxury'). At a political level, the project of 'distinguishing those claims for the recognition of difference that advance the cause of social equality from those that retard or undermine it' (5) is itself deeply enmeshed in a grammar of *ressentiment*.

To but briefly recall the Nietzschean dimensions of this term, classically *ressentiment* is the revenge of the weak, who appeal for acknowledgment and rectification on the basis of the truth of their weakness. As Nietzsche bluntly puts it, 'the weak and the oppressed of every kind [engage in] the sublime self-deception that interprets weakness as freedom, and their being thus-and-thus as a *merit*' (cited in Schrift, 1995: 31). In Wendy Brown's compelling reformulation, women and other historically minoritarian groups have tended to use the rhetoric of *ressentiment* to appeal to the state to rectify their hurt. Brown's critique of this use of identity politics initially sounds close to Fraser's argument. However, Brown's objective is to develop the ground for an 'engagement in political struggles in which there are no trump cards such as "morality" or "truth" ' (1995: 48), where we can 'contest domination with the strength of an alternative vision of collective life, rather than through moral reproach' (47). While again superficially similar in appeal, Fraser's position gets caught in its own conceptual trap and in the end relies upon a renewed voice of feminist reproach, one that seeks to carry the truth of real social injustice. While condemning certain forms of identity politics, or in her terms demands for recognition, Fraser sets up a moral position based in the truth of real harm. It may well be that we need to reassess political strategies in order to better navigate the thorny path of competing demands. However Fraser's argument plays within the rhetoric of *ressentiment*, and instead of exploring other relations of proximity, is content to differentially weight the demands of the weak.

In sum, Fraser's argument shores up the boundaries and the boundedness of the categories of sexuality and class. Against them are defined the 'bivalent' categories of gender and race. However these are still rigidly defined by the analysis of injury in terms of the distinction between cultural versus socio-economic injustice. In simple terms, it is hard to see how these categories could ever intersect given the iron-clad measure that defines and distinguishes them. It follows that Fraser considers 'coalition politics' to be 'wishful thinking' (1997: 4). While we might agree that some visions of coalition politics have had a theoretically thin vision of both of the constitutive terms (of 'coalition' and of 'politics'), her model is fundamentally at odds with what Brown calls for in the stead of the sanitised versions of the politics of difference and identity: to wit, the exigency of 'developing political conversation among a complex and diverse "we"' (1995: 51).[9] This is not some wishy-washy task, but rather one that calls for passion, purpose, and a political vision devoid of morality; for the creation of political spaces that 'are heterogeneous, roving'. Brown also seeks to 'redress our underdeveloped taste for political argument' (51). Compared to this project, Fraser's seems small-minded with its concern to *tidy up* the mess of the last decade of feminist and queer arguments.[10] It is hard not to read her as wanting to *straighten up* what Chris Philo has called 'the postmodern geography in which details and difference, fragmentation and chaos, substance and heterogeneity, humility and respectfulness feature at every turn' (1992: 159).

Now, given my own predilection for at least domestic tidiness, I should add that this description of the social does not entail a celebration of messiness for its own sake; that as Meaghan Morris (1988) so aptly puts it, the task of theoretical work is as fussy and never-ending as housework. However, returning to Foucault's point that the anxiety of our era is to do with relations of proximity, I want to argue that Fraser is exemplary of a theoretical anxiety about the closeness, the touching and confusing proximity of the categories of sexuality, race, class and gender. She is hardly unique in this, and in fact rethinking conceptual relations of proximity is highly anxiety-making to cultural and social theory in general. In very general terms, we are at a moment when the relation between and amongst our cherished analytic categories is both amorphous, and taken for granted. As sites through which to analyse changing social relations, or what Massey calls 'the real multiplicities of space-time' (158), they now come to us with the stamp of 'knowingness': as in gender is 'about' women, race is 'about' blacks, sexuality is 'about' queers. As Eve Sedgwick and Adam Frank have argued, much contemporary theory is plagued by 'what it knows', betraying the vastness of our continued ignorance (Sedgwick and Frank, 1995). At a very basic level, the concepts that we are all so familiar with are losing their primary purpose as relational terms: e.g., that conceptually, gender is the relational term that generates and allows us to configure proximities between other terms such as masculinity and femininity; that race is the term that marks closeness and distance between white/non-white; that sexuality as the relational term marking homo/heterosexuality produces a welter of 'sexualities'. These sites do more than mark the boundaries; they also generate a promiscuity of proximities, a porous closeness of sites that questions the very basis of the site itself.

Approximating site

It will not have passed unnoticed that I have been using the terms 'category' and 'site' somewhat interchangeably. I now want to outline why it is important to think about concepts in terms of temporal-spatial sites. For a start, thinking of concepts as sites necessarily foregrounds the fact that they exist in differing relations of proximity to each other. Conceptualising concepts as sites compels both a recognition of their genealogical trajectories, and acknowledgment that concepts are not geo-politically neutral. In other words, gender or sexuality are not universal: *how* they are used in different contexts greatly affects *how* they may relate in proximity to other sites. At the very least, this might remind Anglo-American theorists of the context of their own writing.

The fundamental point is to reinvigorate thinking about relations of proximity: the changing nature of the distance between concepts. As I have intimated, in terms of the debates over the last decades, these concepts have come to be placed in set patterns to each other, and have lost some of the potential they have had as sites. As Fraser's argument exemplifies, there are now renewed calls for these sites to be hierarchically differentiated, and ordered as to their relative political importance. Following Foucault, I want to focus on the anxiety that the changing relations of proximity generates, and argue for its productiveness. But first we need to consider more carefully what is meant by site.

In 'Of other spaces', Foucault commences with a definition of site which already carries the whiff of anxiety. After Galileo, we find ourselves in 'an intimate, and infinitely open space', things no longer stand still: 'a thing's place was no longer anything but a point in its movement' (23). He summarily charts a genealogy whereby 'today the site has been substituted for extension which itself had replaced emplacement'. Foucault writes that the site 'is defined by relations of proximity between points or elements' (23). As Philo argues, this is 'a form of *spatial ontology*' which privileges the 'observable relationships between the many things under study', 'their nearness, apartness' (Philo, 1992: 148). The site has two major dimensions. At one level, in a somewhat technical sense, the site indicates something akin to the archive as both the milieu and the rationale supporting the collection and analysis of data: 'the identification of marked or coded elements inside a set'. At another, Foucault raises the problem of human siting. Here the question of physical placement meets the categorical logic of the site. In this way, the 'problem of the human site or living space is not simply that of knowing whether there will be enough space for men in the world … but also of knowing what relations of propinquity, what type of storage, circulation, marking, and classification of human elements should be adopted in a given situation in order to achieve a given end' (23).

Following this description Foucault quickly goes on to say that 'space takes for us the form of relations among sites'. Clearly, we do not live within one site, or as he says 'we do not live in a kind of void, inside of which we could place individuals and things' (23). However, if we do not live within a void, as I have argued, the categories we use to analyse aspects of our lives risk becoming dead matter. Paradoxically, this may be due to their reification in cultural and social theory. Hard on the outside, and empty inside, these categories no longer move to the dynamics of mutual proximity, but have become privileged and singular optics through which to explain the world. But let me substantiate this claim in terms of the site of sexuality. At the mundane level of popular culture, the massive publication of sex over the last decade or so has had the effect of constituting sex as no longer very special. If much of this runs on institutionalised confession, the effect is mostly boredom. This situation contrasts with Foucault's description in *The History of Sexuality* (vol. 1), where following the discussion of Charcot's clinic he states: 'The essential point is that sex was not only a matter of sensation and pleasure, of law and taboo, but also of truth and falsehood, that the truth of sex became something fundamental, useful, or dangerous, precious or formidable: in short, sex was constituted as a problem of truth' (1978: 56). Inspired by Foucault's comments, an avalanche of academic claims has accompanied the public airing of sex. Two decades on, however, the truth of sex (itself now accepted to be multiple) is no longer seen as fundamental nor, in the majority of cases, is it dangerous. To take but an obvious example, in the light of the Clinton saga, and the culture of talkshows that supported it, everyone 'knows' about Bill's sexual tastes, at the same time that we can no longer be sure of what 'sex' is. In Britain, recent scandals involving government ministers reveal that trauma is not caused by being openly gay, but rather is still to be found in the closet.[11] A unifocal optic on sex would hardly come close to understanding the issues that are thrown into relief during these episodic events. Beyond the putative centrality of sex, what is of interest are the ways in which these events bring together and highlight the dizzying proximities of truth and lying, private and public, and the fact that the *vox populi* is represented as being more interested in issues of conduct than in sex *per se*. Implicitly, there seems to be an emergent shift from the morality of categories to a concern with the ethical scenarios of how public figures live their lives.

Disperse

While these examples operate more as illustration than as hard empirical proof, I am interested in what the category of sex can and cannot do. Phrased more positively, how we can return to the use of analytic categories some necessary movement and a sense of timing. As Rosalyn Diprose astutely points out, even as sophisticated a theorist as Butler may be overly weighting the category of sex. In terms of Butler's *The Psychic Life of Power*, Diprose argues that 'insofar as individual sexuality becomes her focus as the exemplary product and vehicle of power, [Butler] misses too much ... about the dynamics of affective and social life' (Diprose, forthcoming). Beyond individualised sex, or the sexuality of the individual, this is precisely Foucault's point about the possibilities of using sexuality 'to arrive at a multiplicity of relationships', to use homosexuality 'as an historic occasion to reopen affective and relational virtualities' (1989c: 204; 207). Central to Foucault's theoretical preoccupation with sexuality, especially in his later work, was the will to find ways of figuring the proximity of relationships in terms of an ethical culture (1989c: 207). Equally, in his final work on the self and pleasure, relations of proximity are crucial to thinking the contours of this ethics. To extrapolate from his 'Preface to transgression', the point is to break down the hermetic nature of sex as a category in order to focus on 'its dispersion in a language that dispossesses it while multiplying it within the space created by its absence' (1977b; cited in Schrift, 1995: 28). In other words, we may come closer to the positivity of sex if we actively disperse it. The point is to put sexuality into play with other categories in such a way as to highlight and encourage different relations of proximity. As Deleuze and Guattari remind us, it is not the sexual *qua* sexuality, but rather 'What regulates the obligatory, necessary, or permitted interminglings of bodies is above all an alimentary regime and a sexual regime' (1987: 90).

Given all this, it is hardly surprising that Foucault turned to the ancient Greeks and in particular to the notion of 'the dietetic regimen'; it is, however, surprising that few theorists have followed him in this direction. *The Use of Pleasure* takes us into the intricacies of the regimen, in part so that we can 'dwell on that quite recent and banal notion of "sexuality": to stand detached from it, bracketing its familiarity' (1986b: 3). The regimen was not concerned with the *form* of the practices it brought together; no distinction was made between, for instance, 'natural' or 'unseemly' sexual practices. Rather, 'the usefulness of a regimen lay precisely in the possibility it gave individuals to face different situations' (105). In addition to being situational, it was also to be circumstantial: 'Dietetics was a strategic art in the sense that it ought to permit one to respond to circumstances in a reasonable, and hence useful, manner' (106). As opposed to a moral doxa, the objective of the regimen was to allow men through quotidian activities to place themselves in relation to themselves, and to each other in the wider world. Activities were not judged on their intrinsic merit, but rather on their manner of articulation and enactment: what followed or preceded an activity was of far greater import than the form or content of the activity. In clear contradistinction to conventional morality based on prohibition, 'Regimen should not be understood as a corpus of universal and uniform rules; it was more ... a treatise for adjusting one's behavior to fit the circumstances' (106).

Lest this be understood as mere relativism or consequentialism, central to the workings of regimen was the need to establish the right measure. Quoting from the Platonic dialogue *The Lovers*, Foucault reminds us that 'even a pig would know ... in everything connected with the body ... what is useful is the right measure' (1986b: 102). The injunction to find the 'right measure' was to be found in the manner in which the articulation of categories of behaviour were corporeally enacted. The focus was on 'the relationships between the *aphrodisia*, health, life, and death' (97). In Foucault's description, 'The main objective of this reflection was to define the use of pleasure – which conditions were favorable, which practice was recommended, which rarefaction was necessary – in terms of a certain way of caring for one's body' (97). This brought together the realms of both the corporeal and the ethical, making the reflection upon the different aspects of the body into the ground of ethical conduct. The result was however not a precursor to our own 'body beautiful' cult. Rather the lesson lay in how different aspects of caring for the self interrelated. In other words, this is a reflection on the differing relations of proximity amongst categories that one inhabits. To use Nikolas Rose's (1996) terms, caring for the self in terms of 'small ethical scenarios' forms the fabric of an elastic notion of ethical behaviour. It is the imperative to know oneself and others through how the categories relate to each other, and in their context. This then is the basis for an ethical practice that is informed by how one situates oneself in the time and space of categories of comportment.

Thus the regimen can be seen as the principle that brought together diverse practices, such as exercise, eating, sleeping, sexual activity. Through the advice of practical texts, individuals constituted themselves as healthy and ethical subjects. The regimen provided guidance in terms of how to relate to particular situations and circumstances, and this both in regard to oneself and to others. It is clear that these were not abstract notions, but were indeed practices to be governed by the sense of 'the right measure'. This entailed then an attitude 'rooted in an ethics and not a morality, a practice rather than a vantage point, an active experience rather than a passive waiting' (Rabinow, 1997: xix). While many have queried that Greek society should be taken as the model, and its systems of exclusion are well known, it has also been stressed that for Foucault there was no question of 'returning' to the ancient Greeks. The point is not to repeat the Greek search for self-stylisation, but rather to take from already existing tendencies in order to elaborate an etho-poetics, a worldly ethics of living.

The space-time of ethics

But what is the present day relevance of the regimen, and how does it help us in thinking through the multiplicities of space-time now? For a start, it underlines the fact that how we are placed within sites of difference, and how those sites relate to each other is what orientates us: allows us to reflect on how we stand in relation to our surroundings. Of course, as Foucault puts it, proximity brings anxiety, but it is an anxiety that can fuel reflection, render the sites we inhabit more critical. In Foucault's sense, this reflection upon one's conduct is the basis for an ethical engagement with life. However, this reflection must not circle around itself, but rather should send off lines that connect with other sites, and other reflections.

There are then important links to be made between the ethos embodied in the dietetic regimen, and Foucault's thinking about attitude, or a 'limit-attitude'. It is worth repeating Foucault's definition of 'attitude', remembering that in his essay 'What is Enlightenment' modernity is understood not as an historical period but as an attitude to the space-time of context. That is to say, 'a mode of relating to contemporary reality; a voluntary choice made by certain people; in the end, a way of thinking and feeling; a way, too, of acting and behaving that at one and the same time marks a relation of belonging and presents itself as task' (1997: 308). As Juha Heikkala has argued, this forcefully emphasises the 'contingency of subjectivity': that as Foucault puts it, 'People know what they do; they frequently know why they do what they do; but what they don't know is what what they do does' (Heikkala, 1993: 405). Of course, this is a difficult proposition: how does or how can one know what one does does? But then, one of the 'tasks' of attitude is precisely the motto that Foucault takes from the Enlightenment: '*Aude Sapere*' – dare to know (1997: 306). In more extended terms, attitude 'must be considered both as a process in which men participate collectively and as an act of courage to be accomplished personally' (306).

The emphasis here is on the necessity of embodying a self-reflexive and critical stance in the world. Foucault draws out the essential difference that being modern marks: 'To be modern is not to accept oneself as one is in the flux of the passing moment' (311). Being modern thus entails that (now unfashionable) trait of being deeply interested; of being attached to both the present and to the form of one's relation to it. Crucially, 'modern man is not the man who goes off to discover himself, his secrets, his hidden truth' (312). This is emphatically not liberation, but rather production, attitude 'compels him to face the task of producing himself' (312). This is then no whimsical performance; it is difficult elaboration. Foucault's description captures the way in which attitude is revealed in the force of desire to be within relations of proximity (man's relation to the present, and the constitution of the self as an autonomous subject). As it reworks the relation to self, the tangibility of proximities reorganises the sense of self to

183

self within the limits of the present. In other words, it works with the complexities of space-time, which continually remind us to be 'oriented toward the 'contemporary limits of the necessary'. That is to say, toward what is not or is no longer indispensable for the constitution of ourselves as autonomous subjects' (313). Turning the question into a positive one, Foucault asks that we 'transform the critique conducted in the form of necessary limitation into a practical critique that takes the form of a possible crossing-over' (315).

Only cross-over

To bring the disparate and dense parts of my argument together, it seems to me that we desperately need to reanimate the sites of our analysis. More importantly, given that such calls are commonplace, we need to acknowledge the full weight of Foucault's challenge: to transform the critique of limitations, and turn it into a practical critique. One way of putting this is to think of how we can cross over from the solitary space-time of individual categories in order to renew a critical emphasis on the proximity of sites. The importance of Foucault's description of the regimen is that it brings forward another model of how we might live within proximity. 'The right measure' reminds us that an overprivileging of one site will inevitably refigure the relations of proximity to others. Here ethics would be understood as an ongoing and active practice; one that confronts its own limits. It also questions the limits of our theorising. Living in the space-time of a world rendered local, means that we have the capacity to be intimately confronted with the implications of our actions. That how and in what combination we eat, think, sleep and live will have concrete consequences that render 'far-off' parts of the world closer to home. In contrast to Diderot's statement, we no longer have the comfort of distance in time and space to assuage 'our guilty conscience'. This also applies to how we conceptualise the sites and the categories of identity. In some ways, living in Massey's model of four-dimensionality is both dizzying with possibility, and uncomfortable. However, if we are to take Foucault's challenge, '*Aude sapere*', we must live with the anxiety of proximity, and dare to let go of our familiar conceptual habits, venturing into other space-time configurations of thought.

Acknowledgements

With thanks to Ros Diprose for her insightful comments.

NOTES

1 Here two examples, amongst the plethora of others, come to mind. As with much of her work, Meaghan Morris' biography of the Australian journalist Ernestine Hill takes one point of entry and presents a rich tableau of the four-dimensionality of politics,

history and the present (forthcoming). In the realm of philosophy, Moira Gatens' work on the time-space of gender in terms of genealogy and ethology is ground-breaking (Gatens, 1997). I have also argued elsewhere that 'the surface' is in fact one way of figuring the historical and spatial proximity of social relations: that it allows us to get closer to the 'in-betweenness' of sites or zones of difference (Probyn, 1996).

2 See Probyn *Visceral Proximities: Eating, Sex & Ethics* (forthcoming).

3 Racist campaigns against land claims have inflamed the issue by suggesting that Aboriginal land claims would extend to individuals' back yards. The real point of contention is not individual blocks of land, but the refusal of co-existence on the part of graziers. In their defence, white farmers have interestingly used the idea of prox-imity to argue that most urban Australians, which is to say the majority of the popu-lation, rarely encounter Aboriginal people or culture, whereas living on the land, they are closer in terms of contact and, at times, philosophy with their Aboriginal neigh-bours. Few are those who are willing to also genealogically connect the history of genocide and indentured labour that has enabled their 'living on the land'. But they do exist; they remind and remember as do more acutely the Aboriginal people dispossessed of the land.

4 The question about 'the material' and who is a 'materialist' resurfaced in the American feminist agenda with Teresa Ebert's argument against 'ludic feminism' (1996) and Rosemary Hennessy's claiming of materialist feminism (1993; Hennessy and Ingraham, 1997). For feminists outside of America, the vehemence of these debates may seem rather strange. I certainly find Ebert's division of 'ludic feminism' and 'materialist feminism' to be peculiar, and theoretically facile. The strict claiming of categories in these arguments has much to do with the disciplinary locations and battles around the institutionalisation of feminism (and then cultural studies, followed by queer theory). Again, this is not a new argument, for instance, see the special issue of *Critical Quarterly* on 'the state of the subject' (1987). For my own part, teaching in a Francophone sociology department meant that the boundaries between 'materialist feminism' and 'postmodernist feminism' were constantly debated. In Australia, the 'outside' of feminism is very much present, and I would wager that feminists within academe and those working in public spheres outside the university are in closer proximity than may be the case in the US (for instance, see the special issue of *Australian Feminist Studies* (1998) on the past and present of 'women's studies').

5 In very basic terms, within the area of social movements analysis, 'new social movements' are those that are not primarily constituted through class. This would include a range of contemporary social movements, including peace, environment, some feminist, some ethics-nationalist, but most especially and contentiously gay, lesbian and queer movements. In Laclau and Mouffe's definition, 'The common denominator of all of them would be their differentiation from workers' struggles, considered as "class" struggles, together with an expansion of social conflict' (1985: 159, cited in Weir, 1993: 74). In Lorna Weir's critique, the opposition between 'new' and 'old' social movements has produced 'trite binarisms', and 'fictive historical sequences' (1993: 73). And it is true that the great names of social movements theory (Touraine, Wieviorka, Melucci, Epstein, etc.) have all argued for ways of extending the traditional class bias to include ethnicity, sexuality and environmentalism.

6 As I write the question of big business involvement with gay and lesbian matters is very much in the fore. In 1999 the Sydney Gay and Lesbian Mardi Gras has attracted a record $A 800,000 in corporate sponsorship, including multinationals like Qantas,

Coca-Cola and Stolichnaya vodka. The selling of the Mardi Gras (which is now a registered trade mark) has been contentious as the graffiti around town testifies: 'My sexuality is not sponsored by Telstra' (Telstra is Australia's previously state owned, now privatised telephone and now cable company). However, that corporations sponsor the event does not in itself prove that for the rest of the year hetero-normativity is not the more comfortable assumption.

7 For instance, while the Blairite agenda is exciting, especially for those of us living with a government whose reference point is an idealised 1950s suburbia, none-the-less the seeming inability to think categories simultaneously and differently is striking. In the outline published by Fabian, *The Third Way*, Blair's only mention of women is in the context of the impact of women's gains on 'the family'. For a cogent feminist critique of 'the family' that has the potential to deeply challenge such assumptions, see Smart and Neale (1999).

8 In this respect Susan Boyd's engagement with the Butler/Fraser debate is refreshing in that she works through these questions from a background of feminist legal scholarship, and in relation to a Canadian court case on lesbian spousal support. As she argues, cases like these complicate questions about redistribution and recognition (see Boyd, 1999).

9 This isn't to say that there are not vibrant forms of coalition. The bitter and public resurgence of racism in Australia fronted by Pauline Hanson's 'One Nation' party, and supported by John Howard's government has produced vital groupings like the 'Queers for Reconciliation', combating both racism and homophobia. Given the intertwining of (in Fraser's terms) battles for recognition and for redistribution that are both at stake in Aboriginal land rights, one wonders how her model would understand such political formations.

10 However compared to Martha Nussbaum's recent attack on Butler ('The professor of parody'), Fraser is both measured and mild. In addition to getting several points just plain wrong (that, for instance, Foucault and Butler are against 'normative notions like human dignity' because 'they are inherently dictatorial' (1999: 42)), Nussbaum's text is littered with the deployment of images of women starving, being beaten and raped. These images are devoid of any argument about how feminism can actually serve these women, and in general her own argument sounds like a parody of *ressentiment*. The acute resentment is clear with her concluding statement that Butler 'collaborates with evil' (1999: 45). At the very least it is an example of the sorry and shoddy state of American feminist so-called debate.

11 Here I am thinking of the scandal surrounding the former Minister for Wales, Ron Davies who, following a misadventure on Clapham Common, was unfavourably compared in the press to Chris Smith, Minister for Culture. It was clear that Smith as an out gay man was fine, but Davies as a man who had sex with men, was not.

10

RHYTHMS OF THE CITY

Temporalised space and motion

Mike Crang

This essay is concerned with the intersection of lived time, time as represented and urban space – especially around everyday practice. As such it follows in a long pedigree of works addressing time and space in the city. However, I want to try and rethink some of the better known approaches to offer a less stable version of the everyday, and through this a sense of practice as an activity creating time-space not time-space as some matrix within which activity occurs. The essay thus addresses the paradox that Stewart identifies where the 'temporality of everyday life is marked by an irony which is its own creation, for this temporality is held to be ongoing and non-reversible and, at the same time characterized by repetition and predictability' (1984: 14). I want to look both at stability but also at the emergence of new possibilities through everyday temporality.

To do this I proceed through four circuits, each picking up and expanding upon the previous, developing and transforming it. The first circuit serves to locate the everyday through the study of temporality. The study of the chrono-politics and regulation of daily life serves as an entrée into why 'the everyday' matters. The multiple rhythms and temporalities of urban life thus form the back-cloth for the essay – what Lefebvre evoked, but hardly explained, as a rhythm-manalysis. The second circuit picks up on this but adds to the insights of time-geography in the paths and trajectories that individuals and groups make through the city. Introducing a sense of human action and motility into the experience of time offers a new step while the combination of time-space routines serves to link the everyday to the reproduction of social regularities. However, the sense of time-space created through time-geography is rather rarefied, so the third circuit seeks to develop a critique and step sidewise through a concern with the differences between lived and represented times – a focus on experiential time-space that will lead to considering phenomenological accounts. Here time and space cease to be simply containers of action and I work with a sense of space-time as Becoming, a sense of temporality as action, as performance and practice, of difference as well as repetition; the possibility, as Grosz (1999) argues, for not

187

merely the novel, but the unforeseen. However, the fourth circuit suggests that these actions still share an idea of the self-presence of everyday experience, and will open up ideas of events as problematising the everyday. The attempt is to keep a sense of fecundity in the everyday without it becoming a recourse to ground thinking in an 'ultimate non-negotiable reality' (Felski, 2000: 15). The essay therefore argues for a sense of greater instability – or perhaps better, fragility – within the everyday, focusing upon the flow of experience for the social subject. It is also important, of course, to think through the topology and texture of temporality in the urban fabric, the city as well as its people. But that is a task for a different occasion (see Crang and Travlou, 2000).

Chronotopes of the city

Mikhail Bakhtin introduced the idea of a chronotope as a unifying or typifying relationship of time and space in novels (Holquist, 1984; Holloway and Kneale, 2000). But the creation of distinctive urban time-space registers has also been one of the ways urban life has been characterised. In accounts of urbanisation, time has played a long and important – if often implicit – part. Indeed one story of the city told in terms of space and time could be the conquest of time through space, and the creation of conditions of co-presence. This is a story then of density, proximity, planned and unplanned contact that create a civil society. And yet the moment we think of these terms they surely lead us to others – proximity and density to hustle and bustle. The popular account of metropolitan life is of one of increasing pace. It is a recurrent motif that we can read repeatedly in modernisation theories; there were cold societies of slow change, now there are hot ones; oral culture of recurrent time moves to print capitalism's open linear expansive time; horse gives way to steam, steam to internal combustion, to jet or telephony. A teleological story leads us to current urban nightmares of simultaneity, of real-time connections and interactions overwhelming the city and individuals (e.g. Virilio, 1997). Sociological work offers some support with analyses of the 'time bind' as increasing demands of work, commuting, domestic tasks and social expectations stress people's, and especially women's, time (e.g. Jurcyk, 1998). To suggest how much of a trope this has become, we need only look to the science fiction city, or indeed the heralded information technology 'datascape' written through and through with fantasies of fast straight lines and hurtling people or information (Robins, 1999). The clichéd picture of the city, perhaps its chronotope par excellence, has become the long exposure shot of headlights forming blurred streaking lights (see Thrift, 1997). But we can also trace the trope back to key modernisation theorists like Simmel and Tönnies, whose accounts of dense urban life were written in terms of overload, speed up and the bombardment of increasingly isolated individuals by signs and information (Bouchet, 1998; Friedberg, 1993).

Like many overarching modernisation theses, these accounts work by suppressing other urban temporalities. Typically the dismissing of cyclical time, as outside history or a residue to the main story. Equally typically this risks mobilising gendered assumptions about time that have linked women's experience with cyclical time (be that biorhythmical determinism or a critique of the gendering of household tasks, see for example Leccardi (1996) and Kristeva (1981)) or pitting Western, modern urban time against 'traditional', rural time. The tendency of these narratives of acceleration is, then, to replicate the association of feminised, cyclical time with immanence, place and the everyday while constructing a heroic, Western masculine narrative of time's transcendence over place.

Taking inspiration from Lefebvre (1995) we might instead listen to the rhythms of the city. Lefebvre offered a means of discriminating between different sorts of cities – in his analysis Mediterranean and Northern European – in terms not of a singular tempo or its quickening, but as an assemblage of different beats. Indeed it may be we need to refigure the idea of the urban not as a singular abstract temporality but as the site where multiple temporalities collide, as in Bombay where there is an 'intertwining of times, of attitudes, of the coming together and moving apart of past and present, [which] has historically created Bombay's urban kaleidoscope. It is an urban phenomenon that does not lend itself to simplistic readings of its form, which is pluralistic in nature and does not make explicit its origins, intention or rationale' (Mehrotra, 1999: 65–6). The ordered temporalities of glass, concrete and steel and the bazaar co-exist in the same space. Like Lefebvre's evocative view from his Parisian apartment, Mehrotra's Bombay demands that we think through everyday rhythms as a multiplicity, forming distinctive concordances. Following his lead we might think not just of the one-way story of speed up, but also the circadian beat of commuting, of the changing shifts punctuating the industrial city – the sort of beat that so struck Engels in the tramp of the feet of masses of workers heading to factories in Manchester before dawn – but also the school run and annual cycle.

Beyond these striking moments Lefebvre draws our attention to the overlain multiplicity of rhythms; dominant and quieter, cycles on daily, weekly, annual rhythms that continue to structure the everyday as much as 'linear time'. Indeed Felski (2000: 18) suggests that 'Everyday life is above all a temporal term. As such it conveys the fact of repetition; it refers not to the singular or unique but to that which happens "day after day".' The everyday cycle must include the metronomic beat of official time, the rituals of the life course and rites of passage. A multiplicity of temporalities, some long run, some short term, some frequent, some rare, some collective, some personal, some large-scale, some hardly noticed

– the urban place or site is composed and characterised through patterns of these multiple beats. It is the urban space offered by Rutman's film *Berlin: Symphony of the City* which traces the ebb and flow of a day in the city, the pulsing movement and flows of people and things that make up the daily round (Natter, 1993). Instead of a being a solid thing, the city is a becoming, through circulation, combination and recombination of people and things. This is a seductive vision where the urban field becomes an object in motion, or rather an object with time (Lefebvre, 1995: 223). Lefebvre's influence spread thus to Derrida's vision of temporised places (see Quick, 1998), to de Certeau's (1984) vision of spatialisation as practised place – by which he means the inscription of time on to place, the appropriation of urban places through temporary use – to Harvey's (1985) programmatic move from a study of urban form to the urbanisation of time. Or as Lefebvre suggested:

> Space is nothing but the inscription of time in the world, spaces are the realizations, inscriptions in the simultaneity of the external world of a series of times, the rhythms of the city, the rhythms of urban population ... the city will only be rethought and reconstructed on its current ruins when we have properly understood that the city is the deployment of time.
>
> (Kofman and Lebas, 1995: 16)

The city is then, as Bachelard noted, the poetics of multiple *durées* coming together (Kofman and Lebas, 1995: 29), not necessarily as unified wholes but as sometimes fragmentary and ragged patterns (Réda, 1997). Sassen (1999) argues that cities are places where different temporalities of action come into friction. It is in this sense that we might take Bakhtin's (1981) notion of the chronotope as a unity of time and place, and perhaps adapt it as the sense of temporalised place. A place not necessarily of singular time but a particular constellation of temporalities, coming together in a concrete place (cf. Rämö, 1999).

While Virilio (1997) argues for a premodern unity of time and place suggesting that 'place' as a category becomes shattered by the multiple temporalities that modern life brings, more discriminating analyses undercut the starting point of 'fragmentation' stories by suggesting that the lived city has for a long time been a polyrhythmic ensemble. Thrift (1988) traced examples of this in terms of the public chronology and time experience of the mediaeval period. While many stories talk of capitalism as an encroaching clock time, or as Le Goff (1980) put it the transition from church time to merchant time, or what Castoriadis (1991) called 'public time', Thrift (1988: 66–9) points to the multiple roles of church bells as devices of time between 1100 and 1300. Far from a modern phenomenon, here there were already multiple bells, with the Foucauldian disciplines of the monastic rules of St Benedict dividing the day into eight canonical offices (though not of uniform hours), compounded through a plurality of churches operating on different hours ringing in different services. With public ringing in of curfews and policing the streets, and civic functions marked by church bells, or their own calls, the secular and sacred times of the city were neither distinct nor opposed. Meanwhile careful accounts of the Ottoman empire's adoption and use of the clock tell a different story from the clash of modern and Byzantine times – where clocks were status symbols which were regularly reset to accurately reflect the passing of the solar day. Epochal accounts with bold narratives of a singular commodified time need to be inflected then by a sense of complex starting points. But equally, they produce a time-space that is far from monolithic but produces uneven and varied rhythms in urban space. Contrary to Virilio and other theorists, we have not lost a unity of place with a unity of time, rather places have always had different temporalities orchestrated through them.

In this version of a city with plural rhythms long extant, we can follow other trends than acceleration. We might, for instance, follow Melbin (1987) and look at the colonisation of the night – the steady movement of social life into the dark. There are, inevitably, more stories to tell – with the regulation of nightlife being accompanied by public street-lighting and technologies of policing and surveil-lance (Alvarez, 1995; Schlör, 1998). Night-time revellers and night workers – and the public demons of the criminal classes – follow a different rhythm, often seen as disordering the regular times of regular folk. This story we can extend with current concerns not just of acceleration but the non-stop city as breaking down family time. Every now and again the media return to the demise of 'meal times' as collective times with increasing numbers of people eating ready meals at disparate times – without a collective rhythm holding families together. Shift work and night work increasing in the service sector undermine the solidarities and rhythms of 9–5 lives. There is a sense that the rhythms structuring urban life have shifted and with them the structuration of wider patterns in society (Shapcott and Steadman, 1978). Meanwhile at the same time, new rhythmic groupings may be emerging, not mapping on to the image of stable, permanent (residential) communities, but transient, episodic affinities and comings together – what Maffesoli (1996) termed neo-tribes. This reworking of Durkheim's collective temporalities allows us to include a refashioned notion of the sacred and shared without locking it in the solidarities of tradition (Watts-Miller, 2000). The rhythms of the city thus include the pulsing formation of these intensities and affinities as collective groups – and their dissolution, fragmentation and reformation.

Amid this appealing version of the city as rhythms and urban living as rhythmic composition, though, there is the critique of the polychronic city as a realm of shattered and fragmented times where we see less rhythms than disjointed

191

tempos – as principally women, juggle work rhythms, biological rhythms, school rhythms and domestic expectations (Paolucci, 1998). The fragmentation of household time budgets perhaps brings a different, less harmonic, sense to Lefebvre's rhythmanalysis. When we consider the tangles of people's lives with their different points of intersection with different times and other people, marching to different beats, it seems that we might perhaps take the time-soaked place of Lefebvre rather more as the constrained time-space interaction locale derived from the work of time-geography.

Routes, routines and paths

Lefebvre's rhythmanalysis focuses on the music made by diverse beats forming the experience of place. It offers some purchase on the sense of localities as marked by both their own temporalities or better the conjunction of tempos within them. It certainly seems a more fruitful sense of city life than say Tuan's definition of places as 'pauses'. Rhythms of time suggest activity rather than rumination as defining place. The attempt to dynamise representations of the city and open them to the itinerant and tactile knowledges of immersed participants has indeed been a theme in twentieth-century urban art (Hollevoet, 1992). But this sense of rhythm is perhaps more usually evoked in geography through the legacy of Torsten Hägerstrand. The pulsing of the city, the flows of people through its networks, the constraints of time and motion have long been the subject of time-geographic studies. As Hägerstrand put it in a reflective piece 'We need to rise up from the flat map with its static patterns and think in terms of a world on the move, a world of incessant permutations' (1982: 323). Yet the result is rather different than Lefebvre's vision. Where Lefebvre sought to change our under-standing of the city by unpacking the phenomenology of the place as object, time-geography too often ended up dealing with the measurable and evident – indeed, the mappable.

These maps are expanded through the inclusion of time – typically as a vertical dimension. In this way waves of motion and activity can be traced as an overlay on space. The institutions and temporal parameters they create form specific time-space locales for types of activities. These, then, are the envelopes of space-time, through which people must pass in order to accomplish their daily business. These constraints synchronise (and *synchorise*, bring together in space) actions by diverse people, forming the time-space anchors, or stations, of behavioural patterns. Thus regular envelopes of time-space would be the 9–5 workplace, the school, the nursery, the shops, clubs and of course the domestic arena. Around these anchors are Hägerstrand's prisms of time-space opportunity showing a constrained range of possible activities given the limits of time and mobility – lots of time could be spent on activities near at hand, the further one wishes to travel the less time will remain for any activity. The time-scale might vary to chart daily, weekly rhythms and shifts, seasonal ones, through to life-course changes.

Out of this mapping of the banal, comes something of a ballet of lines of motion. The sense of rhythm and repetition connects provocatively with ideas of routinisation – and the suggestion then of the relationship between societal pressures and individual life. Indeed Lefebvre suggests that 'everyday life' only became visible as urbanisation allowed the observation of uniform and repetitive aspects of social existence (Felski, 2000: 16). This is the sense Hägerstrand labelled as the 'diorama', alluding not to the visual perception but the 'thereness' of rules and regulations. Cullen (1978: 31) suggests the imperatives of routines mean that in quantitative terms, deliberative choices:

> are swamped by a dominant pattern of repetition and routine. We spend very little time each day either deliberating some future action or exe-cuting a previously deliberated one. Most of our time is devoted to living out a fairly sophisticated pattern of well ordered and nearly integrated routine.

If these routinised constraints are one dimension, the opposite of this is Hägerstrand's provocative term of 'project'. The sense of trajectory also means we have a space-time – or a 'ForceSpaceTime' of kinaesthetics (Stewart 1998) – that seeks to include physical and social vectors and velocities (de Certeau, 1984). And with this we seem to add a rather different dimension than Lefebvre offers. Clearly, one sense of project is the intentional and planned courses of action people can undertake. Pred (1981b) shows the idea of a project can encompass a dialectic of life course and daily life with different scales of projects intersecting, and thus meshing longer-term power relations and positions in society with small-scale events. This structuration brings institutional projects into contact with lived experience through routine structures. Pred develops this to suggest that we do not then need to see a possibilist world of constrained action, but can also look at which projects are rendered thinkable, and how the intersection of institutional and individual projects makes available (or not) means to accomplish various goals. This link of habituation and project also joins the two supposed antinomies of cyclical and linear time – though we might again need more empirical specification of how these rhythms structure linear projects for different people, and the shifting importance of each. For instance, Strathern (1992) argued that it is the middle class who tend to make a project out of life, and we might then see a range of projects designed to accumulate forms of capital marking out a middle-class group through a specific temporality in their habitus (cf. Bourdieu, 1990). In this sense trajectory begins to appeal not only to physical motion but the inherited dispositions and capabilities and the varying abilities to colonise the future. We need to include trajectories that are not just 'the thrust into the future that imposes a dominance on social life', but also non-directional, aleatory trajectories that may be cyclical or echo recurring times but allow a coping with the world through 'the serenity of the Greek *kairos*: what we might call the temporal opportunities of everyday life' (Maffesoli, 1998: 108, 110).

This aspect of time-space imposes limits, some addressed within the 'school' of time-geography but also leading us in to different realms. First, despite the primacy given to corporeal lived experience, the practice of mapping activities tends to produce a cadaverous geography. A geography of traces of actions, rather than the beat of living footfalls. So we need not only phenomenology's notion of consciousness as, by definition, an intentional *corporeal* consciousness, but also phenomenology's concern with describing consciousness in terms of the structure of the sensation of the *moving* body. Such a body opposes mere '*flesh*, the body not infused with life but dragged around' (Stewart, 1998: 44). In other words not bodies moving through space-time but making it. One response is the concept of 'action space' that sees not just a time-space grid, as an accounting medium, but temporality as structuring medium (Cullen, 1978: 37). Or going further, seeing these time-spaces as 'a co-ordinated complex of spatio-temporal rhythms with its own internal organisation and structure' (Shapcott and Steadman, 1978: 55). This is to begin to think through space-time as the fluid not just the container; to think temporality not just tempo. This entails a spatiality that is 'atopical' or, as Miller (1995: 7) suggests, that 'Space is less the already existing setting for such stories, than the production of space through that taking place, through the act of narration' (see also Donald, 1997: 183). Space is an eventful and unique happening. Space is more to do with doing than knowing, practice than representation, less a matter of 'how accurate is this?' than of 'what happens if I do it?'. As de Certeau (1984) would argue, the project and story create a theatre of actions, create frontiers and interactions; a space that is topological and about deformation of places rather than topical and about defining places.

This sense of producing space-time returns us to Lefebvre, the form of space-time created characteristic of particular modes of production and forms of social organisation. Lefebvre (1991a, b) argued that the current city was marked by the increasing dominance of an abstract space, a realm of equivalent points. This is a sense of space as a replicable realm of homogeneous instants that differ only through their location. If we are not careful, applying time to geography simply subjects time to just this sort of space. Thus to an East–West, North–South grid we add hours or days; $t_1, t_2, t_3 \ldots t_n$ and so forth. Activities are then located, and emplaced in the same representational logic. Or as Grosz put it:

Even today the equation of temporal relations with the continuum of numbers assumes that time is isomorphic with space, and that space and time exist as a continuum, a unified totality. Time is capable of representation only through its subordination to space and spatial models.

(1995: 95)

This, I would suggest is actually to apply a particular form of spatial categorisation. Or as Grosz (1999: 22) recently put it, if time has numericised duration then mathematisation has rendered space itself as a kind of abstraction of place. It is, after Aristotle's framework, to create a *chrono-chora* framework of abstract models of space and time, a sequence of 'anywhere-whenevers' in Deleuze's terms (1989), fashioned after each other, rather than a *kairo-topos* model of this event through and in this place (Rämö, 1999). The critique of this spatialised model of time can be taken back to Henri Bergson who in the first quarter of the twentieth century criticised models of time as comprising individual instants succeeding one another. He argued instead time works like motion and that this 'spatialised time' attributes to the moving body the immobility of the point through which it passes. As de Certeau (1984) has it, in an almost exact critique of time-geographic maps, we mistake the map for the path, the act of going by for its trace. Bergson suggests space-based versions of time are impure compounds that fail to see that movement (time) is different from distance covered which as a line may be infinitely divided (space) (Deleuze, 1991). This spatial model of time is thus a 'cinematographic illusion' (Douglass, 1998: 26):

which accompanies and masks the perception of real movement ... your succession of points are at bottom, only so many imaginary halts. You substitute the path for the journey, and because the journey is subtended by the path, you think the two should coincide. But how should *progress* coincide with a *thing*, a movement with an immobility?

(Bergson, 1991: 189–90)

Time is qualitatively different from space, and spatial metaphors for time only obscure this (Boundas, 1996: 94) – although it is perhaps better to say that the spatial metaphors generally used are problematic, that concepts of abstract space tend to be applied to make an abstract time. Later I want to include different notions of space but for now let us confine ourselves to homogeneous, empty space rather than say folded or haunted (see also Crang and Travlou, 2000). In the next section I want to look at how temporality is seen as creating heterogeneity.

Experienced time

The idea of time existing 'like beads on a string' in a brute sequence of isolated and self-contained events can be undermined from several directions. Lefebvre's account of the different experiences of different tempos offered a sense of different p(l)aces for different events. However, here I want to expand the notion of event by drawing upon the phenomenology of time. Tracing its roots back to Augustine's writings is a view of the 'expanded present', that has been reworked phenomenologically through the writings of Husserl, Heidegger and Merleau-Ponty (Alliez, 1996: 129; Ricoeur, 1988: 19). Following these authors, I want to make a very selective reading that seeks to emphasise how the relationship of space to time changes when we open up a sense of motion and 'apprehension' – both in the sense of uncertainty and how we grasp the world. I want to do this through a sense of the individual as motion and flow – not in time, but flow as constituting time. For the sake of simplicity the flow here is fairly linear, but thinking back to the polyvalent rhythms mentioned earlier would in reality include loops, turbulence and so forth. This is rather more fundamental than the concept of 'reach' or projects deployed in time-geography, to suggest the level and range of ability to command the future (but see Parkes and Thrift, 1978). Hägerstrand's 'prisms' of mobility might be translated into a phenomenal reach of the subject in time.

Augustine's meditations introduced a 'big now' of the '*distentio animi*'. This is the human consciousness, or soul in his work, composed through successive graspings of a future and past. The present is always a threefold structure where a person's present disposition can only be made sense of in terms of a future and past, so the present becomes an expanded field (Ricoeur, 1984: 9). It is thus a trajectory or project in that intentional sense – where Augustine says of the mind it expects, it is attentive and it remembers (Alliez, 1996: 131). But the situation is more complicated than this – since we have both the phenomenology of time, the phenomenology of the subject and of the world, and as Ricoeur ruefully adds a mimetic spiral of the various ways the subject attempts to represent this both to themselves and others (for a critical note on the tautology possible here see Currie, 1999). We therefore need to unpack this, first, to clarify the grasping of the world in this temporally distended manner, and second to ask whether events and space-time itself, as opposed to its experience or representation, is shaped as a distended present.

Husserl's phenomenology provides a starting clarification, suggesting the structure of any event comprises this sense of the distended present. Thus Husserl suggested a distended present based on the past and future that are bound into the very instant itself – so every instant also contains a just-pastness, and nearly-nowness, say as a sense of continuing action. For example, a ball in flight has the just-pastness and the towards the future nearly-nowness of its trajectory embedded within every moment of the arc. The moments are implicated one in the other. If this is taken as a more general pattern it suggests that events themselves are not discrete objects or happenings but have a temporal structure. It is not, then, a case that humans are super-imposing or configuring an inchoate world of events into representational forms – but that there is a temporality to experience. As Carr (1986: 25) argues '[t]he reality of our temporal experience is that it is organised and structured; it is the "mere sequence" that has turned out to be fictional'. Extending this into thinking about projects we also need to consider the effect of envisaged trajectories, by which I mean we have to distinguish the actual path from the shifting and provisional, and temporally embedded, perspective of an actor at a given moment. There is a sense of 'future perfect vision', the 'will have been' or perhaps conditional 'would have been', registers, as we think of any conscious actor thinking through the possible outcomes of any given project. The threefold temporal structure is thus replicated in mimetic, representational thought.

I want to make two moves to develop this idea of the threefold structure. The first is to try and reinsert a sense of lived space-time through the work of Merleau-Ponty. The second is to introduce the autonomous subject through the work of Heidegger. Merleau-Ponty offers an important moment in this study since, like Hägerstrand, he focuses on the corporeal experience of time, but uses this to open up a sense of phenomenology. Merleau-Ponty (1962: 268) uses Bergson's insight that movement is not simply a successive occupation of a discontinuous series of instants in order to distinguish a public, objective time from lived, subjective temporality. 'It is objective time that is made up of successive moments. The lived present holds a past and a future within its thickness' (1962: 275). However, his formulation takes this rather further asking us to think through how objects and subjects come to shape each other not just *in* space and time but *through defining* space and time. Thus in his discussion of protention and retention he suggests that we should not so much see time as the subject as being stretched. Thus he reworks the example of the thrown ball so that:

> the impending position is also covered by the present and through it all those that will occur throughout the movement. Each instant of the movement embraces its whole span, and particularly the first, which being the active initiative, institutes the link between a here and a yonder, a now and a future which the remainder of the instants will merely develop. In so far as I have a body through which I act in the world, space and time are not, for me, a collection of adjacent points nor are they a limitless number of relations synthesised by my consciousness, and into

which it draws my body. I am not in space and time; I belong to them,
my body combines with them and includes them.

(1962: 140)

The effect is to project the embodied self temporally and spatially anticipating and
haunting, so that future and past, memory and anticipation form the field through
which objects themselves are discernible – they form a structure of foreground
and horizon against which objects appear. This is not being in space and time but
inhabiting space-time. Objects are not just 'in-themselves' but are directional
being 'for-us'. The protention and retention, as well as conscious planning and
projects, mean that let us say our grasping of a house is such that

> the house itself is not the house seen from nowhere, but the house seen
> from everywhere. The completed object is translucent, being shot
> through from all sides by an infinite number of present scrutinies which
> intersect in its depths leaving nothing hidden. ... Each moment in time
> calls the others to witness ... each present permanently underpins a
> point of time which calls for recognition from all the others, so that the
> object is seen at all times as it is seen from all directions and by the same
> means, the structure imposed by a horizon.
>
> (1962: 69)

Or, if we think of the house as a cube, this depends on our understanding that
there are hidden sides that could/will unfold in space and time as we move
around the object. Importantly this opens up a sense of virtual presence, of a field
which is not 'present' in the conventional sense, but which shapes our current
understanding of space and time. Like Hollywood streets that are actually made of
flat facades, this sense of virtual presence, drawing upon expectation and willing
credulity, shapes the perception of an object. The structure of experience
Merleau-Ponty then offers is one of a continuous unfolding of objects to a moving
observer in space-time. Or, as de Certeau (1983: 25) described it, a combination
of wave and landscape that produces a mobile texture of folding relations of seer
and seen – where seeing is already travelling. It is this engagement that moves us
from Husserl's transcendent and essential subjectivity.

This shift to existence before essence leads us to the commitment to engaged
experience or in Heidegger's terms a structure of Care. This practical engagement with the world is a sense of 'pre-ontological' (Heidegger), 'pre-objective'
(Merleau-Ponty) and pre-representational (Bergson) shaping. This move shifts the
concern from epistemological concerns in classical phenomenology to ontological
concerns about the shaping of experience. Both Heidegger and Merleau-Ponty
have a sense of the human subject as being necessarily part of space and time, or

the 'worldliness' of existence, what Heidegger called the 'thrownness' of Being; that is finding oneself amongst the world as it unfolds, and having to start coping from there. This is of course one of the basic points of Heidegger's 'being-in-the-world' or 'Dasein'. But in this context it means:

> The projective character manifested in particular projects ... outlines a future which in turn organises the present world into interlocking complexes of significance, all accomplished on the background of Dasein's thrownness or facticity, that is, its finding itself in a particular situation. This projective structure which accomplishes, projects and organises the world, is actually the self-projection of Dasein onto time ... Dasein's activity is ultimately its structuring of itself.
>
> Carr (1986: 111)

In other words this is not travelling through space-time, but the organisation of space-time. Dasein is a temporal event or to put in another way, we possess ourselves only as a journey (de Concini 1990: 171). This performative production of space-time means being-in-the-world is an activity; in Heidegger's example of the clearing in the woods formed through its context, 'clearing' is an action. It also moves us round again to Lefebvre's rhythmanalysis, since Heidegger offers a difficult – less charitably we might say confused – inflection of shared practice. That is we might see the activity of clearing as being the individual practice, but the clearing as a collective creation (Dreyfus, 1991: 167). Space-time has become the present participle; it is not pre-existent, it is not present but presencing, a threshold of disclos-ing, and appear-ing; it is how things become apprehensible. This public structuring shapes a relationship of dis-stance, that is objects becoming discernible. Thus instead of Merleau-Ponty's structure of field-horizon framing objects, we can 'distinguish the general opening up of space as the field of presence (dis-stance) that is the condition for things being near and far, from Dasein's pragmatic bringing things near by taking up and using them' (Dreyfus, 1991: 132). In other words, particular forms of life that make certain objects appear.

This sense of publicness and sharedness suggests Heidegger's reshaping of the threefold division of Augustine, here as three '*ekstases*' of Dasein thrown into the future and bringing along the past (Ricoeur, 1988). There is also a sense of fore-having, that is things made available or dis-closed in particular ways through a pre-ontological disposition (Dreyfus, 1991: 199). Yet also there is the less often stressed *ekstase* of being-alongside, or collectivity. Certainly the sense of routinisation and the imposition of rhythms is still weaker here than in Lefebvre or empirical time-geographic studies. Perhaps inevitably given the politico-ontological leanings of Heidegger, where he labels inauthentic life as following the crowd (das Man) (Bourdieu, 1991) – the stress is on a self-realisation or a collective realisation that is profoundly anti-urban. But the sense of creating space-time through a tactile apprehension, seems important to develop, not least since the sense of performatively creating time here seems rather different than Lefebvre – performance as orchestration – and raises issues of seeing time-space as practice rather than a way of representing a given time.

The relationship of events to representation has also been rather differently portrayed. For instance Ricoeur (1984, 1988) leans towards the idea that mimesis does not need to be underpinned by a temporal structure of protention and retention. Rather, Ricoeur argues there is a necessary gap or uncertainty between representational structures and the flow of events and experience. Or, as Lyotard would argue, one of the roots of the foundational crisis is the questioning of the conditions of time and space, in part due to temporality being 'figural' and resistant to representation (Quick, 1998: 65–6). This gap leads him to suggest that the relationship of time and narrative is an aporia that drives poetic rather than metaphysical solutions – that is, the gap impels us to spin stories, to account for it. Yet these can never fill the gap but only circle about it.

Spatialised time to temporalised space

What this essay has developed so far is a sense of time-space not as a framework in which events occur, but as itself an event. Instead of adding time to models of space it has tried to suggest adding time to our notions of space. The essay has leant towards Heidegger's view that the idea of unifying consciousness remaining inviolate as a series of images move before it is an intellectualised reflection, not an accurate description of temporal experience (Dreyfus, 1991: 133). Lefebvre offered a rather more ambivalent position in regards to this. He clearly suggested the need to experience particular tempos was not in itself adequate as an account:

> There is a certain externality which allows the analytical intellect to function. Yet, to capture a rhythm one needs to have been *captured* by it. One has to *let go*, give and abandon oneself to its duration. Just as in music or when learning a language, one only really understands meanings and sequences by *producing* them. ... Therefore in order to *hold* this fleeting object, which is not exactly an *object*, one must be at the same time both inside and out.
>
> (1995: 219)

Except he also quite clearly had a suspicion of the process of abstraction. This can be developed in two ways. Much of his work was an account or variations on a theme of the colonisation of the lifeworld, in which his driving narrative was the appropriation of lived experience (by no means uncontested or universally successful) by progressively more abstract conceptualisations. Famously this is his (1991) triplet of lived experience, imaginative engagement and conceptualisation, or spatial practices, representational space and representations of space respectively. Part of his story of capitalism is the movement from absolute to abstract space as a representation, from sacred to geometric, where spatial dimensions form an axis of equivalence allowing any space to be measured and compared against another. This homogeneous, exchangeable form of space is the one generally used as the public representation of time. This is, it has to be said, an appealing but arguable comment on capitalist time, which is after all inflected by many temporalities from grand historical narrative to the daily routines for workers (even those in the most 24/7 occupations).[1] Connecting these times are an expanding range of technologies (from diaries to calendars) that seek to mediate different times and reflexively allow us to monitor and create time-space. The differential connections we can find in what Frank Kermode called the traditionalist modernism of Joyce's *Ulysses* that represents 'an attempt to eternalize a pedestrian dimension of events in a single, randomly chosen day by superimposing a mythological dimension' (Isozaki and Asada, 1999: 78). Indeed the empty, homogeneous time needs these other times for it to make sense, to be distinct and to function. We must acknowledge that: 'As a whole, time is braided, intertwined, a unity of strands layered over each other; unique, singular, individual, it nevertheless partakes of a more generic and overarching time, which makes possible the relations of earlier or later, relations locating times and durations relative to each other' (Grosz, 1999: 23). Anytime, whatevers and abstract time do not have things simply their own way. If this notion of time is powerful, it is not analytically tenable to end up with a dichotomous idea of time as, on the one hand, representations or 'images in consciousness' and, on the other, movements of bodies in space (Deleuze, 1986: 56). Instead we need a sense of time as motion and transformation. It is the idea of motion and transformation that often seems obscured by representing time through a spatial idiom.

Bergson distinguished a temporal pluralism from a spatial pluralism, suggesting two types of multiplicity – one in terms of space, which is a quantitative change (augmentation or diminution), creating multiple and discontinuous actual objects, the second that of time and 'duration' which is a qualitative heterogeneity (about changing type and kind), a multiplicity of fused and continual states which are virtually co-present (Deleuze, 1991: 31). Deleuze thus distinguishes given diversity and becoming difference : 'Difference is not diversity. Diversity is given, but difference is that by which the given is given' (in de Landa, 1999: 31). Deleuze argues that theory has been too preoccupied with extended magnitudes in space instead of intensities in time (Boundas, 1996: 85, de Landa, 1999) – or by making time an independent variable we end up with a succession of instants rather than a dialectical order, missing the production of singularities through watching the accumulation of banalities (Deleuze, 1986: 6). This picks up the sense of 'virtual presence' mentioned previously in the work of Merleau-Ponty but takes it a step further. Here I want to foreground this problematic axis rather more. Instead of seeing the everyday as an unproblematic grounding for experience, I want to suggest that this sort of analysis also suggests a subject who is not self-present. That is not so much distended or stretched through space-time as seeing the subject as 'abyssally fractured' into future and past, always an unstable becoming, always concerned with events rather than things (see Massumi, 1992; Rajchman, 1997). Going the opposite way from the 'big now', using many of the same arguments, Bergson instead hollowed out the present. He pointed out that the present is better defined by what is *not* than what it is:

> You define the present in an arbitrary manner as *that which is*, whereas the present is simply *what is being made*. Nothing *is* less than the present moment, if you understand by that the indivisible limit that divides the past from the future. When we think this present is going to be, it exists not yet, and when we think of it as existing, it is already past.
>
> (1991: 149–50)

In this sense the present ceases to be self-present, and is haunted by the virtual realm (Grosz, 1999). The emphasis is on the flow of time. Bergson thus criticised spatial concepts of time as impure compounds that fail to see that movement (time) is different from distance covered which as a line may be infinitely divided (space) (Deleuze, 1991). Deleuze (1986: 1) suggests three consequences of this approach

> movement is distinct from the space covered. Space covered is past, movement is the present, the act of covering. The space covered is divisible, indeed infinitely divisible, whilst movement is indivisible, or cannot be divided without changing qualitatively each time it is divided. This already presupposes a more complex idea: the spaces covered all belong to a single, identical, homogeneous space while the movements are heterogeneous, irreducible among themselves.

In Bergson's work space preserves indefinitely things which are juxtaposed while time devours the states which flow into each other within it, where the pure present is the progress of the past gnawing into the future (Bergson, 1991: 150). Thus memory should not be seen as a field in which instances and items (or images of the past) accumulate. It is not a representational field of images or instants to be configured – this is a spatial representation of temporal process. The flow of *durée* may produce 'images' but this is rather like a ray of light (1991: 37) that may produce a virtual and static image – an after-image[2] – but is actually based around motion.

Bergson's idea of time therefore emphasises 'the virtual', which appeals to the continued presence of the past. Although we may reflect back on the past and see it as discrete events and epochs this is to spatialise time. Instead we might see within any 'present' a virtual multiplicity of possible futures, and pasts, existing. Thus, the past and present are not 'additive' like spatially bounded entities stacked end on end, what Deleuze called 'extensive boundaries', but are marked by 'internal thresholds of intensity' – offering not just diversity, but the difference that causes different outcomes (de Landa 1998). This is a very different sense of the real and virtual than we had in time-geography. Here the virtual extends like a prism of associations and possibilities brought to bear on a point in the present. But this is not the realisation of possible outcomes. Rather, the virtual unifies a range of mutually impossible and differing paths. As Grosz puts it 'The movement from a virtual unity to an actual multiplicity requires a certain leap of innovation or creativity, the surprise that the virtual lives within the actual. The movement of realization seems like the concretization of a preexistent plan or program; by contrast, the movement of actualization is the opening up of the virtual to what befalls it' (1999: 27). That is, Bergson sees in temporality a dynamic becoming that is different than the rearrangement of what already exists (Adam, 1990: 24). The effect is twofold in that first, it is not human representation of time-space that is fractured into virtual and actual but rather than these are properties of the objective world itself, and second that the distinction between the two becomes indiscernible in practice, even if the poles remain distinct (Rodowick 1997: 92). Past states are drawn into contact with the present, and made *virtually* present by action (143). Memories are organised and called up by the focus of attention to the present and future – what Bergson called 'attention to life' that prefigures Heidegger's notion of Care.

The present virtually includes the past through a process of attention contracting into the future and dilating into the past (Deleuze, 1991: 49). The present and past co-exist in a virtual order:

> We have great difficulty in understanding a survival of the past in itself because we believe that the past is no longer, that it has ceased to be. We have thus confused Being with being-present. Nevertheless the present *is not*; rather it is pure becoming, always outside itself. It *is* not, but it acts. Its proper element is not being but the active or useful. The past, on the other hand, has ceased to act or be useful. But it has not ceased to be.

Useless, inactive, impassive it IS, in the full sense of the word: It is identical with being in itself.

(Deleuze, 1991: 55; emphasis in original)

In a reverse of how we often think of time, the past does not recede but 'literally moves towards the present' and exerts a pressure to be admitted (Deleuze, 1991: 70). Bergson illustrates these points by drawing crossing temporal and spatial axes. The subject he locates at the crossing point. He notes how readily we accept that those positions and things spatially not present are still existing, Merleau-Ponty's virtual spatial unfolding, but that we do not accord the same ontological reality to those temporally not present. This is then the virtual dimension whose continued reality Deleuze insists upon – ontologically different – but there.

The upshot of this is a concern with not spatialising time but temporalising places. As Lefebvre had it in his 'rights to the city' what was required was an assemblage of difference through multiple temporalities:

opportunity for rhythms and use of time that would permit full usage of moments and places … the ludic in its fullest sense of theatre, sport and games of all sorts, fairs, more than any other activity restores the sense of oeuvre conferred by art and philosophy and prioritizes time over space, appropriation over domination [which would oppose] the New Masters [who] possess this privileged space [of urban planning through the], axis of a strict spatial policy. What they especially have is the privilege to possess time. … There is only for the masses carefully measured space. Times eludes them.

(in Kofman and Lebas, 1995: 19–20)

It is clear there are echoes here of de Certeau's (1984, 1997) call for a practice that turns the places of command into spaces through their temporal use – appropriation without possession. It is a sense of time as plurality and difference from the time of the rulers that is crucial in these terms. Seeing time in terms of space is to submit it to the uniformity of a ruling view, to allow its administration. We can thus combine the idea of plural kinds of time and time as plurality with a sense of social vibrancy. But not just through how people arrange themselves in time-space but through their creation of different sorts of time-space. That is there is an 'ontological heterogenesis' (Boyne, 1998: 56) where groups are constituted and defined through inhabiting different corporeal and incorporeal time-spaces. So Felix Guattari suggests we need to see heterogeneity as a 'refrain' explicitly linking uniform time with social domination:

The polyphony of modes of subjectivation actually corresponds to a multiplicity of ways of keeping time. ... In archaic societies, it is through rhythms, chants, dances, masks, marks on the body ... on ritual occasions and with mythical references, that other kinds of collective existential territories are circumscribed. What we are aiming at with this concept of refrain [is] hyper-complex refrains, catalyzing the emergence of incorporeal universes such as those of music or mathematics. ... This type of refrain evades strict spatio-temporal delimitation. With it time ceases to be exterior. ... From this perspective, universal time appears to be no more than a hypothetical projection, a time of generalized equivalence, a 'flattened' capitalistic time; what is important are these partial modules of temporalization, operating in diverse domains (biological, ethological, socio-cultural, machinic).

(1992: 15–16)

These dynamic accounts of time often come at the expense of using one model of spatiality as their foil (Massey, 1998: 30–2) – space that is as simple and bleak as the modellers of spatial science could have wished. We need care not to imply that space is indeed necessarily the opposite of time, and thus all forms of spatiality are conservative (Massey, 1992b).[3] Running the analysis backwards suggests we need senses of space that are not about stasis, that are not the homogeneous ordered realm set up as the villain in these theories. We might well see time-spaces as not closed systems where pre-given variables are related to everything else, but as fields of emergent potentialities, with connections and happenings that may or may not come to fruition but are spheres offering the possibility of multiplicity (Massey, 1998: 26). Seeing spatiality as a becoming seems better able to grasp the rhythmic city, and offers an important pointer beyond the ways geographers have too often ended up inscribing time on to place through a spatial idiom. The sense of dynamism of temporality, written into the inspiration and 'projects' of time-geography should offer us not a way of adding 'time' but rethinking space. Following this side of the argument, not just looking to the time of the subject moving through the world but following that encoded in the world would mean thinking beyond time flowing like simple 'lines' and trajectories to look at loops and recursivity, and fractures and folds in the space-time fabric of the city (Serres, 1995: 57; Crang and Travlou, 2000).[4] Thus instead of a simple direction, flows and projects, we might take Bergson's idea of expanding cycles and circuits of memory and reality, where each recollection deforms and deepens the memory with the memory of its own remembering (Rodowick, 1997: 90).

Conclusion

The time-space that I have been trying to outline here is, then, one of temporal-ised space. That is, it is not simply a matter of mapping time on to space – particularly not when time becomes visible as though it were space. Thinking of the rhythms of particular locales begins to offer a better grasp on the linking of space and time. But this seems a stationary account of motion in places, not actually following the motion itself. In this sense the 'projects' of the Lund school time-geography offer an anchor that can be refashioned so that the paths retain the sense of expectation and memory suggested by Merleau-Ponty and Heidegger. A sense of space-time that brings the virtual into the experience of space, that thinks of space as connected to time seems a fuller account. It is an account of the connectedness of future and past, binding them into the present instant. Rather, then, than offering everyday life and places as self-contained self-present activities, the inclusion of the virtual suggests they always contain other possibilities, that different forms can emerge. The temporalisation of place removes a sense of self-contained moments and acts linked by external logics to open possibilities of immanent and emergent orders. In this sense I have tried to suggest a time-space practice that is neither a stitching together of pre-given points – constituting place – nor as the weaving of a fixed and closed urban fabric. Rather it is a becoming of velocities, directions, turnings, detours, exits and entries (cf. Grosz, 1995: 126–8). The balance of the city in motion between repetitive rhythmical activity making place and passage through place is a matter that will vary empirically. Attention to cycles, to flux and repetition in and through places serves as a useful counter to the rather simple 'flows' I have used here to illustrate ideas of time-space as an event. But in both cases the events created suggest the generative properties of time-space while recognising that everyday acts are inevitably distantiated. Instead of assuming 'a bounded or framed space in which discrete elements may be associated, more or less ambiguously' which subordinates diversity to unity (Rajchman, 1997: 17), it is to see unity as a contingent operation holding together the fractured and virtual diversity of time-space.

Time-space is not a container for images of the past, nor a storehouse of prior experiences in that sense. We should be wary of attempts to suggest time offers a set of representations of future and past. This is an intellectualisation and abstraction from the phenomenological orchestration of time-space. Time is an experience of flow rather than being a series of static images enchained in a sequence. The criticism about models of space and time thus works not only at the level of experience but also that of representation. We need a sense of the event and process of time, rather than letting thinking be dominated by static representations. It may be that we can develop representations that within them encode the forces and movement of time (Deleuze, 1991), not an image added to movement but a sense of dynamic space-time. Crucially time is not an external measure but intrinsic. To create a useful idea of the combination time-space perhaps what we need is to apply the dynamism theory gives to time into space, rather than allow the stasis theory ascribes to space to frame time. Crucially then this uses time to deny the 'self-presence' of the everyday, indeed inserting virtual times to move it from brute actuality. The present is decentred into a set of fractured virtualities. This haunting and opening offers the possibility of difference rather than just repetition. Within space-time there is more than simultaneity, space is more than the synchronic foil to various diachronic models. As Grosz suggests (1999: 25) 'what duration, memory and consciousness bring to the world is the possibility of unfolding, hesitation, uncertainty. Not everything is presented in simultaneity'. Out of the virtual can come the new, that is new forms that do not correspond or fit within dominant logics and expectations. It offers an anarchisation of time away from predictable and projected actions – and projects. Inserting the virtual offers the possibility, and only the possibility, of emergent non-determined forms. Then, perhaps, we can think through a pluralised and eventful sense of lived time-space.

NOTES

1   For instance, alongside modernity goes the rise of the textualised, autobiographical, storied self, as well as many grand narratives of nation, class and even stories of capital itself – be they liberalism or globalisation.

2   This analogy was indeed from where Benjamin took his notion of the after-image as the appearance of history. We might also notice that the study of optical effects like this at the end of the nineteenth century had radically destabilised notions of time and the subject.

3   Specifically, structuralism rejected the dominance of temporal narratives in favour of internally coherent self-standing structures which were a-temporal, which were then labelled 'spatial' (see Massey, 1998: 32; Crang and Thrift, 2000).

4   To think this through we could follow Latour's observation that Serres has 'a topologically bizarre space as [his] reference for understanding time' (in Serres with Latour, 1995: 59).

11

TIME-GEOGRAPHY MATTERS

Martin Gren

Preamble

As time goes by bodies grow older, shoes wear down, things degenerate. As time goes by human corporeal Observers fail to re-cognise practices and places when they re-turn in exactly the same manner as when they left them. As time goes by imagined-and-real things-and-relations change. As time goes by abstract ideologies become ontologically transformed and confirm or alter the concrete morphology of spatio-temporal materialities on earth. As time goes by space-and-time make a difference which makes a difference. As time goes by how do we re-present the meaning-and-matter of life in specific timespaces

in other timespaces?

So it is that what used to be called 'representation' is at heart a problem of *geographical translation*. The task for any earth writer is to speak about whatever phenomena in other spacetimes in the bodily (con)text of the Observer's own timespace. As Olsson puts it:

> Boiled down to its essentials, telling the truth is to claim that something is something else and be believed when you do it. Others trust me, when I say that this is thus or that a = b.[1]

Telling 'the truth' involves the operation of avoiding the pure self-reference of $a = a$. It includes 'telling', to 'be believed', and to be trusted by somebody else. This implies that human embodied knowledges are not things that can be moved from one timespace to another, but activities *performed* in specific contexts. Geographical translations can be brought forth by human Observers in many different domains of observation, one of those inside the discipline of geography is known as 'time-geography'.

Time-geography?

As time has passed, time-geography has been interpreted in different ways. It has been admired for its representational potentiality and for offering a solid conceptual and methodological departure for empirical research, not the least in a transdisciplinary context. It has also been criticised for being too 'physicalistic', too 'reductionistic', too 'objectivistic', too 'masculinistic', and for having a too 'narrow or naive conception of power and human thought-and-action'.[2]

But wait a second! What kind of 'quasiobject' or 'hybrid' do we really have in front of us? Is 'time-geography' a philosophy, an ideology, an ontology, or an epistemology? A research programme, a discourse of practices and power-relations, a method, or simply time-spatial distributions of matter (like texts, diagrams, physical bodies co-ordinated in certain ways in certain places)? Something else or in-between? Should one distinguish Allan Pred or Häger-strand's contemporary Swedish followers as in practice doing the most represen-tative time-geography of today?[3] Who should be allowed the power to define, to include and exclude?

End of rhetorical questions. To what that relational thing called 'time-geography' really refers is in principle undecidable in the age of 'hyperinter-textuality'. In theoretical practice there is no longer, if there ever was, one time-geography but many time-geographies.

Lines (of reasoning) and points (of departure)

Although acknowledging this ambiguity, I have nevertheless used some of the writings of Torsten Hägerstrand as my own prime source of inspiration.[4] The reason is that Hägerstrand is usually recognised as 'the father of time-geography', and I strongly believe that his attempts to outline a basic metaphysical framework will be central also for future time-geographic projects. Given that his own original more explicit formulations of 'time-geography' date back to at least the early 1970s, the time is now perhaps ripe to cast them in the light of some current themes.

I will, however, *not* be offering a detailed step-by-step dissection of Häger-strand's metaphysics.[5] Instead I will *circle around some central ontological and epistemological ingredients* as I understand them. My intention is to make a sympathetic contribution to 'time-geography' (if there ever were to be such a 'thing') from the *inside* rather than providing yet another critique from the outside.

The lines of reasoning that run through the rest of this chapter could be boiled down to two main points. *First*, and I presume contrary to a number of other Observers, I feel that Hägerstrand's metaphysics are not 'too physicalistic', but rather *not physical enough*. Or, to put it differently: the consequences of a corporeal ontology have not yet been fully drawn within time-geography. *Second*, I also believe that the conception of human thought-and-action in Hägerstrand's time-geography could be extended much further somewhere in the direction of what Nigel Thrift recently has referred to as 'non-representational theory'.[6]

The prospect is one of moving time-geography beyond traditional epistemological approaches that invoke transcendental and mentalistic theories of representation, and therefore too readily reproduce various dualistic Cartesian cuts like; 'material-social', 'inner-outer', 'practice-thought', 'fiction-fact', 'reality-representation', 'sign-object', 'body-mind', 'culture-nature', and so on and so forth.

Time-geography and the corporeality of representation

The development of a method for representation which would make it possible to put together (and hold on to) observations in such a way that data was not pulled out from its place- and time-dependent context has been central from the outset in the time-geography project. The following four general requirements for such a representational system were formulated in the early 1970s:

1   It should be easy to realise what the representation corresponds to in the reality.
2   The representation should have a wide scope of applicability.
3   The representation should have the ability to generate questions which one could not pose without it.
4   The representation should admit conclusions and calculations whose correspondence with reality does not have to be verified by observations.[7]

An ordinary map re-presents aspects of the world as one moment frozen in time, as a static snapshot. To overcome the shortcomings of the map's ability to re-present space 'as time goes by' became one of Hägerstrand's most profound professional and disciplinary challenges. In his own words:

> the great adventure of my scholarly life has been to transcend the map. I see, almost literally, the opulence of the world as a *moiré* of processes in conversation.[8]

His attempts to 'transcend the map' have resulted in the now famous time-geographic diagrams. These notational (representational) systems often form the implicit basis for analysis and interpretation by time-geograpers, and they are also

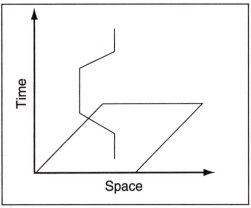

Figure 11.1   A time-space diagram representing a human body in space through time

what tend to be equated with time-geography by outside Observers. The diagrams usually have two axes, one of time and one of space. Figure 11.1 represents a human body in the form of a trajectory that maps the physical location and movement of the body in timespace (for example from home to work and back home again).

According to the ontological point of departure in Hägerstrand's metaphysics human bodies are corporeal beings. This means that they will always be situated in relation to their *own specific and particular embodied positions in timespace*. Fair enough, but at the same time the diagram shows a de-materialised abstraction that appears in the context where the Observer is situated. The notation does not re-present the human body in question from its own *topos*, where it is physically embedded in its own successive movements from within its own actual context(s) of timespace(s), and where its path is 'laid out in walking'.

'In reality' (beware of the rhetoric!) human beings will experience themselves and their environments from within different points, as a consequence of their own corporeal nature. However, in the notation system this myriad of concrete embodied individual timespaces of everyday human life are transformed into an abstract timespace, where they are put in relation to a common external yardstick: one singular space with one time that varies within a dominant time's arrow. What appears is the end-product in the form of a drawn trajectory in an abstract linguistic de-materialised notation system that is the *representational timespace of an outside external Observer*.

It could be argued that this poses quite a *serious representational problem* in time-geography, in so far as the aim is to understand situated bodily practices in everyday life (time-spatial) contexts and move away from de-materialised abstractions. This problematises the standard usage of the co-ordinates of time and space within time-geographic representation itself. What in a corporeal sense exists is in fact many times-and-spaces, and to represent those in relation to only one common timespace frame is therefore in a strict sense *a departure from the corporeal world view of Hägerstrand* as I understand it. Put in other words, the centrality of corporeality is *not* reflected in the way it appears in the de-materialised representation as a trajectory in a time-geographic diagram. By such a mapping, situated embodied *multi*verses are being translated into the *uni*verse of the Observer.

If corporeality is taken towards its own limits, it also implies that we are in fact always dealing with *multiple* corporealities. To put these under the umbrella of one external time-space frame of reference according to the notion of 'objective representation' is to subjectively choose and construct only one out of many possible timespace 'corporealities': the corporeality of the Observer. Being truly faithful to the physical ontology suggests that every trajectory should in fact be placed in its own timespace contexts, and represented in such a way.[9] According to the embodied time-geographical imagination of everyday corporeal life, we should, as Observers of the diagrams, place ourselves *inside the line of each trajectory* instead of observing them from the outside in relation to the coordinates of time and space signified in the diagram.[10]

This is certainly not a call for a more 'realistic' representation, but neither is it to argue that representation is only 'fictional'. It is instead to take seriously and as far as possible the representation of corporeality (the physical ontology in an everyday life context): *representation of corporeality is itself a corporeal activity*.

Corporeality and the subjective character of objectivity

In my mind Hägerstrand was way ahead of his time by insisting on the importance of the material realm. It is however also reasonable to assume that his emphasis on corporeality originally had more to do with a vision of constructing a neutral and objective system of representation, rather than confronting the subjectivity of embodied practices. Through such a procedure, the time-spatial context in which the time-geography Observer is embedded and embodied is not problematised. Therefore the *question of where the embodied corporeal Observer comes from is not being addressed*.

This comes as no surprise because, according to the 'Modern Constitution' of scientific practice, a standard trick has been to generate representational systems of meaning that hide or obscure the subjectivity of the physical body. Objectivity has been socio-materially constructed by placing the situated and embodied character of all knowledge-and-practice, including the Observer, in the background. As Haraway puts it:

Objectivity, which at root has been about crafting comparative knowledge (how to name things to be stable and to be like each other), becomes a question of the politics of redrawing boundaries in order to have non-innocent conversations and connections.[11]

The notion of objective representation has, for a long time now, been seriously challenged by various social constructivisms. This confrontation has focused particularly on epistemological grounds where it is argued that the belief in a possible symmetry between language and reality can no longer be easily sustained. What is important here, however, is that the corporeal ontology and epistemology of human beings points in the same direction.

Today an insistence on the importance of corporeality appears in various discourses that puts the body at centre stage. There it arises not as a neutral material anchorage for bringing forth objective representations, but as an active and constitutive participant. *When 'physicalism' hits bottom it re-enters not as objectivity and neutrality but as an invitation to the subjective, situated and embodied character of all human knowledges.* Limits to representation are not at all only a consequence of 'language', but also stems from the material constraints of the corporeal world. The notion of an unproblematic objectivity must be rejected not only for 'linguistic reasons' but also for 'corporeal reasons'.

This suggests that Hägerstrand's own original ambitions of a neutral system of notation in the form of the space-time diagrams as objective representations of an independent 'reality' out there cannot be sustained. Paradoxically, this conclusion arises not from 'de-spatialised abstract thought' from the outside, *but precisely from the internal corporeal ontology of his own metaphysics.*

One could then make the observation that this also calls for a *critical reconsideration of some of the ontological and epistemological premises in time-geography itself.* It could be argued that a ballast of a somehow naive realism and objectivity, to give its metaphysics a rhetorical steadiness, is no longer adequate. This is not, however, primarily because the tides have now been turned in the direction of social constructivism, but rather as a consequence of the internal corporeal construction of the ship itself. In my understanding this implies *not* more social constructivism but instead 'a turn away from attempts to establish foundations in transcendental theory and a return to natural epistemologies'.[12]

Timespaces and corporeal ontologies

So far human beings have been put on the line in the time-space diagrams in an attempt to problematise notions of representation from within the corporeal metaphysics of time-geography itself. But the reduction by translation of the multiverses to the universe of an Observer could also be related to questions about what appear and do not appear as trajectories in the diagrams. Not only do human beings live in different timespaces, and therefore experience their worlds from their own individual *topos*, but one could also question the common denominators of time and space in the diagrams in relation to the variety of spatio-temporalities that different material objects may have.

The corporealities on my desk at this particular moment could all be mapped in relation to the axes of a time and space, but the objects themselves – the sandwich, the computer, a book by Harries-Jones, the desk itself – are all of different ages. On the more aggregated level of corporeal phenomena in society – the houses, the roads, and so on – we find a similar situation. All the objects listed could be said to exist in the same timespace, and they could all be represented in a singular time-geographic diagram. At the same time, this procedure does not represent the fact that *different entities may have different temporalities*. They are all part of networks in which constituent entities ('local pockets of order') exhibit variable durations and time-spans. Furthermore, one may also take into account that corporeal phenomena may undergo qualitative changes. They may not merely remain the same in place, or move in time over a static space.

Again, what appears as a single timespace in a time-geographic diagram is in fact likely to be many timespaces. Or put differently: as soon as one seriously considers the fundamentals of corporeality, then conceptions of both space and time also turn out to be problematic. One conclusion is that *the conception of both space and time used in the space-time diagrams do not adequately represent the variety of different physical ontologies that exist in the corporeal world*.

The eye of vision/the touch of the tactile

The way I understand it the time-geography of Hägerstrand has sometimes been partly (mis)understood as *only* being a matter of reducing the physicality of life to lines on a paper, whereas the interpretative dimension has been neglected.[13] But just because the lines in your diagram do not vibrate erotically, or tremble with fear, does not necessarily imply that you are ignorant of the importance of sexuality, or of violence.[14] Hägerstrand, and other skilled time-geographers, are of course very well aware of the fact that there are descriptive limits to time-geographic notation, and that the lines need to be interpreted.[15] In that sense, time-geographic representation is not to be equated with physicalism in a reductionist sense, but rather as a common language of corporeality to which different Observers can relate their own systems of meanings. Such abstractions make visible what *otherwise would have remained invisible or taken-for-granted viewed from the inside of the mapped trajectories*.

Another aspect is that a critique of time-geography too easily seems to have come to assume that its particular Gaze is necessarily based on Hägerstrand's supposed privileging of *vision*. But just because the lines do not smell does not necessarily imply that you are ignorant of the importance of sense organs other than the eye. I would instead strongly insist that the basic sense in Hägerstrand's metaphysics is in fact not vision but *touch*. In the corporeal world the most essential form of 'communication' occurs through direct physical contact, and this is an intelligibility clearly not based on the primacy of vision.[16]

In that sense the time-geographic diagram is an *attempt to develop a non-verbal depiction of what physically happens in the space of matter as time goes by*. As Thrift points out:

> Yet, as a written description, it precisely misses Hägerstrand's main aim, which was to find a geographical vocabulary that could describe these pre-linguistic movements pre-linguistically. That was the purpose of his now famous time-space diagrams.[17]

One could therefore distinguish Hägerstrand's project as a search for a *tangible geography of the tangible*. The development of some kind of pre-linguistic representational system based on touch, that would resonate better with Hägerstrand's own corporeal conception of timespace than what is made visible through the traditional time-space diagrams, is perhaps still a challenge for time-geography.[18]

One problem is that we do not intersubjectively share our bodies, but the signs we communicate. It could, for example, be argued that the concept of space used in time-geography appears as 'absolute', 'a-social', as a 'neutral' medium for objectivity, and that it is based on Euclidian geometry. It is also possible to distinguish it as being based on the linear perspective, which is not the outcome of an innocent eye but stems from a particular Gaze situated and embodied in a particular timespace.[19] Furthermore, this Gaze may operate in a masculine space 'from which projective geometry maps out its phallogocentric field of vision (thereby ensuring that the subject or actant who occupies such a position master its space)'.[20] This masculine point of view ensures that unbounded, differential, infinite, open spaces 'remain centred – on an observer – and distanciated – in terms of an object that is "outside" the self-enclosed interiority of a subject'.[21]

Yes, the traditional time-space diagrams seem to privilege vision, but how could the tactile be publicly communicated without being translated into a language that refers to things that everybody can see? Is not geometry the language of the tactile *par excellence*?

Be that as it may. There is also the question of what kind of room there is for meaning inside the metaphysics of Hägerstrand? Is it possible to hold on to a physical ontology when analysing social phenomena, or must such a move necessarily imply an incorporation of some kind of de-materialised social theorising that does not encounter the constraints of corporeality?

Meaning and matter: an ontological dead-end?

Critiques of the 'physicalism' of Hägerstrand's ontology have often implicitly assumed that the material world is separate from the social, and that the important thing to do is to move beyond the material in order to capture the social. One could, for example, argue that the trajectories in time-geographic diagrams are silent lines in relation to class, gender, ethnicity, and the like.[22] As a result, time-geographic notation does not acknowledge the intimate and recursive relationship between a number of different social attributes and the materiality in/of timespace(s).

It must be kept in mind that in Hägerstrand's imagination there is a suspicion towards the realm of meaning and language because of its inabilities to capture the 'time-spatialness' of corporeal life.[23] Consequently, those who would distinguish themselves as (old)time-geographers often quite strongly place themselves as the Other of de-materialised social discourse. Time-geography is, and should be, different from abstract social theorising. In that sense, time-geographers are 'Flatlanders' in a world where matter matters more than meaning.

One could also argue that the ontology and the epistemology in time-geography is in fact not based on a monolistic physicalism, but deeply rooted in a dualistic way of dividing and categorising the world into two distinct and incommensurable halves. In Hägerstrand's own words: 'we as human beings live in two "realities". One consists of patterns of meaning. ... The other is the material reality.'[24] This suggests that time-geography does not necessarily imply a retreat to 'physical reductionism', or that the world of meaning is or should be excluded. In fact Hägerstrand has himself repeatedly insisted that the relationship in-between the two 'realities' should be carefully studied.[25] But how could this really be done on the basis of a Cartesian ontological cut so sharp and deep that no epistemological band-aid will be able to stop the bleeding? As the old saying goes; 'they say that mind is no matter, and matter never mind'.

If it is so that the spatio-temporalities of physical phenomena in a space of matter 'out there' need to be understood by importing explanations from the world of meaning, then the line of distinction in-between the two ontological realms is not erased but confirmed. Therefore, it seems to me, that a 'Flatlander-scepticism' towards social theorising simply seems to *reproduce the dualistic division*. Or put differently: the physical ontology is not being carried all the way through. How, then, could the walls of the social realm be further incorporated within time-geography where one dances on the floorboards of corporeality?

At the limit: the physical becomes the social bond(age)

In Hägerstrand's geographical imagination society is not inside our heads, but rather outside. According to himself:

> Society is not only a set of minds and intangible roles and institutions in interaction. Even if we leave out the entourage of things, society has *corporeality*, as is clearly expressed in such ancient words as some*body* or any-*body*. In other words, meaning and matter come together in the human person. Action in the landscape whatever the meaning is, is also matter acting on matter. *Seen in this perspective actions become space-time trajectories of matter*.[26]

It seems to me that current attempts to seriously consider the fundamental importance of the material realm resonates with at least aspects of Hägerstrand's metaphysics. In some ways there are possibilities to clarify these thoughts through actor-network theory. The latter tries to move beyond dualistic thinking by, amongst other things, accepting non-human material artefacts as members of our human social community.[27]

In actor-network theory, as I understand it, there are no such things as purely material objects that only have an extensionality in timespace, and whose only function is to provide a passive 'coulisse' for de-materialised social relations. Material artefacts, in whatever form they may appear, are constitutive parts of actor-networks in timespace, not merely some things 'out there' that humans ascribe meaning to. Paradoxically the conclusion now becomes that *it is the human corporeal world that is the essence of the human social world!* The physical becomes the social bondage, or as Serres puts it: 'humanity begins with things'.[28]

An implication for time-geography is that objects of the corporeal world are of such significance and importance for the social that they *cannot be reduced to mere matter*. Corporeal objects are not merely entities that occupy space and function as passive constraints for human life. 'Non-humans', or in principle all the corporeality of life that Hägerstrand has insisted upon time after time, must also be regarded as *active stabilisers of 'social relations' in time-space*.[29]

The corporeal approach of time-geography is then not only about the world of pure matter in relation to, for example, ecological issues in a spacetime context, or that activities must be co-ordinated in relation to 'coupling constraints'. Neither is it only about the 'limited packing capacity of timespace'. Material objects are also *actants* in the human social world. This suggests that time-geography could benefit from seriously engaging a lot more with issues about *non-human agency*. If the material world is so important, then why not go all the way and grant it a fair amount of agency? An 'artefactish' is a made object, a bit of fiction, yet also both fact and fetish.[30]

Yet, if one wants to add subsistence to the existence of material 'artefactishes', this implies that time-geography must somehow make a move beyond a language of geometry based on the tactile. The importance of phenomena are not only their physical ontological status, their position or movement in timespace, but what they do. Amongst other things, this calls for a consideration of the conception of (time)space in time-geography, and also geography more generally – Hägerstrand's time in relation to the metaphor of 'network'.

Topologically complex enough?

Thinking in terms of networks may at first glance appear as quite commensurate with the notion of 'webs of trajectories' used in time-geography. A closer look, however, seems to suggest that the importance of, for example, touch as an elementary form of 'communication' and spatial proximity is but only one aspect of networks. And, which would surely disappoint many (time-)geographers, according to Latour not a particularly important one either:

> the first advantage of thinking in terms of networks is that we get rid of 'the tyranny of distance' or proximity; elements which are close when disconnected may be infinitely remote if their connections are analysed; conversely, elements which would appear as infinitely distant may be close when their connections are brought back into the picture. ... The difficulty we have in defining all associations in terms of networks is due to the prevalence of geography. ... [Geographers'] definition of proximity and distance is useless for ANT – or it should be included as one type of connections, one type of networks. ... Out of geographers and geography, 'in-between' their own networks, there is no such thing as proximity or a distance which would not be defined by connectability. ... The notion of network helps us to lift the tyranny of geographers in defining space and offers us a notion which is neither social nor 'real' space, but associations.[31]

So it is that the 'sociology of translation' is not the same as the time-geography of translation. Perhaps, then, time-geographers may need to think through much more seriously their own conceptions of (time)space. Is it *topologically complex enough* for investigating the braided relationships in-between meaning and matter (what Latour calls 'the associations') that Hägerstrand has emphasised? Or, should the time-geographic project be *delimited* to the study of only 'one type of connections, one type of networks', in but one particular type of conception of timespace?

So much for networks, now back to the 'cracked' actor. Who and what is that 'some*body* or any*body*' who creates 'space-time trajectories of matter'? In particular, how should that body be understood as an Observer within time-geography?

The body and the end of representation

In Hägerstrand's metaphysics human Observers are corporeal beings. A conception of 'mind' within Hägerstrand's time-geography must therefore somehow be *embodied*. This ontological point of departure makes it somehow problematic to simply incorporate some kind of external system of meaning into the corporeal body, in order to 'solve' the problem of human agency within time-geography. Taken-for-granted notions about an abstract de-corporealised mind, a 'transcendental subject' that is separate from the body, may easily come with it. The ghost of Descartes would appear again, and the notion that the human body can build up an 'inner' representation of an 'outer world' is still a strong cultural belief in many places.[32] Therefore, before one jumps into the frying pan of linguistic representation, it could be instructive to ask a more fundamental question: *Do human bodies really operate with representation at all?*

In everyday language we say, for example, that we 'take in information from the outside world'. From the point of biological cognition it seems to be rather difficult to pin down any mechanism that is responsible for the transfer of the 'information' from the outside environment to the inside of an organism. The nervous system is operationally closed, and 'information' cannot be 'taken in', it is something actively and physically produced within the body itself.[33] This means that the notion of representation based on possible identity between the 'internal human mind' and its environment, the metaphor of the mirror, cannot be sustained. The corporeal basis for representation is instead based on the *difference between the embodied mind and its environment*. In that sense, the overtly complex environment is reduced to the material complexity of the corporeal Observer.

One could also make the observation that 'information' cannot be represented as trajectories in a time-space diagram because it has no physical extensionality.[34] The everyday expression of 'taking in information' is a de-corporealised semantic description, which does not reflect what is happening in strictly physical terms. Certain parallels exist with time-geography, since a basic proposition within Hägerstrand's metaphysics is that verbal descriptions should be treated with suspicion. According to him verbal language:

is made for telling stories or expressing feelings and wants ... *not* to keep track of what is coexistent. Because of that it takes a lot of effort to think in patterns which are not supported by language.[35]

Words are so often de-materialised abstractions, and therefore they are not able to picture the 'real' material bindings within timespaces. Yet, it is in the domain of language that the human body arises as an Observer that is able to make linguistic descriptions (and linguistic descriptions of these descriptions, and ...).

The Observer and language

One of the consequences of the 'linguistic turn' within the social sciences has been that representation has been problematised. This is true to such a degree that Observers have distinguished a 'crisis of representation', in itself a term that bears witness to a deeply rooted cultural belief in representation. Although now considered problematic, representation is still so often framed by the distinction; language – reality. But according to Maturana and Varela:

> Language was never invented by anyone only to take in an outside world. Therefore, it cannot be used to reveal that world. Rather, it is by languaging that the act of knowing, in the behavioral coordination which is language, brings forth a world. We work our lives in a mutual linguistic coupling, not because language permits us to reveal ourselves but because we are constituted in language in a continuous becoming that we bring forth with others. We find ourselves in this co-ontogenic coupling, not as a preexisting reference to an origin, but as an ongoing transformation of the linguistic world that we build with other human beings.[36]

The 'crisis of representation' arises as long as we use words as if they were designators of objects or situations in the 'real world out there'. As Observers we tend to think of words as standing for something, and we do not easily recognise the co-ordinative role they play in our timespace performances. One reason is that the descriptions that we make as Observers in language do *not reflect how words are used as linguistic co-ordinations of our thoughts-and-actions in timespace*. In the words of Maturana and Varela:

> when we describe words as designators of objects or situations in the world, as observers we are making a description that does not reflect the condition of structural coupling in which words are ontologically established coordinations of behavior.[37]

Another way is therefore to think of language in terms of 'non-representation'.[38] Human Observers are 'languaging beings', which means that the co-ordination of human thought-and-action in timespace becomes central. Language plays a performative and co-ordinative role, and not primarily one of representation. 'Often language's function is simply to set up the intersubjective spaces for these common actions, rather than to represent them', as Thrift observes.[39] The important thing about language, in a non-representational context, is not what words 'mean', 'stand for' or 're-present', whether we use 'table', 'távola', or 'bord', but the thoughts-and-actions that words co-ordinate in timespace.

This suggests a change of perspective from nouns to verbs, and a focus on the bringing forth of the relation rather than the end-product of the related. Consequently, epistemology should no longer be understood as being about what knowledge *is*, but instead about how knowledge is *done*.[40]

> If we reflect a moment on what criterion we are using to say whether someone *has* knowledge, we will see that what we are seeking is an effective action in the realm where an answer is expected. That is, we are expecting an effective behavior in a context that we specify with our question.[41]

A non-representational conception of language could, I believe, open up a quite extraordinary and radical research agenda within time-geography. The whole vocabulary of simple mentalistic concepts that human Observers use in everyday practice, for example; 'nice', 'aggressive', 'clever', 'caring', 'stupid', could all be seen *not* as reflections of an inner mental space (which can be further linguistically chopped up by using more formal concepts like the 'unconscious'), but as *ways of co-ordinating human bodies and relations in timespace*. Of particular importance here is that a non-representational understanding of language would mean that time-geography can *both* hold on to its scepticism towards 'abstract thought', *and* at the same time make substantial contributions to a time-geographical understanding of that thought.

In the perspective of time-geography, language (as a system of signs) may, together with other populations of artefacts, be considered as part of the *environment to which the human body must adapt*. The human body becomes a human being, an Observer, when this adaptation is successfully carried through in timespace.

This, however, certainly does not imply that 'languaging' is the only thing that counts, and neither does it mean that there are *a priori* reasons to accept ready-made Cartesian *a posteriori* divisions. On the contrary, just think of the ways that mobile telephones nowadays co-ordinate human behaviour in certain timespaces. Are 'artefactishes' part of the environment, or, extensions of the human body?

What are the reasons for delimiting 'the Self' by the boundary of the skin? If there is anything really worthy of topological complexity, then it is the in/visible choreography of the human mind.

Languaging and emotioning

Escape from pure self-reference, linguistic reformulation of biologically rooted experiences in everyday corporeal social life. In that domain of existence *a* is not *a* but *b*, so back to basic geo(carto)graphic reason.[42] No Leviathan and no high social ivory towers. In Flatland there are no untouchables, only embodied users and spokes(wo)men. And they are never ever out of, or beyond, their senses!

For corporeal beings all experiences must somehow be biologically rooted in the human body, in the epistemology of the five senses. 'The heart has reasons of its own', as the saying goes. When corporeality is taken towards its own limits, there is no objective experience. Do not look for the smell of invisible ideology behind the touch of appearances, feel instead the taste in front of your own silent eyes.

In a sense, all human understanding, all linguistic reformulations of biologically rooted experiences of happenings in timespace are superfluous. The actual performances in timespace theatres will continue, no matter what kind of meanings are signified by observing Observers. All linguistic explanations arrive to the scene *after* the event. Yet, they are part of the events in the stream of conduct and therefore *not* at all innocent. If they get a hold on somebody, then life can change.

No dirty (God)tricks about everything from nowhere as if the emotionings within the body in timespace did not make any difference. Instead, a *performance where the effects triggered in another Observer are much more important than whatever kind of relationship is being talked about*. So grab a cold tonic for the Foucauldian troops who do not have bodies, but are bodies!

Perhaps, in order to be consistent with its corporeal metaphysics, time-geography also needs to seriously address various kinds of emotionings?[43]

The ethics and politics of time-geography?

This 'non-representational' take on human emotional thought-and-action also suggests that 'time-geography' itself cannot be understood as an innocent neutral transparent medium of objective representation, but rather as a discursive *choreography of performative behaviour*. As Maturana and Varela put it: 'everything we do is a structural dance in the choreography of coexistence'.[44] The space-time diagrams as materialised social relations, as 'immutable mobiles' appearing in 'centers of calculation' for the socio-spatial coordination of everyday corporeal life and practice by the 'magic' of power performed over distance.[45] Bootstrap deconstructed, ontological transformations on display.

Then what kind of human thoughts-and-actions, emotionings, and human relations in timespaces does 'time-geography' privilege? What kind of effects on the coordination of human emotional thoughts-and-actions and relationships of power in time-spaces does 'it' breed and foster? Does 'it' have a room for all those relations we so often signify in language as 'things', or put away in some kind of obsolete and strange 'inner' territory of representation called 'the mental'? What about those thoughts-and-actions that are linguistically co-ordinated in timespaces by words like 'love', 'hate', 'respect', 'revenge', 'fear', 'joy', 'sex', 'forgiveness', 'trust', 'humour', 'anxiety', 'happiness', 'despair', 'play', 'desire'?

Does 'time-geography' have a warm space of feelings for a non-representational and non-dualistic epistemology of doing? Will 'it' welcome heretic jesters? Angels? Blank figures without passports?[46] Those Others who fail to remain the Same? Those topologically complex beings in-between, or beyond, masculine and feminine identities who are not interested in policing boundaries any more?

Coda

In the meantime I sit with closed eyes staring in front of my own small private crystal ball. Although made in Sweden, I nevertheless smell a bunch of 'time-geographers' gathering together outside the Citadel of the Modern Cartesian Confinement. Some of them are writing on the disciplinary Wall; 'We want a new Constitution!' 'There is no Great Divide!' 'A new geography of the Flatland, now!'

Others are in a state of constant trespassing making fleshy translations and carvings sensitive both to the sublime and the mundane. Imaginative geographies of important matter rooted in the meaningful escape from pure self-reference. No more and no less. Welcome to the land of the Observer and second-order observations!

You must remember this; outside the walls of representative confinement the relationship always *precedes*. The simple facts of life cannot be proved, only *probed*.

Acknowledgements

I would like to first of all express my deep gratitude to Steve Hinchliffe for all of his very constructive comments on the first draft. Then there were Eric Clark, PO Hallin, Mikael Jonasson, Gunnar Olsson, Nigel Thrift and Wolfgang Zierhofer, who all have made contributions to the text.

NOTES

1 Olsson, 1991: 167.

2 I make absolutely no attempt here to cover the literature on the pros and cons of time-geography. A few examples of my own sources are; Bladh, 1995; Buttimer, 1983; Carlstein *et al.*, 1978; Giddens, 1986; Gregory (in Gregory and Urry), 1985 Gregory 1994a; Gren, 1994; Hallin, 1988, 1992; Harvey, 1989; Olsson, 1998; Pred, 1977, 1981d; Rose, 1993a; Thrift and Pred, 1981; Åquist, 1992.

3 If one considers the works by Allan Pred as central for contemporary 'time-geography', then Gidden's critical comments (that time-geography 'operates with a naive and defective conception of the human agent', 'tends to treat "individuals" as constituted independently of the social settings which they confront in their day-to-day activities', and 'involves only a weakly developed theory of power', Giddens, 1986: 116–19) become problematic since they cannot readily be applied to Pred's work (see Pred, 1977, 1981d, 1984, 1990a, 1990b, 1991a, 1991b, 1992, 1995).

4 Hägerstrand ,1953, 1972, 1976, 1977, 1985c, 1992a, 1992b, 1993a, 1993b, 1996; Hägerstrand and Lenntorp,1974; the collection of Hägerstrand's writings in Carlestam and Solbe, 1991.

5 I have tried to do so more elsewhere (Gren, 1994).

6 For an introduction to 'non-representational theories', see the first chapter in Thrift, 1996, which has also very much inspired and informed my own writing here (see also the chapters by Pile and Thrift in Pile and Thrift, 1995, and Hinchliffe, 1998).

7 Hägerstrand, 1974: 88; my translation.

8 Hägerstrand (in Buttimer, 1983: 239).

9 However difficult and even impossible that may be, not the least because there are certainly and always material limits to and of representation.

10 To be a Swedish human geographer is quite a challenge indeed. If one is to follow Hägerstrand one has to place oneself 'inside a line', and if one is to follow Olsson one is supposed to be 'lodged inside an American football' (Olsson, 1993, and also Gren, 1994).

11 Haraway 1996: 111 (see also Maturana, 1988; Maturana *et al.*, 1988).

12 Luhmann, 1995: 479. In relation to Hägerstrand, 'natural epistemologies' would point in the direction of 'ecology' and 'evolutionary epistemology'.

13 This may be partly explained by the fact that only parts of Hägerstrand's extensive writings are available in English.

14 I do not want to challenge Gillian Rose's reading of time-geography (Rose, 1993a), but only suggest that there are also other possible readings. Being aware of various masculinist fallacies raised by Rose could, for example, easily enable feminist interpretations of trajectories in space-time diagrams. The instrument of observation should not be too much confused with what some Observers have chosen to do with it.

15 A recent example is Ellegård, 1998, who very strongly insists on the importance of going beyond the surface of 'the lines' in order to reach into the taken-for-granted world behind. For her, time-geography means much more than a simple mapping of trajectories (see also Lenntorp, 1998).

16 See Hägerstrand in Carlestam and Sollbe, 1991: 137–8.

17 Thrift, 1996: 8.

18 Somehow paradoxically this would include a recorporealisation of vision.

19 See Law and Benschop, 1997.

20 Doel, 1998: ch. 2: 15.

21 Doel, 1998: ch. 2: 15. For a detailed analysis of time-geography in relation to feminist perspectives, see Rose, 1993a (ch. 2).

22 See Rose, 1993a.

23 A detailed analysis of the difference between representing phenomena that subsist ('meaning') and phenomena that exist ('matter') is provided by Olsson, 1980.

24 Hägerstrand (in Carlestam and Sollbe, 1991: 197; my translation.

25 See for example the collection of Hägerstrand's writings in Carlestam and Sollbe, 1991.

26 Hägerstrand, 1989: 3.

27 This is not the place to give a systematic account for ANT, see for example; Callon *et al.*, 1986; Hetherington and Munro, 1997; Hinchliffe, 1996, 1997, 1998; Latour, 1993, 1998; Law, 1994; Michael, 1996; Murdoch, 1997a, 1997b; Thrift, 1996.

28 In Serres and Latour, 1995: 166.

29 And note that 'social relations' always also have material ingredients.

30 'Fact + fetish = factish' was the title of a lecture given by Bruno Latour at the university of Göteborg, December 1997.

31 Latour, 1997b.

32 Attempts in human geography to introduce, for example, variants of psychoanalytic theory could be read as an example of this. It is worth noting that Capra makes the observation that there are striking parallels between Newtonian absolute Euclidian space, and the psychological space of Freud's psychoanalytic theory. For example, the Id, the Ego, and the Superego are conceived of as elementary particles moved by forces in a psychological space that evokes the image of billiard ball physics. This space exists independently of the Observer, that is, it is not merely a a semantic way of describing and coordinating human thought-and-action, but is granted an ontological existence apart from those descriptions (Capra, 1982: 186–90). Spatial science in the guise of contemporary cultural geography? See also Duncan, 1996, and Longhurst, 1997.

33 Maturana and Varela, 1980, 1987; Varela *et al.*, 1991; Varela and Dupuy, 1992.

34 For Bateson the 'stuff' of mental process was *difference*, and that is no thing that can not be located in timespace. Bateson even argued that humans cannot perceive anything but difference, for example, the difference between ink and paper is the signal (see Bateson, 1972, 1980; Harries-Jones, 1995).

35 Hägerstrand in Benko and Strohmayer, 1995: 90.

36 Maturana and Varela, 1987: 234.

37 Ibid.: 208.

38 'Words take on a non-representational role; they take on a meaning and persuasive force even though they cannot be directly matched with objects in the world', Barnes and Gregory, 1997: 4. See also the first chapter in Thrift, 1996, the chapters by Pile and Thrift in Pile and Thrift, 1995, and Hinchliffe, 1996, 1997, 1998.

39 Pile and Thrift, 1995: 28–9.

40 See Hinchliffe, 1998, and Rose, 1997.

41 Maturana and Varela, 1987: 173.

42 Olsson, 1998.

43 In the contexts of sociology, see Game and Metcalfe, 1996.

44 Maturana and Varela, 1987: 247.

45 Latour, 1993, 1998; Law, 1994.

46 Hetherington and Lee, 1997.

12

BELONGING: EXPERIENCE IN SACRED TIME AND SPACE

Ann Game

In this chapter I pursue my longstanding interest in the time of lived experience. 'Time unhinged' (Game, 1997) took up the question of living in the now and the status of 'the moment'. There I argued that despite his critique of linear time and causality, Bergson remained too attached to a temporal logic of past-present-future and to the notion of a flow of time towards a future, to be able to recognise the significance of the moment. Here I take this further by considering experiences of the moment that might be regarded as extra-temporal, as outside familiar temporal structures of past-present-future, of either linear time or Bergson's duration. Such moments are moments in eternity. In living these, we experience a now and then all at once, we experience a temporal connectedness. In other words, *living* in the now always implies a now-and-then. Corresponding to this time is a space that, in contrast to familiar notions of an empty space comprised of terms in separation, is full and alive in its connectedness. Developing an understanding of what I shall refer to as sacred time-space contributes to a broader project of understanding relationality, sociality, and the quality of experiences of living 'in-between'. 'Belonging', I will argue, is an experience of living in-between.

When we say that we want to belong we typically locate the realisation of this desire in an identity, in a self or a place or a time past that we would have again in the future. The most common expression of which is a longing for 'home'. But the more we search for the place of belonging, the more it eludes. Destinations reached disappoint, and send us searching again. What I want to suggest here is that we *can* experience a sense of belonging *if* we let go of these quests, and that if we do so, we can experience a belonging here, now. For in the searching itself we are looking in the wrong place. In searching we make the assumption that there is a fixed place and time of belonging, an end point when we will arrive at home; we are mistakenly looking for belonging in Euclidean space and linear time. We are looking elsewhere. But it is now, in sacred time and space, where losing self-identity we feel connected, that we belong. Belonging is experienced in eternity.

226

By reflecting on the quality of everyday experiences of belonging – and, when we are *not* searching, we *do* have such moments – this chapter hopes to develop an understanding of this sacred, mythic time-space of belonging. My view is that recognition of the living reality of the mythic realm, the realm of gods or spirit, opens up the possibility of revitalising a sense of belonging. Letting go of nostalgic longings for a future past, we will be able to experience belonging, with all its spiritual and ethical implications, more fully *now* in our everyday lives.

Childhood

> Ah! How solid we would be within ourselves if we could live, live again without nostalgia and in complete ardor, in our primitive world …
>
> The cosmicity of our childhood remains within us …
>
> … childhood remains within us a principle of deep life, of life always in harmony with the possibilities of new beginnings.
>
> (Bachelard, 1971: 103, 108, 124)

There are moments of sheer joyful pleasure when we want to say 'I feel like a child'. Moments when we feel wide-eyed, wide open, in love with the world. Running into the waves, the salt-smell spray in my face, or feeling the sand between my toes, it happens to me then. And around horses. Smelling them, smelling their steaming bodies, pressing my face into a warm neck, mixing feeds, being with the peaceful munching of their eating and chaff blowing, feeling the soft nuzzling-nibbling of a mouth and nose in my hand, but above all, smelling, breathing horse. I lose myself in sea, sand, chaff and horses' breath. These are moments of feeling 'this is it', 'this is what it was-is about', 'this is right', 'now I have found what I have been looking for, what I have always known'. I get that 'coming home' feeling that we all know so well, a feeling that might best be described as a sense of belonging. And one way or another we associate this feeling of belonging with childhood.

But what are we to understand by the 'childhood' of these adult childhood experiences? For, as Bachelard suggests, we are not talking so much about actual childhood as a *principle* of childhood – a way of being that has a 'primitive' quality to it. This principle 'remains' within us, a changeless state co-existing with irreversible and chronological times. 'I grow old *and* I feel young' only partially captures this experience. 'It is only now, as I grow old, that I feel young' comes closer. For the youth I feel is in the realm of the primitive, a primitive knowledge, beyond the actuality of years. The paradoxical nature of experiences of belonging

– the combination of 'I know this already' *and* 'this feels new' – indicates an extra-temporal quality of this way of being. And thus the childhood of these experiences is a doubled state of now and then, old and new at once. Without causality, it has an eternal quality to it: we simultaneously live 'in our primitive world' *and* experience 'new beginnings'. And in these instants of eternity we feel alive.

Thus the form of selfhood involved in these experiences defies developmental or singular understandings of the self. To find our self, as we might typically say of belonging, is to find a self that is not a singular separate identity progressing through life's stages, but a self in connection. A sense of 'feeling solid within ourselves' comes with neither a turning in nor shutting out of the world, but with openness. In contrast to a commonly presumed fixity in 'being at home', belonging is an in-between state of being within and without our selves. Our inner depths are without. 'Being at home' in this sense is a state of forgetting our self in living our relationality. In feeling 'this is right' we experience a sense of being in connection, securely in connection, with the world – with the sea, sky, hills, horses or whatever it may be. We also experience a temporal connectedness, living both now and then at once, connected with the world across time: I was with the sea, I am with the sea.

What evokes childhood in these experiences is their particular temporal quality. We say we feel childlike and we make an association with a past. But what is the connection between 'this feels childlike' and 'I've had this experience before'. When I am with-in horse, sea, or in a place, arriving – it happened in Rome, for instance – or in a text or in the garden or eating, in a particular way (immediately deadened by purpose, achievement, duty or nostalgia) I feel childlike joy. And it feels right, I belong, my body is comfortable here, it fits. But the 'it' that feels right is both the specific experience and the childlikeness – I am at home in childlikeness or childhood, I am at home in this space, I belong. I belong with horse, with being child. *And* I recognise this from 'the past'. But to describe this as reliving childhood doesn't do justice to the experience. The power of the affect suggests something that runs much deeper. Indeed, although I say I feel like a child, I can't say where this association comes from, I can't connect it with emotions in my childhood. The *reliving* is no weakened version of what we might take to be a real past: reliving is a *living,* and a living heightened by the presence of a living past. In referring to this past as childhood we are not referring to 'the facts of childhood' but something that is 'greater than reality' (Bachelard, 1969: 16). Elusive and powerful: yes, 'greater than reality' is right. As Jung put it, the 'real child' is not 'the cause or pre-condition' of the archetypal child (Jung quoted in Hillman, 1975: 11). There is no causal connection between our empirical childhood and our adult living of the mythic or archetypal child.

At the primitive – archaic or mythic – level the feeling of 'I know this already' refers to a beyond of the real past, as Bachelard would say. I did in fact know horses when I was five, fed, groomed and learned to ride them. But I knew them already and forever. The experience now of 'I know this already' doesn't feel like a mere repetition of that past – it is much more and more alive than that. And thus to have this experience one doesn't have to have fed horses when a child. We know the smell already. As children do too. Where else does the longing for horses come from, or the horse in children, and in adults, to which stories speak, if not the mythic realm? What P.L. Travers (author of *Mary Poppins*) describes as 'a fund of ancient knowledge in man's very bloodstream' (1993: 82).

Involuntary memory

In the light of this discussion about the archaic and mythic in experiences of the past, I wonder now if there might not be something about the temporality of involuntary memory that particularly distinguishes it from other forms? Involuntary memory is just that – involuntary, rather than a wilful recollection of the past, or the memory of dates. And it is alive. Involuntary memory is a living past. Thus, we might assume that its magical enlivening quality comes of a mixing with the immemorial, the extra-temporal. Consider for example one of the most striking of these experiences, when towards the end of the work, Proust's narrator stumbles on cobblestones on his way to the Guermantes party:

> again the dazzling and indistinct vision fluttered near me, as if to say: 'Seize me as I pass if you can, and try to solve the riddle of happiness which I set you.' And almost at once I recognised the vision: it was Venice, of which my efforts to describe it and the supposed snapshots taken by my memory had never told me anything, but which the sensation which I had once experienced as I stood upon two uneven stones in the baptistery of St Mark's had, recurring a moment ago, restored to me.
>
> (Proust 1983: 899–900)

What I want to suggest is that it is not simply the Venice previously experienced that is stumbled upon, but mythic Venice, something much bigger than a particular historical event. The rapturous effect – 'a profound azure intoxicated my eyes, impressions of coolness, of dazzling light, swirled around me' (Proust, 1983: 899) – suggests that the spirit of Venice has worked its magic. It is in the realm and time of the mythic that Venice and the narrator come to life (in a moment, anxiety and intellectual doubts vanish as he is swept away with happiness). And in fact Proust attributes a mythic temporal structure to these memories: they are extra-temporal moments when, through 'the miracle of an analogy', we experience a 'fragment of existence freed from time' (1983: 904, 906):

> A moment in the past, did I say? Was it not perhaps very much more: something that, common both to the past and to the present, is much more essential that either of them?
>
> (Proust, 1983: 905)

These doubled moments of past and present at once are archetypal, mythic moments. The description of them as something more than past or present, something more essential might be taken to refer to the primitive.

Involuntary memories have the same extra-temporality as mythic-archetypal experiences; they also have a mythic quality, they involve an archetypal calling up. Beckett says that involuntary memory is a magician (1965: 34–9), a worker of miracles, looking for a fortuitous moment in order to come to life and to bring to life. Involuntary memory is inhabited by and calls up spirits; it lives in-between. Which suggests that paradoxically what is most particular about memory (would the madeleines work for you?) is also most social. We come to life, our past comes to life, when our particular memories mix with the immemorial. Our living past (and present) is archetypal or trans-subjective. And so it is that when, living relationally, we lose our self, that we are most alive. And we belong, as Proust's narrator belongs in an eternal Venice.

Nostalgia

The at-homeness of belonging needs to be distinguished from nostalgia (as indeed also, from the belonging of 'joining up', the sort of belonging that, far from implying an opening to the world, is associated with conformity and closure). Proust's narrator has no nostalgia for Venice, or at least his nostalgic thoughts about Venice are as lifeless and unmoving as snapshots. Whereas nostalgia is a desire for a re-presentation of a 'real' past as it was (or the childhood I *didn't* have), a return to an original point of departure or home, belonging is a living in our primitive world, the timelessness or eternal return of mythic being. And we live the primitive or archaic *now*, rather than in nostalgia's future past. Or more precisely, it lives us: Proust's narrator is swept off his feet by Venice, when he forgets nostalgia.

Freud's famous quotation on homesickness in 'The "uncanny" ' has contributed to a conflation of these states of being:

> There is a joking saying that 'Love is homesickness'; and whenever a man dreams of a place or a country and says to himself, while he is still dreaming: 'this place is familiar to me, I've been here before', we may interpret the place as being his mother's genitals or her body.
>
> (Freud, 1985: 368)

Freud interprets the feeling 'this place is familiar to me, I've been here before' as a masculine desire to return to origins, to the maternal body. Readings of this passage, particularly by feminists such as Irigaray and Cixous, invariably take it to refer to a nostalgic temporal logic, a desire to return to the past as it was, a desire for an end in a beginning. Their concern is with the masculinity in this structure of desire, the appropriation of the feminine in a quest for masculine identity. But this *could be* an experience in mythic time, that time which suspends profane time, duration and history (Eliade, 1971: 35). The archetype of mother would then be one possible point of mythical connection. When, without nostalgia, we live 'the archetypal power of childhood':

> The father is there, also immobile. The mother is there, also immobile. Both escape time. Both live with us in another time.
>
> (Bachelard, 1971: 125)

This points to what is the real issue for Freud: it is not the desire to appropriate the mother in a return to self-sameness that is pathological for him, so much as an 'infantile', 'outmoded' form of thought that accepts 'the reality' of magic and myth (Freud, 1985: 367–8). Living 'the archetypal power of childhood', living in 'another time', a non-developmental time, is what is problematic for the rationalist Freud. His form of thought has reached an adult stage, grown out of the childishly mythic. The uncanny has to be explained (away). And thus the wonderfully mysterious experience of 'I know this already' is reduced to nostalgia. As Bachelard suggests, we need to let go of nostalgia, together with its reverse of adulthood, in order to be able to live with ardour our connectedness, to feel alive. And, with Serres and Latour, in contrast to 'modern' thought, we might affirm the archaic: 'the past is no longer out-of-date' (1995: 48–62).

The ecstasy of the new

I began this chapter with quotations from Bachelard. There is something about his writing that works for me again and again. Like no other writer. He works in his own terms, performing the creative process of which he writes. Bachelard doesn't simply write *about* newness, aliveness, archetypal living, dreaming and imagining; he performs these, bringing them to life. Every time I read him I want

231

to say yes, yes, that's it; I hadn't seen that before, *now* I get it. Now I get what I have always known. Reliving it 'in a manner that is new' (1969: xxix). Just as he says of the poetic image, his images possess me, and with a 'naïve', childlike, joyousness, they become my own, I feel that I could have, should have created them (1969: xviii–xix). Bachelard's images prompt a feeling of belonging. I feel that I belong in his texts, inhabiting his texts like his dreaming spaces. His words 'take root' in me, they grow, they grow me. 'The image ... becomes a new being in our language, expressing us by making us what it expresses' (1969: xix). His images offer themselves for quotation: 'it has been given us by another, but we begin to have the impression that we could have created it' (1969: xix). They open up to the reader and open us.

In my current readings I am having this experience in connection with the way in which newness and the archetypal or immemorial presuppose each other. In other words, the temporality of creative experiences: the suspension of time in moments of newness (Bachelard, 1971: 173). From every page these ideas leap out at me, speak to me now (1969: xxix). *The Poetics of Space* opens with 'the ecstasy of the newness of the image' – an image that itself has this newness about it, inviting abandonment (1969: xxiv), an image that is connected, and connects us, with the archaic. We don't know quite what is going on, but it feels like a *connection* with something more, something beyond. And to attempt representation would be to lose the quality. For what is communicable is precisely that which defies representation – the trans-subjective is in the domain of the archetypal (1969: xx). As Bachelard's texts are.

It is only in the *experience* that we can really get this connection between the new and the archaic – it *is* a feeling, the elusive is palpable. And hence Bachelard's insistence that only phenomenology – a philosophy based in our experience of the world – with its principle of participation, will work as a method for understanding imagination and creativity. But however elusive and difficult it might be to speak about this connection, there are important implications for understandings of the logic and temporality of creative processes and thus, in turn, for an appreciation of our experiences of creativity.

Bachelard says that the relation between a poetic image and archetype is not causal: the poetic image 'is not an echo of the past'. Rather, 'through the brilliance of an image, the distant past resounds with echoes' (1969: xi). Reverberation rather than causality is at work here, or what we have described as 'originating the origin'. (The original or primal implied in 'archaic' (Jung, 1933: 125) is thus to be distinguished from the logic of a nostalgic origin.) The now and then are experienced at once, they need each other, there is no before and after. Thus moments of the new are in suspended time, outside linear time, but also outside the irreversible time of duration in which every moment is different. They are a-temporal or extra-temporal. An experience of newness and surprise is

232

simultaneously an experience of the archaic. New and old are experienced, *now*, together. Through a receptivity to the new, we are in touch with archaic depths. And we wouldn't experience newness without the archaic. (So although every moment of irreversible time or duration is new, we don't necessarily experience it as such: we don't connect with or open to newness if not connected with or open to the archaic.) Proust says this of the inspirational quality of involuntary memory, that it 'causes us to breathe a new air, an air which is new precisely because we have breathed it in the past' (1983: 903). Newness is always accompanied by 'I know this already', but I didn't know this, or that I knew it, till now. The old and new bring each other into being, mutually enliven, in these creative moments, moments which suspend time. It is in these moments that we feel connected, connected with something bigger than us, with the cosmos and the divine; we feel that we belong.

Sacred time

The archetypal realm has an extra-temporal quality: a present past, eternally mythically present (Eliade, 1959: 70). But presumably we only have access to this realm in an *experience* of a doubled time, an experience of living now and then, new and old, at once: in moments when simultaneously we come to life and the archetypal comes to life. Eliade puts it like this: sacrifice, for example, not only exactly reproduces the initial sacrifice revealed by a god *ab origine*, at the beginning of time, it also takes place at that same primordial mythical moment (Eliade, 1971: 35)

Sacred or divine time abolishes time, suspends profane time and duration. The same primordial mythical moment is *now,* eternally now and then, eternity in the now. Or as John Berger puts it 'the instant and its eternity', reminding us that such experiences of suspended time are momentary (1997: 75). Through archetypal imitation we are transported into the mythical realm. In this '"primitive" ontological conception' (the sort of ontology that Freud took issue with, thought we should have grown out of) 'an act becomes real only insofar as it imitates or repeats an archetype' (Eliade, 1971: 34). In other words, aliveness depends upon receptivity to the mythical; it consists in a mutually mimetic bringing to life. We cannot be really alive unless in tune with the archaic.

Eliade draws attention to the paradox in this, a paradox that is central to my understanding of a relational way of being. He says that archetypal man 'sees himself as real only to the extent that he ceases to be himself', he becomes real through mimetic gestures (1971: 34). We are most alive, feel most 'at home', when we forget our self, lose our selves in our connection with the archetypal, with the universe. Which is precisely what Bachelard says about communication:

when images come to life for us, they do so through archetypal connection: it is that which is *universal* that 'speaks to me' and enlivens.

> Then I may hope that my page will possess a sonority that will ring true
> – a voice so remote within me, that it will be the voice we all hear when
> we listen as far back as memory reaches, on the very limits of memory,
> beyond memory perhaps, in the field of the immemorial. All we com-
> municate to others is an orientation towards what is secret.
>
> (Bachelard, 1969: 13)

Communication takes place from soul to soul (Bachelard, 1969: xvi, xx, 17) – in a mythical realm and time beyond my particularity, my history (see also Hillman, 1975: 12–17). It is in this realm that I feel alive, that I feel that I belong.

In creative experiences of newness – my page ringing true for you and be-coming your page – memory and the immemorial work together, mutually deepening each other (Bachelard, 1969: 5), like Proust's involuntary memory. *The Poetics of Space* brings to life the connection between memory and the immemorial in the reader. Reading this book evokes specific memories and we start dreaming of places in our past, and then this dreaming moves to a different domain where memory and the immemorial mix and our specific memories lose their importance (or perhaps take on a new depth through their connection with something bigger). Bachelard's concern in *The Poetics of Space* is with memories of houses, with identifying an intimate essence 'transcending our memories of all the houses in which we have found shelter' (1969: 3). For it is in living universal images of protected intimacy (spaces of centred, secure openness), that imagination comes to life. In other words, in spaces of belonging, spaces where the old and new come to life, together.

> And the daydream deepens to the point where an immemorial domain
> opens up for the dreamer of a home beyond man's earliest memory.
>
> (Bachelard, 1969: 5)

> ... there exists for each one of us an oneiric house, a house of dream-
> memory, that is lost in the shadow of a beyond of the real past.
>
> (Bachelard, 1969: 15)

When we live 'house' it is always something more than the houses we have 'experi-enced': 'great images' are a blend of 'history and prehistory', 'legend and memory' (Bachelard 1969: 32–33). The hut is surely one such image (1969: 29–37). Bachelard makes a reference to the hut and immediately I can feel hut deep in me, a desire to be in a hut. And although I can recall experiences of huts and hut-making in my childhood (looking for hut opportunities under the clothes' horse, in a tree hollow, in a corner of the garden, amongst furniture, in stables) the sense of knowing huts extends way beyond these memories. It is something much more powerful. 'Hut' has worked for me, even the word, the way Crites says that mundane stories 'take soundings' in sacred stories, the unutterable stories *in* which people live (1989: 69–71), which live in our bloodstream. I am back there (where?) in my hut, now. It is ever so close. I am a hut creature, like my cats climbing into boxes. I am in childhood.

Permanent childhood

Primitive images are lived in childhood. When I am in hut, I am in childhood. If it is in the realm of the archetypal or the mythic that we feel alive, that we have a sense of belonging, then mythic childhood has a particular significance to this way of being. Again with Bachelard, I want to pursue the question of childhood.

When we dream 'we travel to the land of Motionless Childhood, motionless the way all Immemorial things are' (Bachelard, 1969: 5–6; see also 1971: 111, 116). Motionless can be taken to refer to the extra-temporal quality of the experience of living archetypal childhood. 'All the summers of our childhood bear witness to "the eternal summer" ' (Bachelard, 1971: 118). Bachelard also calls this eternally present childhood, 'permanent childhood'. This childhood remains alive in the realm where image and memory unify, where we find '*the real being* of our childhood' (1969: 16). 'In the realm of absolute imagination we remain young late in life' (1969: 33). In states of dreaming-imagining we live in childhood, mythic childhood. That is, a childhood of imagination. Remaining young means letting go of our history and our youth, having the capacity to live in 'pre-history', to live mythically.

> Primal images ... are but so many invitations to start imagining again. They give us back areas of being, houses in which the human being's certainty of being is concentrated, and we have the impression that, by living in such images as these, in images as stabilising as these are, we could start a new life.
>
> (Bachelard, 1969: 33)

When we live in the secure intimacy of these primitive house images, feeling centred, we open ourselves to the new, we open to new beginnings. Receptivity to the primitive is receptivity to newness. It means living in permanent childhood: the primitive world of new beginnings. Which implies that childhood is an archetype of archetypal or mythic being, having the archetypal qualities of both being archaic and opening us to newness. 'Childhood is under the sign of wonder' (Bachelard, 1971: 127).

> In our reveries toward childhood, all the archetypes which link man to the world, which provide a poetic harmony between man and the universe, are ... revitalized.
>
> (Bachelard, 1971: 124)

If it is through archetypes that we are connected to the cosmos, these connections come to life when we live childhood, an 'anonymous childhood' beyond all 'mirages of nostalgia' (Bachelard, 1971: 125). 'Without childhood, there is no real cosmicity' (Bachelard ,1971: 126). Hillman says that archetypal childhood is the carrier of our need for 'the imaginal realm', the realm of 'primitivity' and 'beginnings' (1975: 8–12). And for Bachelard, there is something universal in this because childhood is communicable: a soul is never deaf to this archetype: 'if [a feature] has the sign of childhood primitiveness, it awakens within us the archetype of childhood' (1975: 127). (As Bachelard's images do.) So, in those moments of joyful aliveness, of 'yes, this is it', when I say, I feel like a child, but I don't know where that comes from, it is from deep 'within' me, a primitive childhood we all share and, in some way, can't avoid either.

Sacred space

I have been drawing attention to the temporal quality of experiences of belonging, the place of the archaic in these experiences. But of course Bachelard's concern with images of the house suggests something about the quality of spaces of belonging. So let me turn now to the question of the connection between the space and time of these experiences. For Bachelard, the condition of possibility of creative imagining, of openness to the cosmos, is intimate space – protective, sheltering, inhabited space. This space has the same doubled quality as sacred space in Eliade's account of religious experience: it is both centring and opening. Sacred spaces mark out a centre in the world, fix a point in the homogeneity of profane space, provide a place where we can inhabit the world, *and* they provide a vertical connection with the gods (Eliade, 1959: 20–67).

Architecture and spatial forms across religious traditions – poles, pillars, spires, steeples, temples, domes and bell towers – give expression to this connection with the gods. A connection evoked too by *'images of an opening'* (Eliade, 1959: 26). Think, for example, of the ubiquitous images of openings-sky-angels-god-mortals-heaven on the ceilings and domes of churches in Italy, the dizzying movement and communication between heaven and earth, of baroque ceilings in particular, and the actual openings to the sky in some church domes

236

and the Pantheon. But it is not only in churches that we experience sacred space. And now I think of architectural forms in dwelling spaces structured around an internal opening. In the Italian palazzo, for example, where one goes inside the walls through imposing doorways in order to be open, in the cloisters and courtyard, to the skies. Or, within the Balinese compound – protected at the point of entry, like all doorways in Bali, with offerings to the spirits – where the structures are airy and open to the gods. If the presence of the religious is more explicit in these forms (even if only historically in some Italian buildings) than our more secular living spaces, they nevertheless point to the sacred qualities in any everyday space. Reading Bachelard with Eliade serves to remind us then of the experiences of sacred space in everyday life and domestic space. And, in a sense, in ourselves, for as both Eliade (1959: 57) and Bachelard suggest, a sense of universal belonging, connectedness to the cosmos, works through the homology we make between cosmos, house and human body. I belong in myself, in my body, I belong in this house, I belong. I feel solid in myself, I am connected with the cosmos.

We now might want to rethink the evocative quotation with which Bachelard opens *The Poetics of Space:* 'The world pulse beats beyond my door' (Bachelard, 1969: 3). For although it refers to the horizontal significance of the house, it maybe does not do justice to the vertical. If by 'the world' we understand the cosmos, that is. For then we might also speak of 'the world pulse that beats inside my door' (See also Berger, 1991: 55–6).

The doubled experience of being centred and secure *and* open emerges in our students' writing about intimate childhood spaces, in which they invariably and unselfconsciously associate feeling secure with a sense of mystery. Intimate spaces, whether they are under the house, at the top of a tree, under the sky or in a box, are secret spaces, spaces that house secrets. Spaces in which the mysterious *lives*, where we are open to presences. In hiding places or hide-outs we are open to the hidden, the unknown. We hide away to be in relation with the hidden: in our huts, the mysterious opens to us and we to it. It calls us.

For Bachelard, intimate spaces allow dreaming, 'the house protects the dreamer' (1969: 6–7). Which is perhaps another way of saying that they connect us with the 'world of the gods'. In sacred spaces, we live sacred – mythic, immemorial – time. But can we specify further the connection between sacred space and sacred time? Bachelard's writing suggests that living a certain spatiality is the precondition for dreaming-memory, for images coming to life. And yet the very living of intimate space requires an imagining way of being, the capacity to live immemorial time. As we would presume, there is no causality here: space becomes sacred through dreaming, imagining. But where Bachelard nevertheless emphasises space as the condition of the possibility of the experience of creative imagining, Eliade reverses this emphasis. In sacred space man imitates the gods,

237

'creates the world he inhabits', centres himself (see also Bachelard, 1969: 31–3). But what makes this experience possible is sacred time, 'reversible', 'primordial mythical time made present': 'the mythical moment of Creation' (Eliade, 1959: 62–5, 68–9). In short, sacred space and sacred time are the condition of each other, and of the centring connectedness that we call belonging.

But I think we can push the implications of this idea of simultaneously being centred and open a little further by making a connection with definitions of god that resonate here. I am thinking of 'he whose centre is everywhere and circumference nowhere' (see Borges, 1970; Poulet, 1966). What I am suggesting is that being connected with the world of gods involves an *experience* of that world. In belonging, we experience the presence of the divine with us, within us – a centre that is everywhere. The connectedness of belonging is a divine experience.

Belonging to the universe

'Belonging to the universe' is how Capra and Steindl-Rast, physicist and Benedictine monk, understand spirituality and religious experiences. Belonging is connectedness – feeling 'connected to the cosmos' – with 'affective colouring', with love. 'Love is saying yes to belonging'. A joyful yes (Capra and Steindl-Rast, 1992: 57–8). The experience of limitless belonging, characteristic of mystical experiences in all spiritual traditions is what they have in mind: 'mysticism … is the experience of communion with Ultimate Reality' (1992: 56). Which resonates with William James' accounts of everyday mystical experiences: he cites, amongst others, Tennyson's description of a common experience of finding his 'individuality dissolve' and 'fade away into boundless being' (1960: 369–70). Boundless, limitless belonging: a belonging to the boundless, limitless, feeling connected to all being. This is the famous oceanic feeling that Freud, denying any experience of, dismissed as a regressive 'melting away' of ego boundary (1962: 11–13). Freud was mistaken in his understanding of this as a merger. Quite the contrary, the ego-less states of religious, mystical, spiritual experiences, whatever you want to call them, are about a being in relation with the ineffable, the Unknowable other (which might tell us something about Freud's fear).

Limitless belonging is the opening to the world of which Bachelard speaks, a cosmic inhabiting (1971: 102–3, 124). The 'world dreamer' 'opens himself to the world, and the world opens itself to him' (1971: 173–7). The inhabiting or centring of belonging is then a process, not of closing, but of opening up and out – a going forth towards the world with love, allowing the world in when it calls. In these experiences we can feel physically an opening of the self, an almost aching expansiveness, maybe a desire to weep, the way some music makes you want to weep. Whilst living here, in a Tuscan mountainside, I have felt this. Here, in these hills, so alive with history, with the birds heralding in spring, I feel in relation with the unknowable, the mysterious (see Dessaix, 1997: 322). For all that I would be always a stranger here, I would describe this feeling as belonging: 'this feels right' (the 'this' of the place, of me, remaining indeterminate).

But 'this feels right' points to the moral issues that Capra and Steindl-Rast insist are integral to belonging to the universe. If we are called, if we listen to the call of the world or God or childhood, we must fulfil certain obligations. (It also means having the capacity to live with, or hold, our devils as well as our angels.) The gift of a calling means you have to 'live accordingly' or 'live right', attune yourself to cosmic harmony (1992: 58–9). (Attune to these hills, for instance.) The 'yes' of belonging is a yes to a call, to responsibility: 'this is me and I should be true to me, become what I am'. Take the house, for example, which I have spoken of as a place of belonging. It is also, as we all know, a site of obligations and rights – they come with belonging. This is true too of our relations with the friends and family and children with whom we belong. And for me, this has been the case also with horses: I have felt a compulsion, from an early age, to care for the horses with and by whom I have felt connected and held. We are always, already responsible to and for the other. And yet, we so often respond to this with a sense of duty, of demands being made on us; burdened, we feel ungenerous and resentful: 'the children have to be fed, the horses have to be fed'. With this response we lose a sense of belonging. But if we can let go 'duty' – and it isn't that difficult, if we just allow ourselves, just surrender – then we can experience our obligations as our connections with the world, with life. And being with those we love, by whom we are loved, can shift from feeling like duty to feeling like paradise.

Yes, I want to end this chapter with paradise. It might be that paradise is a term we use to make sense of this feeling of belonging. Don't we say 'this is paradise' when we feel we belong – in connection with places, people, all sorts of experiences? And it is the elusive quality of belonging that we associate with paradise: I am in touch with something, but I don't know quite what. But we often mistake this elusiveness for meaning an elsewhere: somewhere else, in the future, I will find a paradise of which I have had glimpses. We go searching. And miss it, necessarily. For paradise is here: the mysterious, unknowable with which we are connected, to which we belong (lost as soon as belonging turns into a desire for control). This is not nostalgia's lost paradise, but an eternal paradise, lived now.

NOTE

This chapter is part of the joint work, in teaching and writing, that I am involved in with Andrew Metcalfe.

13

HALF-OPENED BEING

Andrew Metcalfe and Lucinda Ferguson

Doors

> the door represents ... how separating and connecting are only two sides
> of precisely the same act. The human being who first erected a hut ...
> cut a portion of the continuity and infinity of space and arranged this into
> a particular unity in accordance with a single meaning. ... By virtue of
> the fact that the door forms ... a linkage between the space of human
> beings and everything that remains outside it, it transcends the separation
> between inner and outer. Precisely because it can be opened, its closure
> provides the feeling of a stronger isolation against everything outside this
> space than the mere unstructured wall. The latter is mute, but the door
> speaks.
>
> (Simmel, 1994: 7)

Space without doors evokes a terrifying condition of endlessness and pointlessness, but doors make space habitable, giving us the *room* or *capacity* for home and
world and self. This role in cosmogony is recognised in the sense of reverence
that touches people whenever they approach a threshold. The obstacle of the
door not only marks beginning and end, it is the beginning of ends, as evident in
the etymology of 'limit'. In the one moment, the door separates and connects
(see Eliade, 1971; Berger, 1984; Phillips, 1993: 83ff.).

Whereas contracts or interactions are based on the overcoming of distance,
relationships rely on insides and outsides that only feel apart when already
together. In other words, they rely on the Janus-faced door. A door literally
implies two fascias, and its limen animates every experience of the in-between or
liminal (van Gennep, 1960; Turner, 1969; Clément, 1994). A door could even be
defined as the mysterious substance that separates and connects the two sides of
life. Consider, likewise, the ontology of the hinge, that connection that separates

the eloquent door from the mute wall. The wall itself allows no play of inside and outside, but, by suspending sides, the hinge gives us the sense of events and spaces unfolding (Bohm, 1985). However hard it is for logicians to accept, people are, as Bachelard said, the paradox of 'half-open being' (1969: 222). The certainty of things relies on the substance of nothing:

> We cut holes for doors and windows to build our house;
> And the usefulness of the house relies on these holes.
> So we should recognise the usefulness of what is not there, as well as
> what is.
>> (Tao Te Ching, chapter xi; author's translation)

This reflection on doors, paradoxes and liminality addresses the transitional terrain postulated by Winnicott in his accounts of child development and ongoing creativity. Between insides and outsides, is a

> third part of the life of a human being ... an intermediate area of *experiencing*, to which inner reality and external life both contribute. It is an area that is not challenged, because no claim is made on its behalf except that it shall exist as a resting-place for the individual engaged in the perpetual task of keeping inner and outer reality separate yet interrelated.
>> (Winnicott, 1991: 2)

Although explicitly talking of teddy bears, art and religion, Winnicott could be talking of the paradoxical *implications* of a door when he says that transitional phenomena display 'the substance of *illusion*' (1991: 3). His point is that the ability to *hold* paradoxical positions is required for culture, and thus for the development of children and for ordinary adult life:

> We share a respect for *illusory experience*, and if we wish we may collect together and form a group on the basis of the similarity of our illusory experiences. This is a natural root of grouping among human beings.
>> (Winnicott, 1991: 3)

Scrupulous concern with logically resolving the paradoxes of inside and outside is a pathology, leading to psychic defences that stifle play, creativity and the liveliness of life (1991: 14).

If Winnicott is right, poststructuralists are pointing to a pathology when they show how conventional social thought reduces the play of insides and outsides by treating them as binary oppositions. Whereas the phenomenology of the door shows that closure relies on openness as a persistent possibility, a present absence, analysts committed to identity and autonomy use it to refer to an absolute break. They seek to reduce doors to walls. But paradox isn't so easily evaded, for an impermeable wall no longer closes off an outside. Indeed, experience 'within' a doorless and windowless wall requires the language of limitlessness!

A major example of this obsessive concern with closure is found in the Western sentimentality about childhood as a special state of innocence, for this distinction is founded on disgust and abjection (Rose, 1993b; Matthews, 1994). Childhood is presented as a developmental stage that people should leave, and many adults can't leave it often enough, repeating their renunciation at adolescence, adulthood and middle age, at their children's birth and parents' death. Many children grow up feeling forever on the edge of a miraculous rebirth into the freedom and independence of adulthood. They can't wait to put the sticky contamination of their childhoods behind them.

This developmental story is *doubly* obsessed by issues of openness and closure, for it celebrates the child's overcoming of its initial 'absence of differentiation between the world and the self, whence arise the feelings of participation and the magical mentality which results' (Piaget, 1973: 266–7). If sociality is based on interaction, the child's sense of participation and substantial illusion is abhorrent, *lacking*, because it doesn't recognise the distance that identity must overcome. The Piagetian childhood must burn its bridges to produce the distance required for the adult's contract story (cf. Pateman, 1988). Accordingly, after first assuming that everything active is alive and conscious, Piagetian children finally restrict these qualities to animals and perhaps plants, thereby finally cutting themselves off from the child's world (Piaget, 1973: 220–1). In *The Child's Conception of the World*, Piaget describes the child as

a being knowing nothing of the distinction between mind and body. ... Compared with us, he would experience much less the sensation of the thinking self within him, the feeling of a being independent of the external world. ... But, above all, the psychological perceptions of such a being would be entirely different from our own. Dreams, for example, would appear to him as a disturbance breaking in from without. Words would be bound up with things and to speak would mean to act directly on these things. Inversely, external things would be less material and would be endowed with intentions and will.

(1973: 49)

The child's thought is characterised as animist and magical, like the thought of 'primitive cultures' (1973: 193).

Whereas Piaget confidently characterises participation as lack, of differentiation, we share Winnicott's emphasis on the lifelong significance of this sense of immersion and participation. Instead of modelling adult sociality on interaction or contract, as a joining of identities, we argue that it is grounded in, or flows as, a continuation of the child's sense of participation in a living world. Piaget treats adult thinking as a self-sufficient or detached process, but, as Merleau-Ponty insists, any adult position in the world rests on a child's sense of pre-positional participation:

> in reality, it must be the case that the child's outlook is in some way vindicated against the adult's and against Piaget, and that the unsophisticated thinking of our earliest years remains an indispensable acquisition underlying that of maturity, if there is to be for the adult one single intersubjective world. My awareness of constructing an objective truth would never provide me with anything more than an objective truth for me, and my greatest attempt at impartiality would never enable me to prevail over my subjectivity ... if I had not, underlying my judgements, the primordial certainty of being in contact with being itself, if, before any voluntary *adoption of a position* I were not already *situated* in an intersubjective world.
>
> (1962: 355)

In this chapter we show how our senses of home and world and self rely on our magical and animist senses of participation, on the suspension of Piaget's distinctions between mind and body, inside and outside, subjective and objective, and self and other. Piaget's characterisation of adult thought only shows how often adults are alienated from their own ways of knowing and being. If actually lived through a distinct sense of inside and outside, and without a sense of the life *in-between* self and other or subject and object, existence would be an experience of horror.

Secrets

A closed door opens the capacity for closing others out. It opens a space of secrets that is crucial to the development of selfhood. We learn this as children through the beginning of our lifelong fascination with boxes, doors, cupboards, houses, buried treasure chests and caskets. These 'are veritable organs of the secret psychological life. Indeed, without these "objects" ... our intimate life would lack a model of intimacy. They are hybrid objects, subject objects. Like us, through us and for us, they have a quality of intimacy' (Bachelard, 1969: 78). In insisting on the mutual implication of our selves and these living objects, Bachelard is decisively undercutting Piaget's model of the child's separation from the world.

The significance of cryptic spaces is demonstrated by my 3-year-old son who, before going to sleep, lovingly tucks under his pillow a picture he's drawn of himself. Similarly, as a 10-year-old, Jung secretly carved a little manikin, wrapped it and hid it for a year in a box in the forbidden attic. He didn't understand his actions, but knew that his security and life depended on the secret: whatever situation arose, he gained strength by thinking of the manikin secure in its hiding place. His autobiography ascribes the episode a major role in his individuation and character-formation, as 'the climax and conclusion of [his] childhood' (1973: 22).

> According to the standard account of childhood development, the child's discovery of the concept of secrecy ... heralds the birth of an inner world. When the child learns that thoughts and ideas can be kept within him/her and are not accessible to others, he/she realises that there is some kind of demarcation between his/her world, which is 'inner', and that which is 'outer'.
>
> (Meares, 1992: 7; cf. Dunbar, 1996: 80ff.)

This secret place has contradictory implications. While it divides people into inner and outer states, subjective and objective selves, an outside appearance and an inner essence, we will also see that when someone is most inside they feel opened out. The power of the openable door to feel closed allows insiders to feel as if there are no sides, as if they are opened *out*, whole or cosmic. Meares adds another 'paradox': because secrets can be shared, the child's discovery of a distinct inner world opens opportunities for connection with others (1992: 11). This claim, however, shouldn't be taken as an endorsement of Piaget's assumption that connection is a secondary process, occurring after children have properly realised their distance from the world. Secrets aren't simply held back from relationships: they only come to life in the erotic tension of relations.

Had Jung been unable to imagine anyone interested in his secret, it would have died as surely as if it had been discovered. I remember the naked delight on the face of a 4-year old boy when three 7-year-olds announced they'd play hide and seek with him. 'We'll let you hide first. We'll count to ten thousand and then come and look for you'. Going first! Ten thousand! The little boy could hardly believe his luck. He rushed here and there before settling on a place behind a tree, where he hid, and hid, and hid, as his vitality and confidence wilted. Unless sought, there is no secret and no self. Rather than revealing a binary demarcation of inside and outside, secrets exemplify a persistent sense of the life in-between without subject or object, neither one nor the other, inside or outside. They indicate the simultaneity of separation and connection rather than the development of a *distinct* self.

The connection implied in the most intimate or private can be readily observed at boarding schools, where I've worked as both student and 'mistress'. One night recently, for example, the girls were outside their dormitory, brushing their teeth, when I came to put them to bed. The empty dorm was a picture of chaos and spontaneity. The girls' belongings – mainly clothes – bulged every which way from swollen dressing tables. They had been abandoned and were motionless, yet were alive, lunging from the snug privacy of the girls' drawers, reaching for the bags they were to enter, for the holidays. They seemed to represent the states of mind of the girls: they too were already, flagrantly, on their way to the homes where, they imagined, their insides could safely hang out.

Slowly, I began to feel a pattern in the mess. The sight now reminded me of *trompe l'oeil*, for the clothes and belongings which seemed haphazard also seemed placed. Although secrets allow girls to establish a role distance and uniqueness against a background of boarding school seriality, the secrets only work if the unknowing is known (Goffman, 1967b). Accordingly, the exposed intimacies in the dorm were those which the girls might want public. They were gift wrapped, to display, and thereby *generate*, and somehow also *deepen* a mystery. They might be defiant defences, to ensure that the otherwise same knew of their exclusion from the girls' innermost hearts, or they might be membership tokens, to allow the girls to celebrate together their different secrets. Either way, such open secrets demonstrate separation's need for a vital relation with connection. By highlighting the display integral to the intimate, they endorse Tournier's shocking suggestion that clothes are the human soul (1983: 199).

Open a dressing table drawer at boarding school and you'll find the husks of the tantalising secrets that make up a person. The drawers are to the dormitory what the soul is to the student. They hold secrets safely so the dorm can hold the girls safely, and girls hold secrets so that the secrets can return the service. When I was a student my drawers held notebooks: a spelling book of favourite words, a little album with photos of hometown friends, and a sticker book. In fastidiously attending to them I looked after myself. Although I couldn't *guarantee* that others hadn't searched my drawers, I was content that no one knew my notebooks existed. These, then, weren't open secrets: the hiding of the books in which I hid was itself hidden.

Strangely, my trust in my roommates' discretion relied on our indiscreet fascination with one other. They wouldn't strip and rape me, through my intimate possessions, because this would destroy the relationship on whose energy we relied. I trusted them to understand that a closed treasure chest holds more interest than an open one. The former lives imaginatively, the latter sinks into the dispiriting objective externality that Piaget ascribes to reality. Bergson is thinking of the latter's desiccation when criticising the drawer as a model of memory (1913: 5).

There was never a pattern to when I would take out my notebooks and fuss over them. It depended how 'I' was feeling. However, I rarely attended to even these secret secrets in the absence of others. There was something curiously arousing, for example, about taking them to my Prep class and working on them once I had finished my homework. My desk created a private space but anyone walking over to discuss school work would encroach on it. If this occurred while I was tending my books, my hackles would rise, my heart palpitate, and my books would suddenly vanish. So many close shaves! Such stimulating delight! Evidently, the meaning of my secret books was only kept alive by the maintenance of this erotic tension between myself and others. Through this game, I also flirted with myself, teasing my self out, trying to catch it, to discover what was special to me.

By withdrawing aspects of my self from others, I fostered my own little secret: a special part of me, for me. But it did not emerge dutifully whenever I opened a notebook, and it wasn't simply in the books' contents. My ability to feel it depended on a lively allure, on the charged relation I maintained with myself. When the secret self was threatened, when it was on the edge because of the presence of another other, I could sense it strongly. I couldn't grab it, apprehend it, but I could feel it, caress it, smell its perfume:

> [T]he caress does not know what it seeks. This 'not knowing', this fundamental disorganisation, is the essential. It is like a game with something slipping away, a game absolutely without project or plan, not with what can become ours or us, but with something other, always other, always inaccessible, and always to come. And the caress is the anticipation of this pure future without content.
>
> (Levinas, 1985: 69)

At base, my self derives from mystique, from the life of this primitive connection *in-between* the 'I' and the 'me', to use Mead's terms (1934). It doesn't derive from the 'I' or the 'me' but from the *relation* that makes them possible.

According to Bousquet, 'I am my own hiding-place' (cited in Bachelard, 1969: 88). By hiding my manikin books in drawers, I was also hiding my secret from me, so that I could feel it through my desire for it. In this way I kept alive the self-love from which I drew strength. Bachelard observes that 'all positivity makes the superlative fall back upon the comparative. To enter into the domain of the superlative, we must leave the positive for the imaginary. We must listen to poets' (1969: 89). To maintain a sense of uniqueness, the self must remain beyond our own reach and speech. The relation involved in this self-love doesn't fit the Freudian model of narcissism, for, as much as my roommates, I appreciated the inestimable value of the unopened treasure chest. Had I sought to enter the casket and lay claim to its contents, I would have closed myself off from the world of difference, turning the box into a final resting place.

Bedtime stories

For we are where we are not.

(Jouve, cited in Bachelard, 1969: 211)

Ordinarily, bedrooms might be the most intimate and comforting rooms in a Western home, but they aren't where people are most individual or self-certain. Where *am* I, and *where* am I, when I feel at home? When asked where they felt most at home in their childhoods, students in my tutorials nominate beds, tree houses, cupboards under stairs. One student then explains that in the safety of her cupboard, with the door closed, she was never just herself, but could be anywhere and anyone she liked, a princess, an Indian, an astronaut. The 'self' was most at home in the non-place and non-time that allowed it to play with its mysterious reserves without needing to reach a conclusion or give an account. The closure of a door opened a world. In this inner of the inner of the inner there's a vital nothingness that's also the fullness of mystery.

'Every poet of furniture', says Bachelard, 'knows that the inner space of an old wardrobe is deep. A wardrobe's inner space is also *intimate space*, space that is not open to just anybody' (1969: 78). The wardrobe is only deep because it closes. The very solidity of the inner sanctum's boundaries allowed them to dissolve, like the wardrobe walls in *The Lion, The Witch and the Wardrobe*, which unaccountably disappeared to give entry to the magical country of Narnia (Lewis, 1991), like the walls of the locked bedroom to which Max was sent in *Where the Wild Things Are*:

> That very night in Max's room a forest grew and grew – and grew until his ceiling hung with vines and the walls became the world all around and an ocean tumbled by with a private boat for Max and he sailed off through night and day and in and out of weeks and almost over a year to where the wild things are.
>
> (Sendak, 1967)

Safe at home, in reverie, we play games and roles like 'astronaut': more than this, though, we play the world, as someone playing a piano brings it to life by testing its limitless resources, as someone plays the world through ritual performance. In play, in a condition of inspiration and bliss, we're in the world and the world is in us. Our home is the place of dreaming, which means that our faculties are scattered, here *and* there. The self at home is at home *and* away, dispersed *and* contained, all over the place *and* grounded. We don't express ourselves as much as recognise ourselves in the doublings of imagination.

An example can be found in my 5-year-old son's 'reading position', which is his favourite place in the world, his home at home. This position involves being snuggled in a parent's lap, enfolded within the open book, held in the room, protected in the house. But in the core of this Babushka doll, there's nothing. The intense stability of reading position is a form of travel, and my son has disappeared. Look at his fingers, on the page, *in* the page. Caillois notes that the tightrope walker must be hypnotised by the rope (1961: 138), and there's a glazed and rapturous look in my child's eyes as he follows the thread of the story. Reading transports him, to other lands and times, to the no-where and no-time of once upon a time. When you open the book you open a door, to read a book well you must also leave it. As Duras puts it, 'one always reads in the dark. ... Even if one reads in broad daylight, outside, darkness gathers around the book' (cited in de Certeau, 1984: 173). The darkness is not the enemy of light but its guardian angel, bent over its shoulder, whispering possibilities.

On the one hand, my son actively plays the roles, feels the pain, is the hero of our stories, perhaps darting to the toy box to grab the necessary props. On the other hand, reading is an experience of intoxication and passivity, of allowing one's self to be swept away. Likewise, as anyone familiar with this reading position knows, it's not easy to say who is reading to whom, or who is reading through whose eyes, or who is breathing for whom. Parent and child are absorbed into each other and into the book, but on the serious playful condition that they won't be required to identify the boundaries. Reading position is a deliciously monster form, even though it is also my son's most centred and concentrated form. 'When I read I am one thousand men, and I am never more myself than when I do' (C.S. Lewis, cited in Tredinnick, 1997: 48). My son knows how to use this playful subjunctive mood to undermine the indicative mood of my proper Fatherly law (see Le Guin, 1990: 37ff.). My 'No' is met with his 'But Maybe'.

The monstrous forms involved in children's reading are often the subject matter of the books. Take Arnold Lobel's 'Shivers', from *Days with Frog and Toad*. Good friends Frog and Toad are snugly at home before the fire on a cold dark night.

> 'Listen to the wind howling in the trees,' said Frog. 'What a fine time for a ghost story.'
>
> Toad moved deeper into his chair.
>
> 'Toad,' asked Frog, 'don't you like to be scared? Don't you like to feel the shivers?'
>
> 'I am not too sure,' said Toad.
>
> (1979: 28)

And so Frog begins his story, about being a small frog who is lost and separated from his parents after a picnic in the woods. His parents had warned him about the Old Dark Frog, a terrible ghost who comes out at night to eat little frog children for supper, and as night falls, out comes the fearsome Old Dark Frog, announcing that he will eat little Frog. As narrator, Frog draws out the tension, making ever more exquisite the line between the pleasure and pain of the audience, and at each step Toad interjects 'Frog, did this really happen?', to which Frog replies 'Maybe it did and maybe it didn't.' Finally, when Frog has stretched the narrative line as tightly as he can, he describes how he outsmarted the ghost and found his parents. They all got safely home. Toad asks again if the story was true, and again he's given an inconclusive answer.

> Frog and Toad sat close by the fire.
> They were scared.
> The teacups shook in their hands.
> They were having the shivers.
> It was a good, warm feeling.

(1979: 41)

This might be a story about stories about stories. Like other hungry giants and monsters in children's stories, the ghost that threatens to eat Frog could be the awesome story that carries away and consumes the identity of the child. The rapturous passivity of being swallowed up, with its ecstatic suspension of identity and truth, works on such a thrilling edge of bliss and fear that it's always an anticlimax when the child escapes, the story ends, and the book's door closes. The ghost is desperately desired at the same time it is feared; it is feared but it is also as safe (and yet as dangerous) as the encompassing parent reading the story. The strange ghost that threatens to consume the child is in fact utterly familiar, not only because the child knows it's coming and recognises in it other giants, witches and monsters, not only because it's uncannily like the child's parents and stories, but because it's uncannily like the child itself, with its desires to eat the world and the stories and parents it loves and hates. In Sendak's book, Max was locked in his room because, when his mother called him 'WILD THING!', he responded by threatening to eat her up. Later, when he's about to return to his mother, the Wild Things plead with him, 'Oh please don't go – we'll eat you up – we love you so!'

Because outsides and insides imply each other, the surprise of wild things waiting outside your door is always expected. Monsters are there with a message just for you, they are your muse, your fate, your route to becoming, the stranger you are to yourself (Serres, 1995; Steiner, 1989). In this annunciatory encounter, are you inside or outside the door? All doors are reversible, all insides are the outsides of somewhere else. It is never possible to say, as we pass through a door, whether we're coming in or going out, but rather than the door's failing, this is its particular blessing. The door allows us to relate to monsters without necessarily or finally being devoured by them. Perhaps more important, it allows them to be meaningful, to be *our* monsters. This is not to invoke the familiar psychoanalytic reduction that the monsters *are* truly us, that the apparently outward journey of Sendak's Max was *really* an inner one. It is to insist that the inner and outer imply one another, and that the monsters and us, and the world between, belong together (Capra and Steindl-Rast, 1992).

At home, held, we can safely allow ourselves to be possessed by possibilities we might elsewhere refuse to recognise. We can welcome the ghost and learn from it how we belong to a larger world than our ego recognises. Accordingly, Bachelard (1969: 6–7) claims that the chief benefit of the house is that it

> shelters daydreaming ... protects the dreamer, ... allows one to dream in peace. ... [T]he house is one of the greatest powers of integration for the thoughts, memories and dreams of mankind. ... Without it, man would be a dispersed being. ... It is the human being's first world. Before he is 'cast into the world', as claimed by certain hasty metaphysics, man is laid in the cradle of the house. And always, in our daydreams, the house is a large cradle.

The cartoonist Michael Leunig wonderfully describes the magic of these places of safe unintegration:

> I built fairy homes in the garden [as a child] and in the flower petals I would make little beds for the fairies to sleep on, and in the morning I would rush out to see whether they'd visited and sure enough they had, I was sure, I could see the marks. To this day the way I work is probably not too different – when I make a cartoon I'm making a little fairy garden in my mind. It's the same feeling. I'm engrossed in creating this fantasy, believing some magic is going to come into it, and when they get published it's my hope that people enter into it in the way the fairies entered into my garden places.
>
> (quoted in *Good Weekend, Sydney Morning Herald*, 13 June 1998: 30)

The house is a box holding the secret that we're not limited by our enclosures.

The holding environment

The sense of integration that comes from the house or the reading position differs from the clear and distinct self that might be fantasised on the basis of a mirror image. If the cradle of the house is the integrated person's bedrock, people are strangely held together with sinews of reverie, sleep and dream. The forces that integrate and accommodate us are precisely the experiences that *we* don't have, that we don't *have*, that aren't located: they are forces that possess or have us, that carry us away to somewhere else. As Levinas puts it, our conscious thought and our deliberate actions rely on deeper processes of rapturous passivity, of possession, inspiration, entrancement, intoxication (1989: 151). The 'real world', beyond the house and cradle, is accommodated within a living concerned cosmos that is created and found through bedtime reverie: 'cosmic reverie ... gives the dreamer the impression of a *home* in the imagined universe' (Bachelard, 1971: 177). The house and the reading position shift attention from the distinctly located self to the endless possibilities, the nothingness, the mysterious darkness, of the self's pre-position (Serres, 1995).

I find it helpful to understand this process in relation to Winnicott's (1965a, 1965b) concept of unintegration. Winnicott would accept Piaget's claim that the infant doesn't inhabit his body with an 'adult' sense of self/other and sub-ject/object distinctions. He lives in a condition of primary unintegration, 'experiencing' a disparate array of feeling states unintegrated by an ego. This dispersal is tolerable, even pleasurable, as long as the child is periodically brought together through the mother's care. As Winnicott says, 'A baby cannot exist alone, but is essentially part of a relationship' (1965a: 88). By literally gathering her baby in her arms, by holding, feeding, bathing and rocking him, the mother allows him to *feel something* (Phillips, 1988: 79). By putting herself in-between his bits, and in-between him and the world, and by intuitively anticipating and responding to his desires, and thereby upholding his magical ability to realise the fantasy that he and the world are as he desires, the mother affords a viable self and an inhabitable world to the child. His experience of integration is based, therefore, on the mother's purposeless love providing a holding environment within which unintegration is safe. Babies are swaddled in sheets, in the rhythms of rocking, in the caress of a lullaby, in their parents' arms. At first these forms of holding are explicate orders, but as the child develops, the mother–child relation becomes an implicate order that allows, that is enfolded – held – within, the child's self (Bohm, 1985).

As this implicate order becomes a way of life, the exoskeleton becomes less conspicuous than a developing vertebrate structure – the backbone of the ego. The ego alone, however, can never carry someone. More often than adults notice or admit, they're carried by 'external' forces. The mother, for example, couldn't intuit her child's needs were she autonomous. Instead she finds that the hugs she gives her child are returned: she hugs herself, makes herself safe, is given access to childhood pleasures and daydreams. She is the parent to her child, and that child is located outside *and* inside. To give another example, most adults going to bed still rely on the swaddling of routines, rituals, hugs, cuddly blankets, doors and windows. These are cases of the holding *mother* being transformed into a cultural and relational *matrix*. Likewise, creativity is nurtured within the rules of play and the frame around art: these forms of holding protect participants while the rules of vertebrate identity are suspended (see Winnicott, 1991; Bateson, 1972; Huizinga, 1970; Milner, 1987: 79ff.; Simmel, 1994; Wilshire, 1991).

If parents and children hold and imply each other, we need to consider the stuff that connects and separates one from the other. Archetypically, this stuff is air, and the aura, energy, atmosphere or environment we feel in a bedroom is living proof of an implicate relational order. Accordingly, when we're healthy, the air in our home is full of love, holding us like a mother, allowing us to be alone but not lonely, to be unintegrated yet safe. When we're unhealthy, the air may be a thick and suffocating force or we may rattle around our houses like peas in a bowl, unheld, unconnected, disintegrating, unable to find any purchase.

Margaret Wise Brown celebrates the loving space of nothingness in *The Quiet Noisy Book* (Brown and Weisgard, 1993). This hymn to the morning concerns a dog called Muffin.

> Quietly something woke him up.
> A very quiet noise ...
> As quiet as quietness ...
> As quiet as someone eating currant jelly.
> As quiet as a little kitten lapping milk ...
> Quiet as air.
> Quiet as someone whispering a secret to a baby.
> What do you think it was?
> Muffin knew what it was!
> It was —
> the sun coming up.
> It was the morning breeze.
> It was the birds turning over in their nests.
> It was the rooster opening his mouth to crow.
> It was the day.

Muffin is cradled by a cosmos that doesn't crowd or abandon, that holds both specificity and connection.

If Muffin hears the world in silence, another Margaret Wise Brown picture book finds it in the blank page. *Goodnight Moon* (Brown and Hurd, 1947) features an infant itemising the (transitional) objects in its environment as it says goodnight to each in turn, presumably seeking to assure itself of the world's continuing existence and presence during the chasm of sleep. 'Goodnight room/Goodnight moon/ … /Goodnight clocks/And goodnight socks'. Towards the end of the book, a hauntingly 'blank' page unexpectedly appears, with the strange text 'Goodnight nobody'. Then the book ends 'Goodnight stars/ Goodnight air/' – the pictures on these pages are interchangeable – 'Goodnight noises everywhere'. The 'old lady' who was sitting with the infant at the beginning of the book has left by the final page, which is dominated by the moon visible through the window.

The infant in this story can feel the presence in the nothingness, the body or spirit in the no-body. This air of tenderness comforts her in the face of the unthinkable anxiety of abandonment or suffocation. Because of the door, this air can fill the bedroom and also escape from there to produce a cosmos. It is important, then, that the moon is outside the closed room *so that* it can return inside, bringing to life and filling with love the air and world between. The moon and infant are not one, any more than the mother and child are. But they hold each other, regard each other, and belong to each other. It is *her* moon, she is *its* friend, she looks at it with the love that she knows it feels looking at her: we could speak of the infant's love of the moon, using the ambiguity of the possessive to draw attention to the substance of the in-between. This is the love, the reversibility and belonging, that holds us in a concerned cosmos. We live in this cosmos through substantial illusions: sometimes we see air and sometimes stars, sometimes foreground and sometimes background. *No-body* and *nothing* are the infinitely tender horizon that bring every *thing* to life: they are God, the Holy Spirit (Capra and Steindl-Rast, 1992; see also Brown and Charlot, n.d.).

The integrated self described and extolled by Piaget and Meares, therefore, is an important socio-cultural achievement, but not a steady state. Rather than transcending or lessening the significance of 'childish' states, integration and unintegration fold into each other, like doors that open and close. The secret around which we integrate and distinguish ourselves is the nothingness of reverie. The confines of the house lead us out into the world.

Moreover, integration isn't an outcome for which we are personally responsible. Home happens and can't be willed. Rather than arriving home, homes takes you by surprise: it occurs to you that you're really at home when you've found yourself lost in the wider world of imagination. Home is a relation between stability and movement or inside and outside, each condition generated in the same moment, shadows of or reversible grounds for each other, none of the terms having to identify itself or give up its mystery. The self is in suspense, between these terms.

Ghosts

In his poem 'The Snow Man', Wallace Stevens distinguishes the nothing that is not there and the nothing that is. The infant in *Goodnight Moon* can feel the nothing that is there, but how should we understand the nothing that isn't there? The two examples we want to give concern ghosts and insomnia.

Ghost stories are to ghosts what sleep is to insomnia. Ghosts stories rely on and reinforce a sense of being held; they can make you feel secure in openly welcoming the ghost at your door. Ghosts, however, are those restless spirits who can't find the sleep of death, who toss and turn without finding a position, who escape their crypts. They are dispirited because they are not *in-between* anything. Without a limit, they are nothingness that can't hold meaning, that can't become full. Ghosts are renowned for their ability to walk through doors and walls, but their misery is that they can't *not* walk through every obstacle. And because they can't hold or be held, or find a limit, they're lost souls, reduced to inarticulate placeless endless background wailing. If you listen at night, you can hear them. We have one at my boarding school. You can't *meet* a ghost, you become ghost, as I did when the girls left for their holidays.

Sharing a boarding house with the tumult of adolescent girls, I often feel drained, physically and emotionally, and at such times I often want the house to myself. I want their omnipotence dissolved. So when the last girl leaves for her holidays, the front door of the boarding house slams with more conviction than the girl alone could have produced. I can almost hear a reverberant *And stay out!* from the tired and impatient old house. I exhale dramatically. The house is now empty. But I do not feel a lifting of the weight I imagined was upon me: the house is heavy. And I suddenly feel out of place. The girls' absence unsettles me and renders the house unfamiliar.

The girls' absence lets me hear the house's silence for the first time. I listen anxiously for the sounds which belong in the house and, when contained within its walls, give it meaning: the girls screeching down the hall, their panics as they get ready for school, the incessant ringing of the telephone. I miss the girls' vitality, which can so effortlessly rope me in with a simple 'Miss Ferguson …?'. During the term I evidently rely on this vitality without realising it, for now I must admit that I'm missing it. I thought they were making demands of me, but now I wonder if I demand their energy to sustain myself. Without it, I feel directionless and pointless. I reside in the boarding house because I'm employed to care for the girls. When the girls are not there, I am not a boarding house mistress, so why should I, and how can I, be there at all? I am a puppet whose strings have dropped to the floor. Without the personal and professional relations between myself and the girls, Miss Ferguson is null and void.

As I walk through the house with no one to greet, no one to avoid, and no one to discipline, I feel meek and hollow. Although the girls appear to believe the authenticity of Miss Ferguson, I do not. I go to strut down the corridor towards my room, as I do many times during the term, and find that my self-assured posture has lost its glory. My inauthenticity is palpable to me. No matter, I console myself, of course this is how it feels when the house is empty; after all, there are usually 62 more people in the house, I would normally make 63; at the moment, I make 1. This confuses me, because despite knowing that I am on my own, I do not feel that I am the only person in the house. There are no other bodies, but I am acutely aware of the presence of the 62 absentees. I can actually feel the life of the absence, the something other than me. I try to assure myself that I will feel better when I reach my room. It's just a matter of pushing through the invisible company I now have within the house.

I pass the dormitories and their wide-flung doors proclaim their innocence. They tease me, daring me to enter and threatening to trap me in a question of the validity of my presence. Go on, have a look, feast your eyes, they mock, making sure I understand that when there is nothing to hide, there is nothing to seek. If I snoop, which is a desire I have, I might just be found out: by the girls, by Miss Ferguson or by my ancestors. I half expect to be surprised by a mirror, or some girls playfully hiding behind a door. Nevertheless, I feel compelled to look and I enter, but when I do, I am startled by the desolation. The austerity in the dormitories is so patently uninviting without the warmth of the girls and their belongings that my prying eyes and earnest body instinctively lock. I feel a shiver down my spine and I feel eyes behind me. I quickly turn around, and simultaneously increase my pace. The echo of laughter I heard as the girls left for their holidays fills my ears and mutates in my imagination. I hear the girls laughing at my pointless voyeurism. What is she looking for ...?

Who or what is playing on my mind?

It is not only the absence but also the space which disconcerts me. All this space with so many objects: basins, beds, tables, towel racks, empty dressing tables, showers. They are functional objects, stoically standing at the ready, proud of the number of human bodies they capacitate, despite the fact that there are no bodies now. These objects must wait, loyally, because they too have been abandoned. I wonder: what if the pipes under the basin were to rust and collapse?, or the bed springs were to snap? ... Well, they might be found at some point, but won't be noticed right now. These possibilities apply to me also and this thought excites and spooks me at the same time. Why, I could sleep outside tonight, under the stars on one of the open balconies. I could take off all my clothes this instant and leave them right where they fall; if I didn't pick them up for a week, no one would know. The thrill of transgression turns to pointlessness ... no one would notice. There are no witnesses.

I have to walk very briskly to my room because these feelings and questions wrap me up through every room I pass. If I walk very quickly and do not look around, perhaps I will not be seen. But this does not make sense: not seen by whom? There are no witnesses. There is no one to walk briskly to, and no one to walk briskly from: there is no one to see me.

I begin to wonder if my fear is that I'll see myself. Is it what my own eyes might see or not see that I am afraid of? Am I afraid of my own growing realisations? Afraid of my realisation that the person I know when the girls are here cannot possibly exist when they're absent. When the girls are away, am I merely part of the furniture? Will I be found too? Or not? I know the answer to this immediately and I feel disconnected – literally, as if I have been taken out of the line. I am left hanging out of place.

The 'Ghost of Miss Wentworth' springs to my mind. Miss Wentworth was a boarding house mistress earlier this century. As the story goes, she was a cranky old spinster who tumbled to her death down the two flights of stairs leading onto the entrance hall of the boarding house. The floor of the entrance hall is marble, with a particularly dark patch at the bottom of the stairs. Anyone doubting the story of Miss Wentworth is escorted to the bottom of the stairs, the corner of the carpet is peeled back and the proof revealed: it is the stain of Miss Wentworth's blood!

The new boarders are introduced to this ancient myth of the boarding house via the older boarders, their older sisters, who were also introduced to her in their early days. This meeting is a 'rite of passage' of sorts. *The Ghost of Miss Wentworth!* I remember being quizzed by the Year 7 girls, just after their introduction to her. My words 'what a load of rubbish' echo in my head. Did I really dismiss her like that, I ask myself, repentant. I hear the girls squeal in protest and ask me to explain the 'blood stain on the marble floor at the bottom of the stairs where she tumbled to her death long ago!' They love this game, as I do. I enjoy their reliance on my cynicism; if it weren't for the safety net my doubts create, there would be no certainty in their fear. There are boundaries for their imaginings and so there is security: I secure them. This is how it's meant to be: carers and students holding each other, like children wrapped up in 'reading position'.

The girls flirt with the idea of Miss Wentworth, as I did at their age. Regardless of whether they really believe in her, she brings to life literature's many stories about boarding schools. More importantly, like Frog's Old Grey Ghost, she connects the listener and the storyteller, and it is through this lively relationship between the new girls and the older girls that Miss Wentworth is able to give the house a life in some ways; her story, or even the idea of her, offers the girls an almost tangible sense of the history that enriches the house, in which they are participating through their storytelling. Miss Wentworth is alive then. And if my own experience is any guide, the girls on the first floor of the boarding house won't ever feel unaccompanied by her if they wake at night and are forced to walk the dark 'vacant' corridor to their bathroom. As a student in this situation I longed for the mistresses who looked after me. I implored the darkness: why can't they catch me out of bed now?!

When the Year 7 girls retold me the story this year I was unperturbed by it. Now, as it runs through my mind, its suggestions make me nervous. The girls are snug in their homes, away from the boarding house, and Miss Ferguson is lost or absent and I am left to face the possibilities the girls have abandoned. I become the target of a mental possession which needs a boundary. I wish I was being told the story of Miss Wentworth by the girls, so I could employ Miss Ferguson's confident denial of ghosts. Is Miss Wentworth with me right now? I'm a self-reliant adult! I'm supposed to have grown out of believing in ghosts. How can I possibly be thinking of her?

I am frightened.

The recollection of my smug dismissal of Miss Wentworth follows me to my room and I feel ashamed for disrespecting her. Suddenly, I realise that because the girls are away, the need for my cynicism has vanished, so that it is me who is now insecure. I have a boundless uncertainty about her existence. But I realise that what I am really afraid of is the elusiveness of the selves I embody. I can't help slowly feeling overwhelmed by malleability: I do not know myself, I do not think I ever have, but I would be willing to submit to any of them now. Who or what does Miss Wentworth represent? Rather than feeling cynical, I sense an eerie identification with her. ... I too could just be one of the stories enriching the house.

My mind flashes back to when I was a student and was picked up and secured by my parents at the end of the term. By comparison, I now feel forgotten, as if I were a student whose parents forgot to come. Have I inhabited my own ghost? Have I given her a body momentarily? Was it *ever* Miss Wentworth? My little room awaits me in this big house. I am a mouse scurrying back to its hole in the wall.

When I arrive there, I'm relieved to find some comfort in my room's familiarity. Aesthetically, it is just as I like it, and, momentarily, I feel recharged, reassured by the reflections I see in my posters, photos, materials and clothes. I look closely at the crowd of posters on my walls, which leave very little wall space free. They are at once windows for my eyes and a wallpaper that serves as a protective barrier against the boarding house and the girls. They enclose me and tuck me into my room.

Today, however, they cannot long protect me from my futility. Without the girls and the structure our relationship creates for me, I feel I'm standing in a room and a house with no walls. I again feel exposed and clingy: shameful. This feeling is not only physical, it also unsettles my sensitivities: what was that sound I'm sure I just heard? 'Hello, is anyone there?' No, no, Lucinda, that was silence you heard. This silence is too quiet and puts me on edge. It is a 'deathly silence' but it is not merely the death of the girls' presence. The house is a heavy weight, it is listless and uncooperative and I can't hold it up any longer. I can not do this on my own. I can't hold myself up. Where are my girls?! At this point I finally realise that the death is my own. It is that of Miss Ferguson. I am Miss Wentworth and this is the utter anguish of a ghost's existence. This is why they wail.

I now understand why Miss Wentworth is described as cranky. She's danger-ously unreliable … life and its shadow, spirit and ghost. Miss Wentworth is alive, smiling benignly, when the senior girls ghoulishly evoke her death. She is the nothing that is there. But there's no lively smile when the community of storytellers has dispersed and she breathes clammy dread down the neck of stragglers. Here (where?) is the nothing that isn't there. Rather than bringing or being life, she insists you attend to the impersonal 'there is' of (your) unheld formlessness. She is – I am – 'there is'. Although you experience horror in the face of this implacable nameless power, there's no distinct 'you' doing the feeling and there's no one who can save you through an act of will. Will you ever get out? 'There is' is what's left when relationships lose their life.

My posters try to defend me, but emptiness pushes into my private space through the gaps between them. I feel its surge, its mass, and then feel swamped and short of breath. Pointlessness empties my room of its secret life. I'm utterly dismayed by recognition of the feeling that I need to get out quickly, to find a door that still works. Agoraphobia and claustrophobia turn each other inside out: I'm crowded and abandoned, I'm smothered by the endlessness of doorless space (Bachelard, 1969: 220–1).

Insomnia

> Night, the essence of night, does not let us sleep. In the night no refuge is to be found in sleep. And if you fail sleep, exhaustion finally sickens you, and this sickness prevents sleeping; it is expressed by insomnia. … In the night one cannot sleep.
>
> (Blanchot, 1989: 267)

If God is the nothingness in-between the infant and the moon, If God is a door that produces the world, and allows us to inhabit and belong in space, doorless-ness is the apparently Godforsaken condition of being unheld. It involves the endlessness and pointlessness that makes the stories of Parzival and the Flying Dutchman so terrifying. In the German story of Parzival, the young hero finds himself at Wild Mountain, where the Holy Grail is kept. Intent on making a good impression and becoming a proper knight, Parzival achieves neither goal, failing his calling as a Christian knight when he fails to ask the tormented Fisher King the cause of his agony. Subsequently unable to forgive himself, yet also angrily quitting God's service because of the suffering God allows, Parzival tries to find his way back to Wild Mountain, to make amends. He wanders the abandoned waste lands for years, but no path is to be found: 'The sun rose, the sun set, but there was no counting of the days. The world seemed a cold and endless wilderness. A man might ride forever and never come out into the sunlight or into sight of a great castle that refused to show its face' (Paterson, 1998: 96).

Like the Flying Dutchman, who vowed to sail forever rather than submit to the sea's power, and whom the Devil kept to his word, Parzival exists in a self-imposed exile, a hell without arrivals and departures, without beginnings or endings, without the passionate urgency of mortality, without home or away or the belonging of lively relation. Parzival and the Dutchman are not in-between anything, and are therefore limitless, unable to hold any meaning, neither empty nor full, neither held nor holding. They subsist in a ghostly undead and unalive condition, unable to be still but unable to go anywhere.

This lifeless undeath can swamp people at any time. It often happens, for example, in the middle of redrafting a piece of writing, when the words won't hold and, no centre or end or source of agency being imaginable, the middleness drops from experience. But the most common form of this condition might be insomnia: 'To sleep badly is precisely to be unable to find one's position. The bad sleeper tosses and turns in search of that genuine place which he knows is unique. He knows that only in that spot will the world give up its errant immensity' (Blanchot, 1989: 265). Involving a self ineffectually seeking a grip in a careless and distant 'real world', the pathology of insomnia indicates how 'life' would be if we fulfilled Piaget's model of adulthood, if interbeing and pre-positional connection became unimaginable and if closed doors lost their limit because they lost their implied outside. Strangely, this independent life becomes existence without distinction, from which no self can separate.

When suffering insomnia, I cannot find comfort or purchase or a door to close off the experience. *oh no now I'm in trouble not this again try on the side no good roll over don't panic stay calm stay calm breathe at least there's no cramp tonight try your hand under the pillow it's so quiet there must be a sound somewhere it might be better on my back and when I get to sleep the children will wake me to take them to the toilet* breathe *you forgot to breathe I wish the clock wasn't so loud oh no what is the time? this is a disaster I've got classes tomorrow no don't think about the time it'll only make it worse how can she sleep? she's stealing the sheet it's so hot kick the sheet off maybe I need a drink I should get up that's what I should do but I'm too tired and I really need to sleep maybe the pillow will be cooler if I turn it over I can't get up I'm too tired I need to sleep now* The more I will myself to sleep, the less I can find it, and while I know this, and know I should break the cycle, I lack the sense of 'I' to get up. My muscles won't move to my will.

I can't sleep and can't wake. My mind churns monotonously, images and phrases coming and going with relentless menace. These aren't my ideas, just ideas I'm condemned to think forever. However nonsensical, I cannot stop them repeating self-importantly, like the cheaply-made advertisements on stale rotation on midnight television. It isn't that I have insomnia, it's that everything is insomniac. Insomnia menaces me with an endless existence without witnesses, relations or vitality, and I cannot escape because, without doors, I'm moved by neither self nor grace. Incapable of leaving insomnia, I can only be inexplicably delivered from it, when some thing reveals a door that breaks the spell.

For all the fear it evokes, however, insomnia is impassive. It waits, never preying or attacking. But, knowing that it awaits, the restless sleeper tries to flee, tries to close off from it, to save the self from dissolving in it. And this attempt to turn away from insomnia delivers the would-be sleeper to it. Insomnia is self-inflicted, a malady of pride, of a self too faithless and fearful to fall. Its doorless-ness is produced by the self's very attempt to stay in control behind doors that simply close off dangers.

The way to separate from insomnia is to let go of the attempt to deny it, even if this appalls our instincts of self-preservation. Victims who toss and turn, looking for an escape, are turning in the wrong direction and prolonging the suffering: they should turn *toward* the waiting insomnia, incline to it, attend to it, just as car drivers must steer *into* skids, must leave the skid by first coming to terms with it. Because if a single relation is formed in the midst of insomnia, a door has miraculously appeared, an in-betweenness has been formed to hold us, and the ghostly spell is broken. This door will found and bless the world, turning the ghosts into spirits who can be met, with whom we can have a relation. Its complexity will hold and heal the space laid waste by the fantasy of pure distance.

Because there can be no doors without implied doorlessness, it is a condition that cannot be denied. It is always with us. Nevertheless, doorlessness loses its horrible ability to possess us when we realise that it remains connected to doors. Rather than an independent condition that pre-existed God chronologically, doorlessness is a pre-positional possibility formed at the moment the door originally made the world. There can be no doorlessness without implied doors. Once the insomniac relates to doorlessness, then, they stumble on the door still implied in and leading out of the condition; they find the God still implied, as an absence, in Godforsakenness. They learn that it isn't possible to be abandoned by God, though we often turn our backs on God in the self-imposed exile of sulking. They learn, too, that while we cannot save ourselves, the miracle of grace is always at hand when we ask for it.

Paradoxical being

The dominant assumptions about the self and sociality reproduce Piaget's. They instruct me to tell students that we – as adults, as progressive Western individuals – have outgrown belief in angels, ghosts and spirits. But angels, spirits and ghosts are the life-form of the paradoxes that are implied in doors and in humans as half-open being. They are substantial illusions, the tangible presence of nobody and nothing. As Winnicott and Bachelard insist, these paradoxes must be held: just as the self is separate and connected at once, angels are here *and* not here, and neither reductive alternative to this claim is as true.

Besides, whether we consciously believe in them or not, we know that ghosts, angels and spirits exist, and it may be this knowledge that leads to terrified denials. We live with them hidden around us, between us. They carry us away, they give us homes and an intersubjective concerned world, they even allow us to be decision-making individuals. The speaker who reacts to the atmosphere in the lecture theatre; the sailor who listens to the wind and reads the clouds; the novelist whose characters create their own lives; the surfer who rides the imperceptible lines of force etched into the water's surface; the reader who enters and breaths the text: adult thinking often derives from immersion in and relation to the world, and not from severed distance. The knowledgeable person hears and touches the world where the ignorant or incompetent person insists on its externality and dumb passivity.

These relational forms of knowing highlight the importance of processes like intuition, revelation and inspiration, and of in-between media like air, angels and spirits. When the headmistress speaks of 'school spirit', she means what she says, even if she doesn't integrate this into her conventional assumptions about rationality and distinct selfhood. If I confidently denied the existence of spirits to students, this spirit would be between us, holding the relation that gave force to this denial.

Existence would be a ghostly horror of alienation if actually lived through a distinct sense of inside and outside. People committed to identity and autonomy seek to reduce the complex implications of doors to the simply closed door, which they treat, in fact, as a wall that simply banishes outside. But we've seen that paradox isn't so easily evaded, for an impermeable wall, having no sides, no longer closes off an outside. Without doors, the world has no horizon to end it, connect it, embrace it. This is why people point to doors when identifying their homes: there is no containment or wholeness without some hole in the house's perimeter. The outside has to enter before it can be apart.

We have therefore tried to continue where poststructuralism usually leaves off, pursuing its criticisms of binary oppositions into a concern with the paradoxes *between* separation and connection, and then into an interest in varieties of religious experience. The hole, the holy and the whole go together.

14

SAVING TIME

A Buddhist perspective on the end

David R. Loy

> Each culture believes that every other space and time is an approximation
> to or perversion of the real space and time in which it lives.
>
> (Mumford, 1934)

What is the 'real space and time' in which our culture lives? A hundred years ago it would have been easier to say, but the twentieth century has complicated things. In its early years relativity theory displaced the Newtonian paradigm of space and time as empty formal containers. At the other end of the century, postmodernism has shattered our linear future-orientation: the self's modernist alienation has given way to a fragmentation which releases our subjectivity from the intentionalities that used to focus it, encouraging a more fluid and multiple sense of personal and social identity.

This collapse of linearity would seem to imply the timelessness of simultaneity, but that has not meant freedom from time: time – or the lack of it – has become more of a problem for most of us. The dissolution of linear temporality is linked with accelerated ephemerality and a widespread increase in insecurity and anxiety, contributing a 'manic' quality to much of public and private life. A 'time-compression' effect means we experience ourselves as having less time to do the things we need or want to do. I am surely not the only contributor to this volume to notice the not-always-amusing irony in wondering when I could find the time to write this essay. A 1992 survey by the National Recreation and Park Association found that 38 per cent of Americans report 'always' feeling rushed, up from 22 per cent in 1971. Lou Harris polls have shown a 37 per cent decrease in Americans' leisure time over a twenty-year period, leading him to assert that 'Time may have become the most precious commodity in the land' (Levine, 1997: 107). But what if commodifying time is itself the problem?

Some see the comparison as nostalgia for a more leisurely past that never was. While Juliet Schor (1992) and A.R. Hochschild (1997) have argued that

262

Americans are working much longer hours, some other academics, most notably John Robinson and Geoffrey Godbey (1997), have responded by claiming that Americans have much more leisure time today than in the 1960s, with the notable exception of those who have graduate education (which may be an important exception, given the background of those most likely to read and write essays such as this one). Then do we actually have less time, or does it only feel that way, because there is something different about the way we 'use' time today? I am in no position to adjudicate the debate, yet it serves to remind us that temporality is one of those social constructions which, once objectified, returns the compliment by objectifying us. For that reason, however, we need to consider not only economic and technological factors (e.g., instantaneous communications) but also ideological ones that address how and why we construct time the way we do, especially the role such constructions play in our larger world-view of what the world is and where we are going within it.

In *The Condition of Postmodernity* David Harvey presents a sophisticated Marxist understanding of the problem. Recognising that there is no single, objective sense of time or space, Harvey emphasises that neither can be assigned meaning independent of material processes: their apparent objectivity is due in each case to the material practices that reproduce them, and they are never socially neutral, usually being the focus of struggle. Marx himself had little to say about how clock-time has transformed people's subjectivities, but his later works examined how capitalist accumulation is based upon 'the annihilation of space by time'. Harvey continues this analysis by understanding our present experience of time-space compression as one of a series generated by the logic of capital accumulation and its tendency toward vertical integration to circulate market commodities more speedily.

Time's arrow may only fly in one direction, yet reductionism is a game that can be played either way. Insofar as Marxist socialism is a capitalist heresy, it shares many of the same assumptions, and by no coincidence the classless utopia of classical Marxism participates in the same linear orientation as the widespread abundance that 'free enterprise' also promises for the faithful – perhaps in the near future provided we do not lose our nerve. But do materialist perspectives on our postmodern spatiotemporality explain enough about it, or do they merely recycle the problem, by assuming and reproducing an orientation that now needs to be questioned?

This chapter will argue that our problem with time-space compression can be better explained from an opposite perspective that privileges the unavoidable *spiritual* role of space and time. From this alternative view, materialist approaches, like the abstract and homogeneous Kantian space-time they are examples of, can be understood as *decayed*, because such hollowed-out space-time is unable to serve a spiritual function.

Our contemporary preoccupation with the tension between linear time and postmodern time serves to distract us from what, historically, has been a far more important understanding of time. In traditional societies, including the premodern West, space and time are religious in their basic structure. Instead of being some neutral, Cartesian-like grid for mapping the affairs that happen in this world, spatiotemporal schemas provide the sacred patterns that give meaning to the affairs of this world. This much is generally known, but I want to go further by suggesting that that is still the case today, for space and time are essentially religious in the functionalist sense that such schemas cannot help embodying our most basic understanding of what the world is and how we should live in it. From this perspective, space-time compression is a late stage (the last stage?) in the failure of 'secular' time, in which its inability to provide an adequate schema of meaning for our lives becomes evident. The future-directedness of linear time provided that meaning by promising fulfilment in the future: originally in the millennarian End of Time with the return of Christ, which was imminent; later, in the Enlightenment promise to remake society and nature until they fulfil our desires. Today our inability to respond adequately to the many environmental crises, among other things, signifies the collapse of our collective faith in the future. If the future cannot save us any more, the solution we seek can no longer be located there. Now our time is truly hollowed out into a meaningless continuum. But if time always has a religious function, such eviscerated temporality is best understood as *nihilistic*, because signifying our inability to discover any meaning to individual or collective life in the world.

This of course challenges the distinction we normally assume between the religious orientation of traditional societies and the secularity of modern Western civilisation. So I shall begin by attempting to deconstruct that duality, by employing a category derived from Buddhism. As is well known, the concept of *anatta* 'no-self' is essential to Buddhist teachings, but to make sense of it we must relate it to an even more important concept: *dukkha*, usually translated as 'suffering' yet better understood as frustration or unhappiness. The four truths into which Sakyamuni often summarised his teachings revolve around this concept: life as *dukkha*, the cause of that *dukkha*, the end of *dukkha*, and the path to end *dukkha*.

Elsewhere (Loy, 1996) I have argued that today we can use the psychoanalytic understanding of repression to help us understand *anatta* and *dukkha*. Then *anatta* implies that our primary repression is not sexual wishes (as Freud thought), nor even death fears (as many existential psychologists think) but awareness of non-self – the intuition that 'I am not real' – which we become conscious of (the 'return of the repressed') as a sense of *lack* infecting our empty core. The death-repression emphasised by existential psychology transforms the Oedipal complex into what Norman Brown calls an Oedipal *project*: the attempt to conquer death by becoming the father of oneself, i.e., the creator and sustainer of one's own life. Buddhism merely shifts the emphasis: the Oedipal project is better understood as the attempt of the developing sense-of-self to attain autonomy, the quest to deny one's groundlessness by becoming one's own ground.

Then the Oedipal project derives from our intuition that self-consciousness is not something 'self-existing' (*svabhava*) but a mental construct. Consciousness is more like the surface of the sea: dependent on unknown depths that it cannot grasp because it is a manifestation of them. The problem arises when this conditioned consciousness wants to ground itself – i.e., to make itself *real*. If the sense-of-self is an always-insecure construct, its efforts to real-ise itself will be attempts to objectify itself in some fashion.

The consequence of this perpetual failure is that the sense-of-self has, as its inescapable shadow, a sense-of-*lack* which it always tries to escape. The return of the repressed in the distorted form of a symptom shows us how to link this basic yet hopeless project with the symbolic ways we try to make ourselves real in the world. We experience this deep sense of *lack* as the persistent feeling that 'there is something wrong with me', but that feeling manifests, and we respond to it, in many different ways.

The problem with our objectifications is that no object can ever satisfy if it is not really an object we want. When we do not understand what is actually motivating us – because what we think we want is only a symptom of something else (our desire to *become real*, which is essentially a spiritual yearning) – we end up compulsive. According to Nietzsche (1986: 331), a Christian who follows the Biblical admonition and plucks out his own eye does not kill his sensuality, for 'it lives on in an uncanny vampire form and torments him in repulsive disguises'. Yet the opposite is also true: if today we think we have killed or escaped such a spiritual drive we are deceiving ourselves, for that drive lives on in uncanny secular forms which obsess us because we do not understand what motivates them.

The point of this Buddhist approach is that such an understanding of *lack* straddles our usual distinction between sacred and secular. Their difference is reduced to where we look to resolve our sense of *lack;* but if that *lack* is a constant, and religion is understood as the way we try to resolve it, we have never escaped a religious interpretation of the world and perhaps never can. Our basic problem is spiritual inasmuch as the sense-of-self's lack of being compels it to seek being one way or another, consciously or unconsciously, whether in religious ways or in 'secular' ones. What we think of as secular projects – in particular, our pre-postmodernist preoccupation with the future as the place where our *lack* will be resolved – are just as symptomatic of our spiritual need. The problem for us is that our grasping at the future rejects the present; we reach for what could be because we feel something lacking in what is. Norman Brown (1961: 277)

265

summarises the matter brilliantly: time is 'a schema for the expiation of guilt', which in my Buddhist terms becomes: time originates from our sense of *lack* and our futile attempts to fill in that *lack*. In sum, the temporality we live in is the canvas we erect before us on which we paint the dreams that fascinate us, because they offer the hope of filling up our sense-of-*lack*. It is not the only way that humans have tried to resolve their *lack*, but it has been a very important part of our way.

Then the collapse of the future ('progress') means our *lack* is losing the projects whereby it hoped to become real, with some drastic consequences. One is our increasing distractability: various addictions both physical and mental become more compulsive. Another, probably more relevant for most of the people who read this essay, is a heightened anxiety that feels compelled to do more things more quickly, to bury itself in its projects. In this latter case, our collective sense of increasing time-pressure reflects the increasing burden of our *lack* without the space-time schema that previously enabled us to objectify it and thus get a handle on it. On this account, space-time compression should be understood as part of a more general social crisis of meaning.

This *lack* approach to time enables us to understand what otherwise seems so peculiar about the temporality of many non-Western cultures and shows how our own neutral space-time continuum evolved 'out of' such a different temporality. It may also help us better understand the contemporary (Western) crisis of time and social meaning and – perhaps – offer ways of resolving it.

Non-Western Time

> In prehistoric times, and in primitive societies until well into this century, the supernatural and the passage of time as represented by the yearly cycle were so closely linked that they were virtually indistinguishable.
>
> (Thompson, 1996: 4)

Not only the yearly cycle, for it is no exaggeration to make the same claim about the measurement of time generally: it is 'inextricably bound up with belief in the supernatural' (Thompson, 1996: 3), since time is not a contentless container but essential to the way that most traditional societies try to bring this world into harmony with a higher one. According to this approach, our *lack* is due to their disharmony (or the threat of it), and the solution is not time as empty continuum but time as pattern to be renewed or re-enacted. In contrast to our linear succession of cause-and-effect events, this is an associative temporality in which history and cosmology are inseparable, for time and what happens in time are not to be distinguished. Rituals such as purificatory sacrifices are performed according to cyclical natural phenomena (especially solstices and moon-phases) because they are needed at those times – which is to say, they do not occur *in* those periods but as a part *of* those periods.

Mircea Eliade's scholarly legacy is controversial, yet I think he had an important insight into this distinction. For Eliade, archaic societies lived in a 'paradise of archetypes' because their time structure was based upon periodic regeneration of the creation observed in nature (in Aveni, 1995: 65). This temporality presupposed that every re-creation repeats the initial act of genesis; by re-enacting the creation myth, participants relived the creation of order out of chaos, of meaning out of meaninglessness. In this manner such events were never allowed to become 'past' in the way that we understand the word, as historically superseded by the present. While we create history by choosing to separate ourselves from past events (a good working definition of 'secular time'), traditional societies deny that duality by reliving the past. For them history is not the burden it is for us, since the past is not a weight to be overcome nor is the future a set of novel possibilities to be realised.

It is dangerous to generalise about 'archaic societies', but the classic Maya, one of the most time-obsessed cultures of all time, provide a good example. They did not measure time by the sun; for them time *is* the sun's cycle (and they were on to something: for time would certainly cease for us if the sun ceased). The sun was not a separate entity that followed a course through the heavens, nor did particular gods symbolise each day of their 20-day cycle; rather, a day *manifested* its god. According to this schema, close attention to the demands of each day was necessary to preserve the parallel relationships between humans and the supernatural world. The past continued to repeat itself in the present because life's events were carefully timed and regulated: only by weaving the two together was the order of the cosmos maintained. In such a world view the future is of little interest, for the important events will continue to recur as long as the same balance of cosmic forces is maintained (Aveni, 1995: 190ff.).

> These people did not react to the flow of natural events by struggling to harness and control them. Nor did they conceive of themselves as totally passive observers in the essentially neutral world of nature. Instead, they believed they were active participants and intermediaries in a great cosmic drama. By participating in the rituals, they helped the gods of nature to carry their burdens along their arduous course, for they believed firmly that the rituals served formally to close time's cycles. Without their life's work the universe could not function properly.
>
> (Aveni, 1995: 252)

The Aztecs were even more devoted to the sun, for all their festivals were preoccupied with paying their 'debt' to it by supplying the essential energy that would keep it on its course. Only by fuelling the sun with enough of life's vital fluid – blood – could they maintain the balance of powers among the cosmic forces that would keep the world from being destroyed again. The primary responsibility of the state was ensuring that there were enough sacrifices, providing a social mission that we can hardly admire but must understand: for in this manner the Aztecs, like the Maya, did not distinguish time from meaning, since the nature of their time determined the nature of their *lack* and what they needed to do to overcome it.

The parallel holds for spatiality. The Aztecs, like most other non-modern civilisations including pre-Copernican Europe, understood themselves to be living at the centre of the universe, and this was no mere conceit for them. Their *axis mundi* was the Templo Major, which faced west towards the setting sun, the most propitious position to receive the sacrifices that would maintain that deity on his course. For these pre-Columbian American cultures, as with many premodern peoples, each of the four cardinal directions had its own sacred meaning; and due to the sun the primary axis was east–west rather than our north–south. One does not need to go back very far to find such resonances in Western culture: many of the earliest New World maps place east at the top; and in addition so too the etymology of 'orientation', one of its old meanings being that part of the church one looks at when viewing the altar from the nave – traditionally east. 'So there are religious roots to our cardinal directions, and each of the compass points is tied to the pagan art of sun worship buried deep in our past' (Aveni, 1995: 263–4).

In contrast, the Chinese emperor, who was believed to be the terrestrial complement of the immovable celestial pole, aligned his palace and city precisely in a north-south direction, which made them symmetrical and harmonious with the heavens; as with the pre-Columbians, however, each of the cardinal directions had its own symbolic colour, bird and season. Although this sacrality of space (or time) is not so prominent a theme in Buddhism, the Tibetan *tangka* of tantric Buddhism emphasises spatial relationships by placing different Buddhas (with different consorts, colours, etc.) in different directions; for example, Amida (and his paradise) are in the west. I am sure that each reader will be able to supplement this with other examples that come to mind. The *prima facie* notion that Chinese *fengshui* must be silly derives directly from our modern presumption that space, like time, is homogeneous and neutral, yet a survey of other cultures reveals this to be a minority opinion. The more common view throughout history is that space, like time, is infused with significance because different places and directions require different responses from us; to fulfil our spiritual role in the cosmos we must discriminate between them. From this perspective, the most interesting thing about 'Western' (*sic!*) space-time is not that it is 'natural' or 'objective', but how our unique understanding of a homogeneous space-time continuum ever evolved out of such spiritually-charged schemas.

Western time

> In our case, the cultural taproot of calendrical organization grew out of economic determinism, though ... the Judeo-Christian religion played a major role. Babylonian counting and timekeeping began with an economic motive, and the idea that time is money grew out of medieval mercantilism. Today the rigid control of human time is still powered largely by the business of making money in a highly industrialized, technological world.
>
> (Aveni, 1995: 334–5)

This is a common enough view, distinguishing the economic determinism of 'our' temporality from the religious approaches of other peoples. But such a materialistic perspective to understanding European calendrical origins presupposes the conventional duality between economics and religion, something that may be questioned if, as Weber argued, our forms of economic organisation also have religious roots, and if, to use my Buddhist term, economic growth too has become our collective effort to overcome *lack*.

In either case, the true story is much more complicated, as Aveni himself shows. 'The calendar we live by is loaded with hidden meaning; and though we use it unceasingly, we seem to have lost all contact with its roots. ... To understand our time is to chart the course of Western Judeo-Christendom' (11–12). Our week has seven days because that re-creates the pattern of creation at the beginning of Genesis; we rest on the Sun's day because on that day the Lord rested and Christ arose from the dead to become the light of the world. The names for the seven days originate from the seven heavenly bodies known before the telescope, and Aveni does not doubt that such a scheme derives from the ancient practice of 'scientific' astrology. Although a later construct, our way of dating history according to *Anno Domini* is perhaps unique in its assumption or creation of one focal point to all time (although some such a focal point is very helpful if there is to be a universal time-medium within which all events can be located; compare the parallel role of Renaissance spatial perspectivism). In Japan, official dating follows the more widespread premodern convention of using the reign of a king; I write this in the year *Heisei* ('promoting peace') 11, because the eleventh of the present emperor's reign.

269

None of the above is very surprising in itself, but the important thing is to read it in the light of what was said earlier about non-Western time: contrary to the linear cause-and-effect continuum that we tend to project back upon our ancient world as well, our origins reveal the same associative temporality which does not distinguish between cosmos and history, between time and what happens in time, for there too time is not a homogeneous continuum but a meaning-providing pattern to be re-enacted. The ancient Mediterranean ('the middle of the earth') world was not very concerned about determining the timing of one historical event by relating it to another; that presupposes our anachronistic belief in an objective and universal time-continuum in which everything can be situated. So we should not be surprised that the first clocks we know of, such as the Tower of the Winds in Athens, were not so much efforts to measure time as models of the cosmos constructed to represent and celebrate the beauty and simplicity of heavenly motion (Aveni, 1995: 92).

Perhaps that helps us understand the extraordinary fascination of the first wave of cathedral clocks in the high middle ages: they symbolised the divine structure of time. The medieval Christian mind too did not yet conceive of history as our continuous chain of cause and effect. As Auerbach puts it, 'the here and now is no longer a mere link in an earthly chain of events, it is *simultaneously* something which has always been, and will be fulfilled in the future; and, strictly, in the eyes of God, it is something eternal, something omnitemporal, something already consummated in the realm of fragmentary earthly event' (1968: 64). Describing the monastic origins of our clock-time, David Landes emphasises how the first clocks had the effect of creating

> 'only one time, that of the group, that of the community. Time of rest, of prayer, of work, of meditation, of reading: signaled by the bell, meas-ured and kept by the sacristan, excluding individual and autonomous time.' Time, in other words, was of the essence because it belonged to the community and to God; and the bells saw to it that this precious, in-extensible resource was not wasted.
>
> (1983: 68, quoting Albert d'Haenens)

But the notion that time is a *resource* is an anachronism that describes our attitude, not theirs, for it detracts from the essential point that time was God's and should be used to glorify Him (which is why usury was immoral: it profited from using something – time – that belonged only to God).

In sum, medieval time largely retains the non-Western non-duality between cosmology and history, between meaning and temporality – with one important difference. For the Maya and the Aztecs, the end of time, which might occur without their laborious efforts, would be the greatest possible catastrophe; for

Christians (except perhaps the most powerful and wealthy) the second coming of Christ, which most believed imminent, was a salvation devoutly to be wished for. That turned out to be a crucial difference.

For our purposes, apocalyptic millenarianism can be characterised as the belief that the tension between this world and the supernatural will be finally resolved in the (near) future when the supernatural manifests itself so completely that this world is purified and transformed. The crucial step toward modern time was the notion of a future golden age not outside history but *within it*, and it seems to have been taken by Joachim of Fiore (*c.* 1135–1202), an extraordinarily influential Calabrian abbot and hermit who had visions of complex patterns that drew together all the different threads of revelation and history. Joachim retained the traditional belief that events on earth correspond to what is happening in another dimension, but envisioned a Christian utopia on earth *in the future*, a perfect society which would require the creation of new social structures. He wrote and preached about the Three Ages of the Father, the Son, and the Spirit; the last, which would arrive by about 1260, would be a supremely happy era during which a renewed Church would regulate all aspects of life. 'Without necessarily meaning to, he had made the crucial connection between apocalyptic change and political reconstruction. From the thirteenth century onwards, the two could never be entirely separated' (Thompson, 1996: 128). The trajectory that would culminate in our cherished faith in progress was set, having grown out of

> the Christian belief that all things are made expressly for the end they fulfill ... faith in the Second Coming pulls the true believer forward through the good life toward the ultimate judgment that gives reason to it all. In modern times, this notion of advancement along a time line still prevails, except that technology has replaced religion as the force that propels events to succeed one another; nonetheless, the doctrine remains teleological.
>
> (Thompson, 1996: 128)

Thus, far from being rational alternatives to religious apocalypticism, 'many of the ideas we consider to be the very opposite of "medieval", such as faith in progress and the promise of utopia, have roots in the religious beliefs of the middle ages – and End-time beliefs at that' (Thompson, 1996: 57). According to Christopher Hill, the most interesting legacy of the millenarian ferment in mid-seventeenth century England is that it 'shifted the golden age to the future. Once it had shaken off its apocalyptic associations, it could easily link up with Bacon's scientific optimism to form a theory of progress' (Hill, 1980: 58).

Having traced the religious roots of modern temporality so far, we may be tempted to locate the rupture between them in the development of mechanistic science; from a *lack* perspective, however, the continuity between them is more important. The medieval search for spiritual wisdom included an attempt to understand the workings of God in the natural world. How does nature manifest God's mind and will? How do its phenomena embody his 'signature'? And, most importantly, what does this understanding of the world reveal about our role in it, i.e., the meaning of our lives? Galileo's insight turned out to be the revolutionary one: 'the Book of Nature is written in mathematical symbols' by 'the great Geometer', so the key to its hidden laws is to be found by discovering the mathematical laws of the cosmos – laws that could not have been discovered without chronographs which 'disassociated time from human events and helped to create the belief in an independent world of mathematically measured sequences' (Mumford, 1934: 15).

The three most consequential scientific discoveries all involved a new understanding of the *meaning* of space and/or time. Copernicus displaced us from the centre of the universe, although without understanding this marginalisation in our 'secular' terms, since he viewed the sun as a spiritual being. Darwin's theory did much the same for our temporality: instead of being at the centre of a cosmic drama of fall and salvation, which provided the teleology of history, humankind was a late and evidently accidental result of the same biological forces that produced every other species.

We usually see these discoveries as crucial to the development of our secular world, but they may also be viewed as simply displacing our understanding of *lack* and therefore our approach to resolving it: instead of being the crux of creation and history, our *lack* too is marginalised and its problematic shifted. We can no longer appeal to the structure of the cosmos for our salvation, but must work it out ourselves. The new space-time was abstract and homogenised, yet from this perspective it is somewhat misleading simply to call it 'secular' or 'objective': rather, it provided us with a different kind of grid on which to work out our *lack*. In place of an intricate spiritual obstacle-course to be followed according to fixed rules, we now had an infinite football field where we had to set up the goalposts and decide which way to run.

This alternative perspective becomes clearer in the case of the third, and in some ways the most important, scientific development: Newton's mechanics, which, by accounting for both terrestrial and heavenly motion with the same gravitational laws, eliminated the need for anything more than a deistic Creator. The Newtonian universe was self-sustaining and completely mechanical – with the remarkable exception that it retained God in the wholly passive role of a privileged observer. 'The Deity endures for ever, and is everywhere present, and by existing always and everywhere, He constitutes duration and space. ... [He is] a being incorporeal, living, intelligent, omnipresent, who in infinite space as it were in his sensory, sees the things themselves intimately, and thoroughly perceives

them, and comprehends them wholly by their immediate presence to himself'
(*The Mathematical Principles of Natural Philosophy*, quoted in Mason, 1944: 86).

Thus Newton's postulation of an absolute space and time, and an ether per-
vading them, was connected with his understanding of this single spiritual
observer in the universe. As this implies, such space-time can hardly be
considered secular, for it still requires a God and in this respect bears comparison
with, e.g., the Mayan world view which also identified the supernatural with time
itself. On this account, the mechanistic world view finally eliminated all traces of
its spiritual origins only with Einstein's relativity theories; but with relativity we
begin a return from objective time to event-time, since Einstein's time is not a
homogeneous continuum but relative to the observer's situation.

To sum up: far from being a non-spiritual alternative to the sacred spatio-
temporalities of premodern cultures, the evacuated and abstract space-time we
have until recently viewed as objective – the 'real space and time' in which we
alone have been living – is religious in its origins and continues to be so in its
function insofar as it remains essentially connected with our social and individual
devices to resolve our *lack*. Those projects have changed because our way of
understanding our *lack* has changed, but what has not changed is the need to
address that *lack*. Inasmuch as the collapse of future-directed linearity leaves us
bereft of a way to do that, our postmodern temporality, now truly hollowed-out,
can be viewed as nihilistic. For perhaps the first time in recorded history, we lack
a socially agreed way to get a handle on our *lack*.

So where does that leave us today? Understanding the relativity of all spatio-
temporal schemas does not in itself help us determine which ones we want to live
'in'. We do not need to worship the sun to keep it on its course, and we can
hardly orient ourselves by constructing some new supernatural world; yet the
fragmentations of postmodernity leave us more anxious and adrift than we usually
like to admit. Although time and space are so constitutive of our experience that
it seems presumptuous to suppose we can alter them at will, the advent of a new
millennium seems a fitting occasion to wonder: what other alternatives, if any, are
available?

Losing time

The odd thing was, no matter how much time he saved, he never had any
to spare; in some mysterious way, it simply vanished. Imperceptibly at
first, but then quite unmistakably, his days grew shorter and shorter.

(Ende, 1984: 65)

In Michael Ende's wonderful children's novel *Momo*, the eponymous heroine is an orphan who discovers the secret of time. She lives alone in a park, but has many friends who learn to love her because she is so good at listening and playing games. The trouble is, their time (like everyone else's) is being stolen by grey-suited time-thieves who persuade people not to waste time but to invest it instead – and then they literally smoke up the profits, leaving everyone else too busy to enjoy their lives. With the help of Professor Hora and Cassiopeia, a slow-moving tortoise, Momo outwits the time-robbers and liberates the time they have stolen.

As this fable implies, our problem with time today may not be very different from our problem with everything else. In fact, temporal compression reveals our more general problem in an especially clear fashion, and therefore hints at how it might be addressed.

That problem is commodification, which tends to convert everything into marketable resources valued according to their exchange value. The whole earth – our mother as well as our home – continues to be commodified in new and ingenious ways, most recently including the genetic codes of biological species and even the tragicomedy of carbon emission trading rights. More prosaically, human life has been commodified into labour (work time), bought and sold according to supply and demand; and today that applies to our understanding of time generally, that most precious of 'resources' because we can never have too much of it.

The commodification of time was made possible, indeed inevitable, by the clock. As clock-time became central to social organisation, so the social world became 'centered around the emptying out of time (and space) and the development of an abstract, divisible and universally measurable calculation of time'. Regardless of whatever we may personally think about its reality, the collective objectification of clock-time means that we must all now live according to it, for the complexities of our social interactions require such a continuum for their co-ordination – despite the fact that 'our mechanical way of repatterning time has led to a way of knowing it that is totally divorced from the real world. We have reduced time to pure number' (Aveni, 1995: 135).

According to Aveni (1995: 100), 'Our quest for the precise time of day may go down in history as the greatest obsession of the twentieth century.' Before doing anything Gulliver looked at his watch; he called it his oracle. The Lilliputians concluded, quite naturally, that it must be his God. Today it is usually only encounter with premodern societies that gives us some perspective on our obsession. Tribal societies, which lack such a precise and abstract reference point, continue to puzzle us because they do not objectify a time apart from the activities which occur 'in' it. Evans-Pritchard's classic study on the Nuer of central Africa rather wistfully concludes (1969: 103):

I do not think that they ever experience the same feeling of fighting against time or having to coordinate activities with an abstract passage of time, because their points of reference are mainly the activities themselves, which are generally of a leisurely character. Events follow in a logical order, but they are not controlled by an abstract system, there being no autonomous points of reference to which activities have to conform with precision.

According to E.R. Leach, the Kachin language of North Burma embodies a similar understanding by treating words for time more like adverbs than nouns. According to Edward Hall, for the Hopi Indians as well time has no objective reality: 'the Hopi cannot talk about summer being hot, because summer is the quality hot, just as an apple has the quality red' (Levine, 1997: 94).

Clock-time or event time: with the former, objectified time is outside the activity and regulating it; with the latter, the time of an activity is integral to the activity itself. We can sometimes hear the difference in the way music is played: the notes march along following the time-signature, or we are so absorbed in the notes that we do not notice the time-signature at all. Toscanini versus Furtwangler.

This suggests that our problem with time today is not so much its compression as what that compression takes for granted: the dualism we experience between an event and its time. For Mahayana Buddhism this is a fundamental delusion which contributes to our *dukkha*. The Japanese Zen master Dogen, who wrote perhaps the most on this dualism, deconstructs their thought-constructed bifurcation by reducing each to the other: by demonstrating that *objects are time* (objects have no self-existence because they are necessarily temporal, in which case they are not objects as usually understood); and, conversely, that *time is objects* (time manifests itself not in but as the ephemera we call objects, in which case time is different than usually understood). 'The time we call spring blossoms directly as an existence called flowers. The flowers, in turn, express the time called spring. This is not existence within time; existence itself is time.' In his *Shobogenzo* Dogen combines subject and predicate in the neologism *uji*, 'time-being':

'Time-being' here means that time itself is being ... and all being is time. Time is not separate from you, and as you are present, time does not go away.

Do not think that time merely flies away. Do not see flying away as the only function of time. If time merely flies away, you would be separated from time. The reason you do not clearly understand time-being is that you think of time as only passing. ... People only see time's coming and going, and do not thoroughly understand that time-being abides in each moment.

275

Time-being has the quality of flowing. ... Because flowing is a quality of
time, moments of past and present do not overlap or line up side by side.

Do not think flowing is like wind and rain moving from east to west.
The entire world is not unchangeable, is not immovable. It flows. Flow-
ing is like spring. Spring with all its numerous aspects is called flowing.
When spring flows there is nothing outside of spring.

(Dogen, from Welch and Tanahashi, 1985)

Such time cannot be saved because we are it. To treat it as a commodity is to be
caught up in a delusion that makes us hurry up in order to have the time to slow
down – the trap that Momo's time-thieves encourage. The commodifying attitude
that tries to save time cannot help but carry over into the rest of our lives.
Understanding time as a resource to be used like any other means we lose the
ability to *be* it. It is another case of being objectified by our own objectifications.

Marx had much to say about objectification and commodification, but here a
materialist approach is less illuminating than a spiritual one, for, as the previous
sections have tried to show, time inevitably raises the most fundamental questions
about the meaning of our lives in the world. One of those questions is: why do we
distinguish between events and clock (absolute) time? Economic determinants are
still the most popular, but are they sufficient to explain our obsession with clock-
time?

For Aveni, our most basic motivation, the common denominator of human
space-time schemas, is a quest for order, which is necessary to secure the cosmos
and the self that inhabits it. 'Temporally speaking, we desire the capacity to
anticipate where things are going, to relieve our anxiety by peeking around
nature's corner as far as it will allow' (1995: 331). Our Newtonian faith in
astronomical regularities gives us an advantage over the Maya, who discovered an
order in the heavens that regulated their society, but one which required their
constant attention and participation. Our knowledge about the solar system is
thus 'time-saving', yet we pay a different price: the more we understand about
the universe, the more pointless it all seems. 'We emerge as the great nonpartici-
pants who have no influence on the outcome. ... We have not the slightest cosmic
duty to perform' (Weinberg, 1988: 154). Since our time apparently is not needed
by the cosmos, it seems devalued.

Insofar as it is our fate to be self-conscious in an objective world, the meaninglessness of the universe seems to be inescapable ... or is it? How much of that
meaninglessness is a consequence of the abstract homogeneity of our objectified
space and time – a search for a similarly abstract and homogeneous life-meaning
that must be the same in all places and all times? What does the presupposition of
such an objective meaning make us miss about *this* time and *this* space, right here
and now? Insofar as we feel alienated from our world (which for Buddhism gets at

the basic problem with the self), our detachment leads us to think that we are more or less the same person here as there, then as now. From a Buddhist perspective, this delusion is resolved in the realisation that I am non-dual with my world, and my responsibilities follow from my embeddedness in this particular space-time situation. Then to seek for the cosmic meaning of life is like a journalist asking a grandmaster: 'Please tell us, what is the best move in chess?'

What prompts us to such a new understanding? Most often, liminal events such as our fear of death and our sense of *lack*. Deeper than the desire for order is what Damian Thompson describes as our 'deep-seated human urge to escape from time which, in the earliest societies, was usually met by dreams of a return to a golden past' (1996: 325). Christianity put an end to that, by situating us toward the end rather than the beginning of time. Yet the difference is less important than the common need to transcend time as we know it, for its ineluctable course carries us all to the same final destination. Thompson sums up his study of apocalyptic time by concluding that the human understanding of time is always distorted by death: 'The belief that mankind has reached the crucial moment in its history reflects an unwillingness to come to terms with the transience of human life and achievements. Our urge to celebrate the passing of time fails to conceal an even deeper urge to escape from it' (1996: 332).

> Rest not! Life is sweeping by;
> Go and dare before you die
> Something mighty and sublime,
> Leave behind to conquer time.
> (Goethe)

Who would not like to conquer time? Because whether or not time conquers all, it conquers us. Goethe had a great fear of death (greater than most of us? or just more conscious?), which the symbolic immortality of his literary success evidently did not allay. The compulsion to accomplish something does not need to be so dramatic. The psychoanalyst Neil Altman wrote in similar terms about his years as a Peace Corps volunteeer in southern India:

> It took a year for me to shed my American, culturally based feeling that I had to make something happen. ... Being an American, and a relatively obsessional American, my first strategy was to find security through getting something done, through feeling worthwhile, accomplishing something. My time was something that had to be filled up with progress toward that goal.
>
> (quoted in Levine, 1997: 204–5)

Individualistic cultures emphasise achievement more than affiliation. In psychoanalytic terms, the pressure we then feel to accomplish something is an introjection of the intentions we project outward into the world. Since the self lacks any being or ground of its own, according to Buddhism, it is best characterised as an ongoing process which in this way seeks perpetually, because in vain, to feel secure, to make itself real. If our more individualised ego-self is that much more of a delusion, it will be all the more unsatisfied; but then it must explain that dissatisfaction: the reason is that I have not attained my goals. Since the goals I do accomplish bring no satisfaction (*lack* still itches), the need develops to construct more ambitious projects. ... Unfortunately, the same dynamic seems to be operating collectively: it explains our preoccupation with economic growth and technological development. As Weber realised, this historical process has become all the more obsessive because it has lost any teleological end-point.

The more objective time is for us, the more subjective (alienated) is the sense of self that is *in* (i.e., other than) time, which therefore *uses* time in order to try to gain something from it – for then the greater, too, is one's awareness of the end of one's own time. Our own sense of separation from time motivates us to try to secure ourselves within it, yet according to Buddhism the only satisfying solution is the essentially religious realisation that we are not other than it.

Earlier we saw how Dogen uses the interdependence of objects and time to demonstrate that objects are unreal, but their relativity also implies the unreality of objective time. If there is only time then there is no time, because there can be no container without a contained, when there are no nouns, there are no referents for temporal predicates. When there are no things that have an existence apart from time, then it makes no sense to speak of things as being young or old. Dogen makes this point using the image of firewood and ashes:

> Firewood becomes ash, and it does not become firewood again. Yet, do not suppose that the ash is future and the firewood past. You should understand that firewood abides in the phenomenal expression of firewood, which fully includes past and future and is independent of past and future. Ash abides in the phenomenal expression of ash, which fully includes future and past. Just as firewood does not become firewood again after it is ash, you do not return to birth after death.
>
> This being so, it is an established way in buddha-dharma to deny that birth turns into death. Accordingly, birth is understood as no-birth. It is an unshakeable teaching in Buddha's discourse that death does not turn into birth. Accordingly, death is understood as no-death.

> Birth is an expression complete this moment. Death is an expression complete this moment. They are like winter and spring. You do not call winter the beginning of spring, nor summer the end of spring.
>
> (Dogen, in Welch and Tanahashi, 1985)

Because our life and death, like spring and summer, are not in time, they are timeless. If there is no one non-temporal who is born and dies, then there are only the events of birth and death. But if there are only those events, with no one *in* them, then there is no real birth and death. Or we may say that there is birth-and-death in every moment, with the arising and passing-away of each thought and act.

If we resolve our own sense of something lacking in us, we can realise that there is nothing lacking in the present which needs to be full-filled in the future. Even as 'I' am neither young nor old, so spring is not an anticipation of summer: it is whole and complete in itself. There is nothing mystical about this, but it can be expressed only paradoxically: as a *nunc* both *stans* and *fluens*, as an eternal now (it is always *now*) which none the less flows (that *now* manifests in different ways).

There is an exact parallel here with space: things (e.g., our bodies) are not in space, for they are irremediably spatial; and space is things, i.e., things are what space is doing right here. Yet if there is only space then there is no objective space, since there can be no container without a contained; one way to put it is that everywhere becomes the centre of the universe. This sheds considerable light on, for example, the many conflicts between indigenous cultures (which understand themselves as belonging to the land, because part of it) and our globalising economic culture (which understands the commodified land as belonging to someone who can sell it).

The end of death implies the end of self implies the end of *lack* implies the end of objectified space-time. ... What implications does this have for our social need to develop a new spatiotemporal schema? I conclude with a reflection on these questions.

A genuine end to objectified, commodified time – as opposed to a postmodern fragmentation that merely heightens anxiety and makes us more compulsive – implies a 'new' understanding of life as play. That is because play is what we are doing when we do not need to gain something *from* a situation – in particular, when we do not need to use a situation to end our sense of *lack*. When we do not need to extract something from this time-place, we will not continue to devalue it by contrasting here-and-now with some other location (e.g., heaven) or time (the future), or with homogenised Newtonian/Kantian space-time generally. That would resacralise the world, not as something objective and abstract, but as this here, this now.

The same would be true for time and space themselves. If we did not dualise between an event and its space/time, we could realise something hitherto overlooked about the nature of events: not only rites of passage such as birth, marriage, death, etc., but everyday wonders such as preparing and eating food; nurturing children, making and appreciating music, dance, etc.; even sleeping and being reborn each morning. To resacralise such events would be to resacralise such space-times, and therefore space-time itself – no longer hollowed out, but always with a particular texture and flavour. If we did not need to extract something from our space-times, something which might survive our deaths or end our *lack*, then we could recover an awareness of how essentially mysterious they are. Such space-time may be as much a god as we can find today; and if our *lack* were truly resolved, it might be as good a god as we need.

BIBLIOGRAPHY

Abercrombie, N. and Longhurst, B. (1998) *Audiences*. London, Sage.

Abram, D. (1997) *The Spell of the Sensuous*. New York, Vintage Books.

Adam, B. (1990) *Time and Social Theory*. Cambridge, Polity Press.

—— (1992) 'Modern times: the technology connection and its implications for social theory', *Time and Society* 1(2): 175–92.

—— (1995) *Timewatch: The Social Analysis of Time*. Cambridge, Polity Press.

—— (1998) *Timescapes of Modernity. The Environment and Invisible Hazards*. Routledge, London.

Adas, M. (1989) *Machines as the Measure of Men: Science, Technology and Ideologies of Western Dominance*. Ithaca, Cornell University Press.

Agamben, G. (1999) *Remnants of Auschwitz. The Witness and the Archive*. New York, Zone Books.

—— (1993) *Infancy and History*. London, Verso.

Aikin, J. (1795) *A Description of the Country from Thirty to Forty Miles Around Manchester*. London, John Stockdale.

Aitken, S. and Carroll, M. (1996) 'Man's place in the home: telecommuting, identity and urban space'. Internet 1998–06–11: ,http://www.ncgia.ucsb.edu/conf/BALTI-MORE/authors/aitken/paper.html..

Alliez, E. (1996) *Capital Times*. Minneapolis, Minnesota University Press.

Alvarez, A. (1995) *Night: NightLife, Night Language, Sleep and Dreams*. New York, Norton & Co.

Amato, J.A. (2000) *Dust: A History of the Small and Invisible*. Berkeley, University of California Press.

Amin, A. and Graham, S. (1997) 'The ordinary city', *Transactions of the Institute of British Geographers* 22(4): 411–29.

Anderson, B. (1983) *Imagined Communities: Reflections on the Origin and Spread of Nationalism*. London, Verso.

Anderson, E. (1990) *Street Wise: Race, Class and Change in an Urban Community*. Chicago, University of Chicago Press.

Andrews, C. (ed.) (1936) *The Torrington Diaries*, volume 3. London, Methuen.

Anon. (1795) 'Untitled obituary for Josiah Wedgwood' *The Gentlemans' Magazine* 65: 84–5.

Ansell Pearson, K. (1997) *Viroid Life: Deleuze on Nietzsche and the Transhuman Condition*. London, Routledge.

—— (1999a) 'Bergson and creative evolution/involution: exploring the transcendental illusion of organic life', in J. Mullarkey (ed.) *The New Bergson*. Manchester, Manchester University Press, 146–67.

—— (1999b) *Germinal Life*. London, Routledge.

Antliff, M. (1999) 'The rhythm of', in J. Mullarkey (ed.) *The New Bergson*. Manchester, Manchester University Press, 184–208.

Åquist, A.-C. (1992) *Tidsgeografi i samspel med samhällsteori*. Meddelanden från Lunds Universitets Geografiska Institutioner, avhandlingar 115.

Armstrong, T. (1998) *Modernism, Technology and the Body: A Cultural Study*. Cambridge, Cambridge University Press.

Ashton, T.S. (1948) *The Industrial Revolution, 1760–1830*. London, Oxford University Press.

Assad, M.L. (1999) *Reading with Michel Serres: An Encounter with Time*. Albany, State University of New York Press.

Auerbach, E. (1968) *Mimesis*. Princeton, NJ, Princeton University Press.

Austin, T. (1997) *Never Trust a Government Man: Northern Territory Aboriginal Policy 1911– 1939*. Darwin, Northern Territory University Press.

Aveni, A. (1995) *Empires of Time*. New York, Kodansha.

Azaryahu, M. (1998) 'It is no fairy tale – Israel at 50', *Political Geography* 18: 131–47.

Bachelard, G. (1969) *The Poetics of Space*, tr. M. Jolas. Boston, Beacon.

—— (1971) *The Poetics of Reverie*, tr. D. Russell. Boston, Beacon.

—— (1932/2000) *The Dialectic of Duration*. Manchester, Clinamen Press.

Bachmann, M. (1991) *Dalcroze Today. An Education Through and Into Music*. New York, Oxford University Press.

Bahktin, M. (1981) *The Dialogical Imagination*. Austin, University of Texas Press.

Barnes, T. and Gregory, D. (eds) (1997) *Reading Human Geography: The Poetics and Politics of Inquiry*. London, Arnold.

Barrell, J. (1982) 'Geographies of Hardy's Wessex', *Journal of Historical Geography* 8(4): 347–61.

Bartky, I.R. (1989) 'The adoption of Standard Time', *Technology and Culture* 30: 25–56.

Bashford, A. (ed.) (1998) Theme Issue: 'Return of the repressed', *Australian Feminist Studies* 13(27): 47–136.

Bateson, G. (1972) *Steps to an Ecology of Mind*. New York, Ballantine Books.

—— (1980) *Mind and Nature*. Glasgow, Fontana/Collins.

Bateson,G. and Bateson, M.C. (1987) *Angels Fear: Towards an Epistemology of the Sacred*. New York, Macmillan.

Bauman, Z. (1995) *Life in Fragments: Essays in Postmodern Morality*. Oxford, Blackwell.

Beck, U. (1994) 'The reinvention of politics: towards a theory of reflexive modernization', in U. Beck, A. Giddens and S. Lash (eds) *Reflexive Modernization: Politics, Tradition and Aesthetics in the Modern Social Order*. Cambridge, Polity Press.

Beckett, S. (1965) *Proust*. London, Calder and Boyars.

Bender, J. and Wellbery, D. (eds) (1991) *Chronotypes: The Construction of Time*. Stanford, Stanford University Press.

Benjamin, W. (1985) *Charles Baudelaire: A Lyric Poet of High Capitalism*. London, Verso.

Benko, G.B. and Strohmayer, U. (1995) *Geography, History and Social Sciences*. Dordrecht, Kluwer Academic Publishers.

Berg, M. (1985) *The Age of Manufactures*. London, Fontana.

Berger, J. (1984) *And Our Faces, My Heart, Brief as Photos*. London, Writers and Readers.

—— (1991) *And Our Faces, My Heart, Brief as Photos*. New York, Vintage.

—— (1997) *Photocopies*. London, Bloomsbury.

Bergmann, W. (1981a) *Die Zeitstrukturen sozialer Systeme: eine systemtheoretische Analyse*. Berlin, Duncker and Humblot.

—— (1981b) 'Zeit, Handlung und Sozialität bei G. H. Mead', *Zeitschrift für Soziologie* 10: 351–63.

—— (1983/1992) 'Das Problem der Zeit in der Soziologie: ein Literaturüberblick zum Stand der "zeitsoziologischen" Theorie und Forschung', *Kölner Zeitschrift für Soziologie und Sozialpsychologie* 35: 462–504; translated by Belinda Cooper, 'The problem of time in sociology', *Time and Society* 1(1): 81–134.

—— (1992) 'The problem of time in sociology: an overview of the literature on the state of theory and research on the "Sociology of Time", 1900–82', *Time and Society* 1(1): 81–134.

Bergson, H. (1910) *Time and Free Will*. London, Swan Sonnenschein.

—— (1913) *Creative Evolution*. London, Macmillan.

—— (1896/1988) *Matter and Memory*. New York, Zone Books.

—— (1932/1979) *The Two Sources of Morality and Religion*. New York, Greenwood Press.

—— (1991) *Matter and Memory*, tr. (from 5th edition of 1908) Paul and Palmer. New York, Zone Books.

—— (1999) *Duration and Simultaneity*. Manchester, Clinamen Press.

Berman, M. (1991) *All That is Solid Melts into Air: The Experience of Modernity*. London, Verso.

Berry, B.L., Harpman, E. and Elliot, E. (1995) 'Long swings in American inequality: the Keznets conjecture revisited', *Journal of the Papers in Regional Science* 74(2): 153–74.

Bertaux-Wiame, I. and Thompson, P. (1997) 'The familial meaning of housing in social rootedness and mobility: Britain and France', in D. Bertaux and P. Thompson (eds) *Pathways to Social Class: A Qualitative Approach to Social Mobility*. Oxford, Oxford University Press.

Biernacki, R. (1995) *The Fabrication of Labor*. Berkeley, University of California Press.

Bird, I.L. (1966) *The Englishwoman in America*. Toronto, University of Toronto Press.

Bischofberger, E., Dahlqvist, G. and Elinder, G. (1991) *Barnets integritet: etik i vårdens vardag*. Stockholm, Almqvist and Wiksell.

Björkqvist, E. and Åhlman, M. (1996) 'Vårdetikens betydelse för omvårdnaden – om faktorer som påverkar sjuksköterskans beslut och handlingssätt vid vårdetiska situationer', unpublished paper, Helsingborg, College for the Health Sciences.

Bladh, G. (1995) *Finnskogens landskap och människor under fyra sekler: en studie av samhälle och natur i förändring*, Forskningsrapport 95:11, Högskolan i Karlstad.

283

Blair, T. (1998) *The Third Way*. London, Fabian Publications.

Blanchot, M. (1989) *The Space of Literature*, tr. A. Smock. Lincoln, University of Nebraska Press.

Block, R.H. (1980) 'Frederick Jackson Turner and American geography', *Annals of the Association of American Geographers* 70: 31–42.

Bohm, D. (1985) *Unfolding Meaning*. London, Routledge.

Boltanski, L. (1999) *Distant Suffering: Morality, Media and Politics*. Cambridge, Cambridge University Press.

Borges, J.L. (1970) 'The fearful sphere of Pascal', in *Labyrinths*. Harmondsworth, Penguin.

Bornstein, M. (1979) 'The pace of life revisited', *International Journal of Psychology* 14.

Bouchet, D. (1998) 'Information technology, the social bond and the city: Georg Simmel updated', *Built Environment* 24 (2/3): 104–33.

Boundas, C. (1996) 'Deleuze-Bergson: an ontology of the virtual', in P. Patton (ed.) *Deleuze: A Critical Reader*. Oxford, Blackwell, pp. 81–106.

Bourdieu, P. (1990) *Photography: A Middlebrow Art*. London, Polity Press.

—— (1991) *The Political Ontology of Martin Heidegger*. London, Polity Press.

Bourke, V.J. (ed.) (1983) *The Essential Augustine*. Indianapolis, Hackett.

Boyd, Susan B. (1999) 'Family, law, and sexuality: feminist engagements', *Social and Legal Studies* 8(3): 369–90.

Boyer, P. (1978) *Urban Masses and Moral Order in America, 1820–1920*. Cambridge, MA, Harvard University Press.

Boyer, R. (1990) *The Regulation School: A Critical Introduction*, tr. C. Charney. New York City, Columbia University Press.

Boyne, R. (1998) 'Angels in the archive: lines into the future in the work of Jacques Derrida and Michel Serres', in S. Lash, A. Quick and R. Roberts (eds) *Time and Value*. London, Sage, 48–64.

Bramwell, A. (1985) *Blood and Soil: Walther Darré and Hitler's Green Party*. Buckinghamshire, Kensal Press.

Brenner, M.H. (1973) *Mental Illness and the Economy*. Cambridge, MA, Harvard University Press.

Bringing Them Home: Report of the National Inquiry into the Separation of Aboriginal and Torres Strait Islander Children from their Families (1997). Canberra, Commonwealth of Australia.

Brown, M.W. and Charlot, J. (n.d.) *A Child's Good Night Book*. New York, Harper & Row.

Brown, M.W. and Hurd, C. (1947) *Goodnight Moon*. New York, Harper & Row.

Brown, M.W. and Weisgard, L. (1993) *The Quiet Noisy Book*. New York, HarperTrophy.

Brown, N.O. (1961) *Life Against Death: The Psychoanalytic Meaning of History*. New York, Vintage.

Brown, S. and Capdevila, R. (1999) 'Perpetuum Mobile: substance, fame and the sociology of translation', in J. Law and J. Hassard (eds) *Actor Network Theory and After*. Oxford, Blackwell, 26–50.

Brown, T. (1981) *Ireland: A Social and Cultural History, 1922–79*. London, Fontana.

Brown, W. (1995) *States of Injury: Freedom and Power in Late Modernity*. Princeton, Princeton University Press.

Bruner, J. (1990) *Acts of Meaning*. Cambridge, Cambridge University Press.

Brunton, R. (1997) 'Shame about Aborigines', *Quadrant* May: 36–9.

Buck, N., Gershuny, J., Rose, D. and Scott, J. (1994) *Changing Households: The British Household Panel Survey 1990–1992*. Essex: ESRC Research Centre on Micro-Social Change.

Buck-Morss, S. (2000) *Dreamworld and Catastrophe: The Passing of Mass Utopia in East and West*. Cambridge, MA, MIT Press.

Butler, J. (1993) *Bodies that Matter: On the Discursive Limits of 'Sex'*. London and New York, Routledge.

—— (1997) *Excitable Speech: A Politics of the Performative*. New York, Routledge.

—— (1998) 'Merely cultural', *New Left Review* 227: 33–44.

Butterfield, F. (1998) 'Prisons replace hospitals for the nation's mentally ill', *The New York Times* 5 March: A1.

Butterfield, H. ([1933]1979) *Writings on Christianity and History*, ed. C.T. MacIntyre. Oxford, Oxford University Press.

Buttimer, A. (1983) *The Practice of Geography*. London, Longman Group Ltd.

Byng, J. (1936) *The Torrington Diaries*, volume 3, ed. C. Andrews. London, Methuen.

Byng-Hall, J. (1990) 'The power of family myths', in P. Thompson and R. Samuel (eds) *The Myths We Live By*. London, Routledge.

Caillois, R. (1961) *Man, Play and Games*, tr. M. Barash. New York, Free Press.

Callon, M. (1986) 'Some elements of a sociology of translation: domestication of the scallops and the fisherman of St Brieuc Bay', in J. Law (ed.) *Power, Action and Belief*, Sociological Review Monograph 32. London, Routledge and Kegan Paul, 196–233.

Callon, M., Law, J. and Rip, A. (eds) (1986) *Mapping the Dynamics of Science and Technology: Sociology in the Real World*. London, The Macmillan Press Ltd.

Canada (1871) *Census of Canada*, Cornwall, Schedule No. 6, Return of industrial establishments, Microfilm (reel C10,006), Public Archives of Canada.

Capek, M. (1971) *Bergson and Modern Physics*. Dordrecht, Holland, Reidel.

Capra, F. (1982) *The Turning Point: Science, Society and the Rising Culture*. Wildwood House.

Capra, F. and Steindl-Rast, D. (1992) *Belonging to the Universe*. San Francisco, Harper-Collins.

Carey, J. (1983) 'Technology and ideology: the case of the telegraph', in J. Salzman (ed.) *Prospects: An Annual of American Cultural Studies*. New York, Cambridge University Press, 303–25.

—— (1989) *Communication as Culture: Essays on Media and Society*. London, Unwin Hyman.

Carlestam, G. and Sollbe, B. (eds) (1991) *Om tidens vidd och tingens ordning: texter av Torsten Hägerstrand*. Byggforskningsrådet. Stockholm.

Carlstein, T., Parkes, D. and Thrift, N. (eds) (1978) *Human Activity and Time Geography*. London, Edward Arnold.

Carr, D. (1986) *Time, Narrative and History*. Bloomington, IN, University of Chicago Press.

Castells, M. (1989) *The Informational City: Economic Restructuring and Urban Development*. Oxford, Blackwell.

—— (1996) *The Rise of the Network Society*, volume 2, *Networks and Identity*. Oxford, Blackwell.

Castoriadis, C. (1991) 'Time and creation' in J. Bender and D. Wellbery (eds) *Chronotypes: The Construction of Time*. California, Stanford University Press, 38–64.

Caygill, H. (1999) *Walter Benjamin: The Colour of Experience*. London, Routledge.

Chimisso, C. (2000) 'Introduction', in G. Bachelard *The Dialectic of Duration*. Manchester, Clinamen Press, 1–16.

Cixous, H. (1997) *Rootprints: Memory and Life Writing*. London, Routledge.

Clark, G. and Dear, M. (1984) *State Apparatus: Structures and Language of Legitimacy*. Boston, Allen & Unwin.

Clark, T.J. (1985) *The Painting of Modern Life*. London, Thames and Hudson.

—— (1999) *Farewell to an Idea: Episodes in a History of Modernism*. New Haven, CT, Yale University Press.

Clément, C. (1994) *Syncope: The Philosophy of Rapture*, tr. S. O'Driscoll and D. Mahoney. Minneapolis, University of Minnesota Press.

Coleman, W. (1966) 'Science and symbol in the Turner frontier hypothesis', *American Historical Review*, 72: 22–49.

Collier, P. (1991) 'The inorganic body and the ambiguity of freedom', *Radical Philosophy* 57, Spring: 3–9.

Commonwealth Submission to the National Inquiry into the Separation of Aboriginal and Torres Strait Islander Children from their Families (1996). Canberra, Commonwealth of Australia.

Conley, V.A. (1994) 'Foreword: East meets West', in C. Clement *Syncope: Philosophy of Rapture*. Minneapolis, University of Minnesota Press, ix–xviii.

Cornwall Freeholder, 19 January, 16 February, 2, 9, 30 March, 6 April, 4, 14, 18 May, 19 October, 16 November 1883 and 23 April 1886.

Cornwall Reporter, 9 February and 16 December 1876.

Cornwall Standard, 12 and 19 August 1886.

Corporation of the Town of Cornwall Town Council Minutes, Simon Fraser Centennial Library, Cornwall, 25 October 1868, 14 August 1871, 14 September 1873.

Crang, M. and Thrift, N. (2000) 'Introduction: species of spaces', in M. Crang and N. Thrift (eds) *Thinking Space*. London, Routledge, 1–37.

Crang, M. and Travlou, S. (2000) 'The city and topologies of memory', *Environment & Planning D: Society & Space* 18.

Crary, J. (1999) *Suspensions of Perception: Attention, Spectacle and Modern Culture*. Cambridge, MA, MIT Press.

Cresswell, T. (1997) 'Imagining the nomad: mobility and the postmodern primitive', in G. Benko and U. Strohmayer (eds) *Social Theory: Interpreting Modernity and Postmodernity*. Oxford, Basil Blackwell, 360–2.

Crites, S. (1989) 'The narrative quality of experience', in S. Hauerwas and L.G. Jones (eds) *Why Narrative?* Grand Rapids, Eerdmans.

Critical Quarterly (Special Issue) (1987) 'The state of the subject' 29: 4.

Cronon, W. (1992) *Nature's Metropolis: Chicago and the Great West*. New York, Norton.

Cullen, I. (1978) 'The treatment of time in the explanation of spatial behaviour', in T. Carlstein, D. Parkes and N. Thrift (eds) *Timing Space and Spacing Time*, volume 2, *Human Activity and Time Geography*. London, Arnold, 27–39.

Cunnison, S. (1986) 'Gender, consent and exploitation among sheltered housing wardens', in K. Purcell *et al.* (eds) *The Changing Experience of Employment*. Basingstoke, Macmillan.

Currie, G. (1999) 'Can there be a literary philosophy of time', in J. Butterfield (ed.) *The Arguments of Time*. Oxford, Oxford University Press, 43–63.

Curtis, B. and Pajaczkowska, C. (1994) 'Getting there: travel, time and narrative', in G. Robertson, G. Mash, M. Ticker. J. Bird, B. Curtis and T. Putnam (eds) *Travellers Tales: Naratives of Home and Displacement*. London, Routledge.

Daniels, S. (1993) *Fields of Vision: Landscape Imagery and National Identity in England and the United States*. Cambridge, Polity.

Davies, K. (1990) *Women, Time and the Weaving of the Strands of Everyday Life*. Aldershot, Avebury.

—— (1994) 'The tensions between process time and clock time in care-work: the example of day nurseries', *Time and Society* 3(3): 277–303.

—— (1996a) *Önskningar och realiteter. Om flexibilitet, tyst kunskap och omsorgsrationalitet i barnomsorgen*. Stockholm, Carlssons.

—— (1996b) 'Capturing women's lives. a discussion of time and methodological issues', *Women's Studies International Forum* 19(6): 579–88.

—— (forthcoming) 'New times at the workplace – opportunity or angst? The example of hospital work', in M. Soulsby and J.T. Fraser (eds) *Time at the Millennium: Changes and Continuities (The Study of Time X)*. New York, Greenwood Publishing Group.

Davison, G. (1992) 'Punctuality and progress: the foundations of Australian standard time', *Australian Historical Studies* 25: 169–91.

de Certeau, M. (1983) 'The madness of vision', *Enclitic* 7(1): 24–31.

—— (1984) *The Practice of Everyday Life*. Berkeley, CA, California University Press.

—— (1986) *Heterologies: Discourse on the Other*, tr. Brian Massumi. Minneapolis, University of Minnesota Press.

—— (1988) *The Writing of History*, tr. Tom Conley. New York, Columbia University Press.

—— (1997) *The Capture of Speech and other Political Writings*. Minneapolis, University of Minnesota Press.

de Landa, M. (1998) 'Extensive borderlines and intensive borderlines', in L. Woods and E. Rehfled (eds) *Borderline*. New York, Sprimnger Wien, 18–24.

—— (1999) 'Deleuze, diagrams and open ended becoming', in E. Grosz (ed.) *Becomings: Explorations in Time, Memory and Futures*. New York, Cornell University Press, 29–41.

Dear, M.J. (1988) 'The postmodern challenge: reconstructing human geography', *Transactions of the Institute of British Geographers* 13: 262–74.

Dear, M.J. and Wolch, J. (1987) *Landscapes of Despair: From Deinstitutionalization to Homelessness*. Princeton, Princeton University Press.

—— (1989) 'How territory shapes social life', in M. Dear and J. Wolch (eds) *The Power of Geography*. Boston, Unwin, 1–19.

de Concini, B. (1990) *Narrative Remembering*. Lanham, MD, University of America Press.

Deem, R. (1996) 'No time for a rest. An exploration of women's work, engendered leisure and holidays', *Time and Society* 5(1): 5–25.

Defoe, D. (1971) *A Tour Through the Whole Island of Great Britain*. Harmondsworth, Penguin.

Deleuze, G. (1986) *Cinema 1: The Movement-Image*. Minneapolis. University of Minnesota Press.

—— (1989) *Cinema 2: The Time-Image*. Minneapolis, University of Minnesota Press.

—— (1991) *Bergsonism*. New York, Zone Books.

—— (1996) *Foucault*. Paris, Minuit.

Deleuze, G. and Guattari, F. (1987) *A Thousand Plateaus*, tr. Brian Massumi. Minneapolis, The University of Minnesota Press.

—— (1998) *A Thousand Plateaus*, tr. Brian Massumi. Minneapolis, The University of Minnesota Press.

Dessaix, R. (1997) 'At last the secret', in D. Modjeska, A. Lohrey, R. Dessaix *Secrets*. Sydney, Pan Macmillan.

di Leonardo, M. (1987) 'The female world of cards and holidays: women, families and the work of kinship', *Signs* 12(3): 440–53.

Dickens, C. (1989) *American Notes for General Circulation*. London, Penguin Books.

Dillon, B (1997) 'The carcass of time', *Oxford Literary Review* 19: 133–48.

Dillon, M.C. (1997) *Merleau-Ponty's Ontology*, second edition. Evanston, IL, Northwestern University Press.

Diprose, R. (forthcoming) ' "The body in its sexual being": Butler and Merleau-Ponty', in A. Munster and E. Probyn (eds) *Body to Body: A Critical Reader in Corporeality*. London and New York, Routledge.

Disturnell, J. (1850) *Disturnell's Railroad, Steamboat and Telegraph Book*. New York, J. Disturnell.

—— (1853) *Disturnell's Railway, Steamship and Telegraph Book*. New York, J. Disturnell.

Doel, M. (1998) 'Poststructuralist geography', unpublished manuscript.

Dominick, R.H. (1992) *The Environmental Movement in Germany: Prophets and Pioneers, 1871–1971*. Indianapolis, Indiana University Press.

Donald, J. (1997) 'This, here, now: imagining the modern city', in S. Westwood and J. Williams (eds) *Imagining Cities: Scripts, Signs, Memory*. London, Routledge, 181–201.

Douglas, M. (1997) 'In defense of shopping', in P. Falk and C. Campbell (eds) *The Shopping Experience*. London, Sage.

Douglass, P. (1998) 'Deleuze, cinema, Bergson', *Social Semiotics* 8(1) 25–36.

Dreyfus, H. (1991) *Being-in-the-World: A Commentary on Heidegger's Being and Time, Division 1*. Cambridge, MIT Press.

du Gay, P. (2000) *In Praise of Bureaucracy*. London, Sage.

Dunbar, R. (1996) *Grooming, Gossip and the Evolution of Language*. London, Faber and Faber.

Duncan, D.E. (1998) *The Calendar*. London, Fourth Estate.

Duncan, N. (ed.) (1996) *BodySpace: Destabilizing Geographies of Gender and Sexuality*. London, Routledge.

Duncan, S. (1993) 'Sites of representation: place, time and the discourse of the other', in S. Duncan and D. Ley (eds) *Place/Culture/Representation*. London, Routledge, 39–56.

Durkheim, E. (1915/1965) *The Elementary Forms of the Religious Life*. London, Free Press.

Ebert, Teresa L. (1996) *Ludic Feminism and After*. Ann Arbour, The University of Michigan Press.

Eigenheer, R. (1973–74) 'The frontier hypothesis and related spatial concepts', *The Californian Geographer* 14: 55–69.

Elchardus, M. and Glorieux, I. (1994) 'The search for the invisible 8 hours: the gendered use of time in a society with a high labour force participation of women', *Time and Society* 3(1): 5–27.

Eliade, M. (1959) *The Sacred and The Profane: The Nature of Religion*, tr. W. Trask. San Diego, NY and London, Harcourt Brace & Company.

—— (1971) *The Myth of Eternal Return*, tr. W.R. Trask. Princeton, Princeton University Press.

—— (1975) *Myth and Reality*. London, Harper Colophon.

Elias, N. (1982a) 'Über die Zeit' (Part 1), tr. H. Fliessbach and M. Schröter. *Merkur* 36(9): 841–56.

—— (1982b) 'Über die Zeit' (Part 2), tr. H. Fliessbach and M. Schröter. *Merkur*, 36(10): 998–1,016.

—— (1984) *Über die Zeit*, tr. H. Fliessbach and M. Schröter. Frankfurt a. M., Suhrkamp.

Ellegård, K. (1998) 'Under ytan – ingångar till det kulturgeografiska äventyret', in M. Gren and P.O. Hallin (eds) *Svensk kulturgeografi: en exkursion inför 2000-talet*.

Ellickson, R.C. (1996) 'Misconduct in public spaces', *Yale Law Journal* 105(5): 1,165–248.

Ellis, H. (1959) *British Railway History: An Outline from the Accession of William IV to the Nationalisation of the Railways 1877–1947*. London, Allen & Unwin.

Ende, M. (1984) *Momo*, tr. J. Maxwell Brownjohn. London, Puffin.

Engels, F. (1987) *The Condition of the Working Class in England*. Harmondsworth, Penguin.

Evans-Pritchard, E.E. (1969) *The Nuer*. New York, Oxford University Press.

Fabian, J. (1983) *Time and the Other: How Anthropology Makes its Other*. New York, Colombia University Press.

Falk, T. and Abler, R. (1980) 'Intercommunications, distance and geographical theory', *Geografiska Annaler* Series B: 59–67.

Felski, R. (2000) 'The invention of everyday life', *New Formations* 39 13–32.

Ferguson, A. (1966) *An Essay on the History of Civil Society*. Edinburgh, Edinburgh University Press.

Ferland, J. (1989) ' "In search of the unbound prometheia": a comparative view of women's activism in two Quebec industries, 1869–1908', *Labour/Le Travail* 7: 113–26.

Field, N. (1997) 'War and apology: Japan, Asia, the Fiftieth, and after', *Positions* 5(1): 1–49.

Fiennes, C. (1949) *The Journeys of Celia Fiennes*, ed. C. Morris. London, Cresset Press.

Fischer, C.S. (1985) 'Studying technology and social life', in M. Castells (ed.) *High Technology, Space and Society*, volume 28, Urban Affairs Annual Reviews. Beverly Hills, Sage Publications, 284–300.

—— (1992) *America Calling: A Social History of the Telephone to 1940*. Berkeley, University of California Press.

Fishman, J. (1972) *Language and Nationalism* Rowley, MA, Newbury House.

Fleming, S. (1889) 'Time-reckoning for the twentieth century', *Annual Report ... Smithsonian Institution ... for the Year Ending June 30, 1886*. Washington, DC, Smithsonian Institution.

Forman, F.J. with Sowton, C. (eds) (1989) *Taking our Time: Feminist Perspectives on Temporality*. Oxford, Pergamon Press.

Forrester, J. (1992) 'In the beginning was repetition: on inversions and reversals in psychoanalytic time', *Time and Society* 1(2): 287–300.

Foster, J.W. (1977) 'Certain set apart: the Western Island in the Irish Renaissance', *Studies* 66: 264–73.

Foucault, M. (1973) *Madness and Civilization: A History of Insanity in the Age of Reason*. New York, Vintage Books.

—— (1977a) *Discipline and Punish*. Harmondsworth, Penguin.

—— (1977b) 'Preface to transgression', in Donald F. Bouchard (ed.) *Language, Counter-memory, Practice*, tr. Bouchard and Sherry Simon. Ithaca, Cornell University Press.

—— (1978) *The History of Sexuality*, volume 1, tr. Robert Hurley. New York, Vintage.

—— (1986a) 'Of other spaces', *Diacritics* 16(1): 22–7.

—— (1986b) *The Use of Pleasure: The History of Sexuality*, volume 2, tr. Robert Hurley. New York, Vintage.

—— (1989a) *The Order of Things: An Archaeology of the Human Sciences*. London, Routledge.

—— (1989b) *The Birth of the Clinic*. London, Routledge.

—— (1989c) 'Friendship as a way of life', in S. Lotringer (ed.) *Foucault Live*. New York, Semiotext(e).

—— (1997) 'What is Enlightenment?', in Paul Rabinow (ed.) *Michel Foucault: Ethics, Subjectivity and Truth*. New York, The New Press.

Frank Zappa & The Mothers of Invention (1968) 'What's the ugliest part of your body' (from *We're Only in it for the Money*). The Zappa Family trust d/b/a Frank Zappa Music (BMI).

Friedberg, A. (1993) *Window Shopping: Cinema and the Postmodern*. Berkeley, University of California Press.

Fukuyama, F. (1992) *The End of History and the Last Man* London, Hamish Hamilton.

Fulford, S. (1997) 'Review of Cixous, Rootprints', *Oxford Literary Review* 19: 149–54.

Fourier, C. (1971) *The Utopian Vision of Charles Fourier*. London, Cape.

Fraser, N. (1997) *Justice Interruptus: Critical Reflections on the 'Postsocialist' Condition*. London and New York, Routledge.

—— (1998) 'Heterosexism, misrecognition and capitalism: a response to Judith Butler', *New Left Review* 228: 140–50.

Franssén, A. (1997) *Omsorg i tanke och handling*. Lund, Arkiv förlag.

Freeman, C. (1982) 'The "understanding" employer', in J. West (ed.) *Work, Women and the Labour Market*. London, Routledge and Kegan Paul, 135–53.

Freud, S. (1962) *Civilization and its Discontents*. New York, Norton.

—— (1985) 'The "uncanny" ', in *Pelican Freud Library*, volume 14, *Art and Literature*. Harmondsworth, Penguin.

Frow, J. (1997) *Time and Commodity Culture*. Oxford, Clarendon Press.

Gadamer, G. (1972) *Wahrheit und Methode*, 3rd edition. Tübingen, Mohr Verlag.

—— (1975) *Truth and Method*. New York, Seabury Press.

Gaita, R. (1997a) 'Not Right', *Quadrant* January/February: 46–51.

—— (1997b) 'Genocide and pedantry', *Quadrant* July/August: 41–5.

Gale, R.M. (ed.) (1978) *The Philosophy of Time*. Brighton, Harvester Press.

Game, A. (1991) *Undoing the Social: Towards a Deconstructive Sociology*. Buckingham, Open University Press.

—— (1997) 'Time unhinged', *Time and Society* 6 (2/3): 115–29.

Game, A. and Metcalfe, A. (1996) *Passionate Sociology*. London, Sage.

Garhammer, M. (1995) 'Changes in working hours in Germany: the resulting impact on everyday life', *Time and Society* 4(2): 167–203.

Gatens, M. (1997) 'Sex, gender, sexuality: can ethologists practice genealogy?', *The Southern Journal of Philosophy* 35: 1–15.

Gell, A. (1992) *The Anthropology of Time: Cultural Constructions of Temporal Maps and Images*. Oxford, Berg.

Gershuny, J. (2000) *Changing Times*. Oxford, Oxford University Press.

Gibelman, M. and Demone, H. (1998) 'Preface,' in M. Gibelman and H. Demone (eds) *The Privatization of Human Services: Policy and Practice Issues, Volume I*, New York, Springer Publishing Company, xi–xxii.

Giddens, A. (1979) *Central Problems in Social Theory: Action, Structure and Contradiction in Social Analysis*. London, Macmillan.

—— (1981) *A Contemporary Critique of Historical Materialism: Power, Property and the State*. London, Macmillan.

—— (1984) *The Constitution of Society: Outline of the Theory of Structuration*. Cambridge, Polity Press.

—— (1986) *The Constitution of Society*. Cambridge, Polity Press.

—— (1987) 'Time and social organisation', in *Social Theory and Modern Sociology*. Cambridge, Polity Press, 140–65.

—— (1990) *The Consequences of Modernity*. Cambridge, Polity Press.

—— (1991) *Modernity and Self-Identity: Self and Society in the Late Modern Age*. Cambridge, Polity Press.

—— (1994) 'Living in a post-traditional society', in U. Beck, A. Giddens and S. Lash (eds) *Reflexive Modernization: Politics, Tradition and Aesthetics in the Modern Social Order*. Cambridge, Polity Press.

Gilligan, C. (1982) *In a Different Voice: Psychological Theory and Women's Development*. Cambridge, Harvard University Press.

Gilroy, P. (2000) *Between Camps: Nations, Cultures and the Allure of Race*. London, Allen Lane.

Ginzburg, C. (1994) 'Killing a Chinese mandarin: the moral implications of distance', *New Left Review* 208: 107–20.

Gladstone, W. (1863) *The Life and Works of Josiah Wedgwood* (an address given at Burslem, 26 October 1863).

Glennie, P. and Thrift, N. (1996) 'Reworking E.P. Thompson's "Time, work-discipline and industrial capitalism" ', *Time and Society* 5(3): 275–99.

—— (1998) 'The spaces of times', paper presented to the ESRC Workshop: Rethinking History and the Human Sciences.

Glover, S. (1998) 'Ex-patient recalls grim life in clinic', *Los Angeles Times* 16 June: B3.

—— (1999) 'Casting a critical eye on church of castoffs', *Los Angeles Times* 1 February: A1.

Goff Le, J. (1980) *Time and Culture in the Middle Ages*. Chicago, University of Chicago Press.

Goffman, E. (1961) *Asylums: Essays on the Social Situation of Mental Patients and Other Inmates*. Garden City, NY, Anchor Books.

—— (1967a) 'Where the action is', in *Interaction Ritual: Essays on Face-to-Face Behaviour*. New York, Doubleday.

—— (1967b) *Interactions*. Chicago, Aldine.

Graham, G. (1997) *The Shape of the Past: A Philosophical Approach to History*. Oxford, Oxford University Press.

Gregory, D. (1994a) *Geographical Imaginations*. London, Blackwell Press.

—— (1994b) 'Time-geography,' in *The Dictionary of Human Geography*. London, Blackwell Reference.

Gregory, D. and Urry, J. (eds) (1985) *Social Relations and Spatial Structures*. London, Macmillan.

Gren, M. (1994) *Earth Writing: Exploring Representation and Social Geography In-Between Meaning / Matter*. Department of Geography, University of Gothenburg, Series No. 85.

Gren,M. and Hallin, P.O. (eds) (1998) *Svensk kulturgeografi: en exkursion inför 2000-talet*. Studentlitteratur.

Gressley, G.M. (1958) 'The Turner thesis – a problem in historiography', *Agricultural History* 32: 227–49.

Greu, M. (1994) *Earth Writing*. Göteborg, University of Goteborg.

Griffith, J. (1995) 'Life of strife in the fast lane', *Guardian* 23 August, 'Society': 4.

Grossberg, L. (2000) 'History, imagination and the politics of biology: Between the death and the fear of history', in P. Gilroy, L. Grossberg and A. McRobbie (eds) *Written Guarantees: In Honour of Stuart Hall*. London, Verso, 148–64.

Grosz, E. (1995) *Space, Time and Perversion: Essays on the Politics of Bodies*. London, Routledge.

—— (1999) 'Thinking the new: of futures yet unthought', in E. Grosz (ed.) *Becomings: Explorations in Time, Memory and Futures*. New York, Cornell University Press, 15–28.

—— (ed.) (1999) *Becomings. Explanations in Time, Memory and Futures*. New York, Columbia University Press.

Guattari, F. (1992) *Chaosmosis: A New Ethico-Aesthetic Paradigm*. Indiana, Indiana University Press.

Gulley, M. (1959) 'The Turner frontier: a study in the migration of ideas', *Tijschrift voor Economische en Sociale Geografie*, 50, Part I 65–72, Part II 81–91.

Gunnarsson, E. and Ressner, U. (1985) *Fruntimmersgöra. Kvinnor på kontor och verkstadsgolv*. Stockholm, Prisma.

Gunnarsson, E. (1994) 'Att våga väga jämt! Om kvalifikationer och kvinnliga förhållning-ssätt i ett tekniskt industriarbete', Ph.D. dissertation. Luleå, Tekniska högskolan.

Gunnell, J.B. (1968) *Political Philosophy and Time*. Middletown, CT, Wesleyan University Press.

Haebich, A. (1992) *For Their Own Good: Aborigines and Government in the South West of Western Australia 1900–1940*. Nedlands, University of Western Australia Press.

Hägerstrand, T. (1953) *Innovationsförloppet ur korologisk synpunkt*. Meddelanden från Lunds Universitets Geografiska Institution. Gleerupska univ.-bokhandeln. Lund.

—— (1972) 'Om en konsistent individorienterad samhällsbeskrivning för framtidsbruk', *Justitiedepartementet Ds Ju 972:25*. Stockholm.

—— (1974): 'Tidsgeografisk beskrivning. Syfte och postulat', *Svensk Geografisk Årsbok*, no. 50.

—— (1975) *Dynamic Allocation of Urban Space*. Farnborough, Saxon House.

—— (1976) 'Geography and the study of interaction between nature and society', *Geoforum* 17: 329–34.

—— (1977) *Culture and Ecology: Four Time-Geographic Essays*. Rapporter och notiser 39. Institutionen for kulturgeografi och ekonmiska geografi vid Lunds universitet. Lund.

—— (1982) 'Diorama, path, project', *Tijdschrift voor Econ. En Soc. Geografie* 73(6): 323–39.

—— (1985a) 'Time and culture', in G. Kirsch, P. Nijkamp and K. Zimmermann (eds) *Time Preferences: An Interdisciplinary Theoretical and Empirical Approach*. Berlin, Wissenschaftszentrum, 1–15.

—— (1985b) 'Den geografiska traditionens kärnområde', Lecture, Geographic Days in Uppsala, 1 June 1985, in *Geografiska notiser*.

—— (1985c) 'Time-geography: focus on the corporeality of man, society and environment', reprint from *The Science and Praxis of Complexity*. The United Nations University, 193–216.

—— (1989) Reflections on 'What about people in regional science?' *Papers of the Regional Science Association* 66: 1–6.

—— (1992a) *Samhälle och natur*. Rapporter och notiser 110. Institutionen för kulturgeografi och ekonomisk geografi vid Lunds Universitet. Lund.

—— (1992b) 'Geografins innehåll och historiska utveckling'. *Svensk geografisk årsbok*: 9–18.

—— (1993a) 'Vår roll är förvaltarens, inte exploatörens', in B. Kullinger and U.-B. Strömberg (eds) *Planera för bärkraftig utveckling*. Byggforskningsrådet. Stockholm.

—— (1993b) 'Gränser – en försummad dimension i kritiskt tänkande', in *Det kritiska uppdraget: den problemorienterade forskningen i framtiden*. Institutionen för tema, Universitetet i Linköping.

—— (1996) 'Att äga rum. utveckling', *Svensk geografisk årsbok*: 105–11.

Hägerstrand, T. and Lenntorp, B. (1974) 'Samhällsorganisation i tidsgeografiskt perspektiv', in *SOU 1974:2 Ortsbundna levnadsvillkor*. Bilaga 2: Ortssystem och levnadsvillkor. Stockholm.

Hall, P. (1988) *Cities of Tomorrow: An Intellectual History of Urban Planning and Design in the Twentieth Century*. Oxford, Basil Blackwell.

—— (1993) 'Policy paradigms, social learning, and the state', *Comparative Politics* 25(3): 275–96.

Hallin, P.O. (1988) *Tid för omställning – om hushållens anpassningsstrategier vid en förändrad energisituation*. Meddelanden från Lunds universitets geografiska institutioner, avhandlingar, no. 105.

—— (1992): 'New paths for time-geography?', *Geografiska Annaler* 73B(3): 199–207.

Halter, A. (1992) 'Homeless in Philadelphia: A qualitative study of the impact of state welfare reform on individuals', *Journal of Sociology and Social Welfare*: 17–19.

Hanson, S. and Pratt, G. (1995) *Gender, Work and Space*. London, Routledge.

Haraway, D. (1996) 'Situated knowledges: the science question in feminism and the privilege of partial perspective', in J. Agnew, D. Livingstone and A. Rogers (eds) *Human Geography: An Essential Anthology*, Oxford, Blackwell.

Hareven, T.K. (1982) *Family Time and Industrial Time: The Relationship Between the Family and Work in a New England Industrial Community*. Cambridge, Cambridge University Press.

Harries-Jones, P. (1995) *A Recursive Vision: Ecological Understanding and Gregory Bateson*. University of Toronto Press.

Harrison, M. (1986) 'The ordering of the environment: time, work and the occurrence of crowds 1790–1835', *Past and Present* 110: 134–68.

Harvey, D. (1985) *Urbanisation of Consciousness*. Oxford, Blackwell.

—— (1989) *The Condition of Postmodernity: An Enquiry into the Origins of Cultural Change*. Oxford, Basil Blackwell.

—— (1990) 'Between space and time: reflections on the geographical imagination', *Annals of the Association of American Geographers* 80(3): 418–34.

—— (1993) 'From space to place and back again: reflections on the condition of postmodernity', in J. Bird, B. Curtis, T. Putnam, G. Roberston, and L. Tickner (eds) *Mapping the Futures: Local Cultures, Global Change*. London, Routledge, 3–29.

Hastings, A. (1997) *The Construction of Nationhood: Ethnicity, Religion and Nationalism*. Cambridge, Cambridge University Press.

Headrick, D.R. (1981) *The Tools of Empire: Technology and European Imperialism in the Nineteenth Century*. New York, Oxford University Press.

Heffernan, M. (1991) 'The desert in French Orientalist painting during the nineteenth century', *Landscape Research* 16(2): 37–42.

Hegel, G.W.F. (1952) *Phänomenologie des Geistes*, first published 1807. Hamburg, Felix Meiner Verlag.

—— (1967) *The Phenomenology of Mind*, tr. J. B. Baillie. New York, Harper and Row.

Heidegger, M. (1980) *Being and Time*, first published 1927, tr. J. Macquarrie and E. Robinson. Oxford, Blackwell.

Heikkala, J. (1993) 'Discipline and excel: technologies of the self and body, and the logic of competing', *Sociology of Sport Journal* 10: 397–412.

Hennessy, R. (1993) *Materialist Feminism and the Politics of Discourse*. London and New York, Routledge.

Hennessy, R. and Ingraham, C. (eds) (1997) *Materialist Feminism: A Reader in Class, Difference, and Women's Lives*. London and New York, Routledge.

Hetherington, K. (1997a) *The Badlands of Modernity: Heterotopia and Social Ordering*. London, Routledge.

—— (1997b) 'Museum topology and the will to connect', *Journal of Material Culture* 2(2): 199–218.

—— (forthcoming) 'The topology of Utopia', mimeo.

Hetherington, K. and Law, J. (1997) 'Allegory and interference: representation in sociology', unpublished manuscript.

Hetherington, K. and Lee, N. (1997) 'Social order and the blank figure', unpublished manuscript.

Hetherington, K. and Munro, R. (eds) (1997) *Ideas of Difference: Social Spaces and the Labour of Division*. Blackwell Publishers/ *The Sociological Review*.

Hill, C. (1980) *Some Intellectual Consequences of the English Revolution*. Madison, University of Wisconsin Press.

Hillier, B. (1966) 'An arcanist at Etruria', *Proceedings of the Wedgwood Society* 6: 91–5.

Hillis, K. (1999) 'Towards the light "within": optical technologies, spatial metaphors and changing subjectivities', in M. Crang, P. Crang and J. May (eds) *Virtual Geographies: Bodies, Space, Relations*. London, Routledge, 23–43.

Hillman, J. (1992) *Re-Visioning Psychology*. New York, Harper Perennial.

—— (1975) *Loose Ends*. Dallas, Spring Publications.

Hinchliffe, S. (1996) 'Technology, power, and space', *Environment and Planning D: Society and Space* 14: 577–585.

—— (1997) 'Home-made space and the will to disconnect', in K. Hetherington and R. Munro (eds) *Ideas of Difference: Social Spaces and the Labour of Division*. Oxford, Blackwell.

—— (1998) 'Fabricating workers – the end of knowing and the injunction to activity', unpublished manuscript.

Hirsch, F. (1977) *The Social Limits to Growth*. London, Routledge and Kegan Paul.

Hobsbawm, E.J. (1975) *The Age of Capital 1848–75*. London, Weidenfeld and Nicolson.

Hochschild, A.R. (1983) *The Managed Heart*. Berkeley, University of California Press.

—— (1997) *The Time Bind*. New York, Metropolitan.

Hofstede, G. (1991) *Cultures and Organisations: Software of the Mind*. London, McGraw-Hill.

Hohn, H.-W. (1984) *Die Zerstörung der Zeit. Wie aus einem göttlichen Gut eine Handelsware wurde*. Frankfurt a. M., Fischer Alternativ.

Hollevoet, C. (1992) 'Wandering in the city *Flânerie* to *Dérive* and after: the cognitive mapping of urban space', in C. Hollevoet (ed.) *The Power of the City/City of Power*. New York, Whitney Museum of Art, 25–56.

Holloway, J. and Kneale, J. (2000) 'Mikhail Bakhtin', in M. Crang and N. Thrift (eds) *Thinking Space*. London, Routledge, 71–88.

Holquist, M. (1990) *Dialogism: Bakhtin and his World*. London, Routledge.

Honour, H. (1961) *Chinoiserie*. London, J. Murray.

—— (1977) *Neo-Classicism*. Harmondsworth, Penguin.

Hopkins, E. (1982) 'Working hours and conditions during the industrial revolution: a reappraisal', *Economic History Review* 35(1): 52–66.

Hopkins, H. (1982) 'The subjective experience of time with particular reference to time-bound institutions', Ph.D. thesis, University of Lancaster.

Hopper, K. *et al.* (1997) 'Homelessness, severe mental illness, and the institutional circuit', *Psychiatric Services* 48: 659–65.

Howard, J. (1997) *Opening Address to the Australian Reconciliation Convention*.

Howell, S. (1992) 'Time past, time present, time future: contrasting temporal values in two Southern Asian societies', in S. Wallman (ed.) *Contemporary Futures: Perspectives from Social Anthropology*. London, Routledge, 124–37.

Hughes, E.C. (1984) *The Sociological Eye: Selected Papers*. New Brunswick, Transaction Books.

Huizinga, J. (1970) *Homo Ludens*. Frogmore, Paladin.

Hupfield, H. (1931, 1973) *As Time Goes By*. Harms Inc, USA/Reedwood Music Ltd, London W1X 2LR.

Husserl, E. (1964) *The Phenomenology of Internal Time Consciousness*, tr. J.S. Churchill. The Hague, Martinus Nuhoff.

Hutchinson, J. (1987) *The Dynamics of Cultural Nationalism: The Gaelic Revival and the Creation of the Irish Nation State*. London, Allen & Unwin.

Ingold, T. (1986) *Evolution and Social Life*. Cambridge, Cambridge University Press.

Isozaki, A. and Asada, A. (1999) 'Simulated origin simulated end', in C. Davidson (ed.) *Anytime*. Cambridge, MA, MIT Press, 76–83.

Jackson, P. and Smith, S. (1984) *Exploring Social Geography*. London, Allen & Unwin.

James, N. (1989) 'Emotional labour: skill and work in the social regulation of feelings', *Sociological Review* 37(1): 15–42.

James, W. (1960) *The Varieties of Religious Experience*. Glasgow, Collins.

Jameson, F. (1984) 'Postmodernism, or the cultural logic of late capitalism', *New Left Review* 146: 53–92.

—— (1991) *Postmodernism: or, the Cultural Logic of Late Capitalism*. London, Verso.

Janelle, D. (1968) 'Central place development in a time-space framework', *The Professional Geographer* 20: 5–10.

Japhet, S. (1998) 'Some Biblical concepts of sacred space', in B.Z. Kedar and R.J. Werblosky (eds) *Sacred Space: Shrine, City, Land*. New York, New York University Press, 55–72.

Jaques, E. (1982) *The Form of Time*. London, Heinemann.

Jay, M. (1992) 'Of plots, witnesses, and judgments', in S. Friedländer (ed.) *Probing the Limits of Representation: Nazism and the 'Final Solution'*. Cambridge, MA, Harvard University Press. 97–107.

Jewitt, L. (1865) *The Wedgwoods: Being a Life of Josiah Wedgwood*. London, Virtue Brothers.

Joas, H. (1985) *G. H. Mead: A Contemporary Re-examination of his Thought*, tr. R. Meyer. Cambridge, Polity Press.

Jones, M.M. (1991) *Gaston Bachelard: Subversive Humanist*. Madison, WI, University of Wisconsin Press.

Jönhill, J.I. (1997) 'Samhället som system och dess ekologiska omvärld: en studie i Niklas Luhmanns sociologiska systemteori', Lund Dissertations in Sociology 17, Sociologiska institutionen, Lunds universitet.

Jung, C.G. (1933) *Modern Man in Search of a Soul*, tr. W.S. Dell and Cary F. Baynes. New York, Harcourt Brace & Jovanovich.

—— (1973) *Memories, Dreams, Reflections*. New York, Vintage.

Jurczyk, K. (1998) 'Time in women's everyday lives: between self-determination and conflicting demands', *Time and Society* 7(2) 283–308.

296

Kasarda, J. (1989) 'Urban industrial transition and the underclass', *Annals of the American Academy of Political and Social Science* 501: 26–47.

Keith, M. and Pile, S. (eds) (1993) *Place and the Politics of Identity*. London, Routledge.

Kenny, J. (1995) 'Making Milwaukee famous: cultural capital, urban image, and the politics of place', *Urban Geography* 16(5): 440–58.

Kern, S. (1983) *The Culture of Time and Space 1880–1918*. Cambridge, MA, Harvard University Press.

Kidd, R. (1997) *The Way We Civilize: Aboriginal Affairs – The Untold Story*. Brisbane, University of Queensland Press.

Kieve, J. (1973) *The Electric Telegraph: A Social and Economic History*. Newton Abbot, David and Charles.

Kingston British Whig, 17 November 1883.

Knight, C. (1845) *Capital and Labour*. London.

Knights, D. and Odih, P. (1995) 'It's about time! The significance of gendered time for financial services consumption', *Time and Society* 4(2): 205–31.

Knox, P. (1991) 'The restless urban landscape: Economic and sociocultural change and the transformation of Metropolitan Washington DC', *Annals of the Association of American Geographers* 81(2): 181–209.

Kodras, J. (1997a) 'The changing map of American poverty in an era of economic restructuring and political realignment', *Economic Geography* 73: 67–93.

—— (1997b) 'Restructuring the state: devolution, privatization, and the geographic redistribution of power and capacity in governance', in L. Staeheli, J. Kodras and C. Flint (eds) *State Devolution in America: Implications for a Diverse Society*. Urban Affairs Annual Reviews 48. Thousand Oaks CA, Sage Publications, 79–96.

Kodras, J. and Jones, J.P. (1990) 'Introduction,' in J. Kodras and J.P. Jones (eds) *Geographic Dimensions of United States Social Policy*, New York, Edward Arnold, 1–16.

Kofman, E. and Lebas, E. (1995) 'Lost in transposition – time, space and the city', in E. Kofman and E. Lebas (eds) *Henri Lefebvre: Writings on Cities*, Oxford, Blackwell,.

Kristeva, J. (1981) 'Women's time', tr. A. Jardine and H. Blake *Signs* 7(1): 17–35.

Krygier, M. (1997) *Between Fear and Hope: Hybrid Thoughts on Public Values*. Sydney, ABC Books.

Kubler, G. (1962) *The Shape of Time*. New Haven: Yale University Press.

Laclau, E. (1990) *New Reflections on the Revolution of our Time*. London, Verso.

Laclau, E. and Mouffe, C. (1985) *Hegemony and Socialist Strategy: Towards a Radical Democratic Politics*. London, Verso.

Laermans, R. (1993) 'Learning to consume: early department stores and the shaping of the modern consumer culture', *Theory, Culture and Society* 10: 79–102.

Lake, R. (1997) 'State restructuring, political opportunism, and capital mobility,' in L. Staeheli, J. Kodras and C. Flint (eds) *State Devolution in America: Implications for a Diverse Society*. Urban Affairs Annual Reviews 48. Thousand Oaks CA, Sage Publications, 3–20.

Lamb, H.R. and Grant, R. (1982) 'The mentally ill in an urban county jail', *Archives of General Psychiatry* 39: 17–22.

Landes, D. (1969) *The Unbound Prometheus*. Cambridge, Cambridge University Press.

—— (1983) *Revolution in Time: Clocks and the Making of the Modern World*. Cambridge, MA, Belknap Press (Harvard University Press).

Langer, L. (1991) *Holocaust Testimonies: The Ruins of Memory*. New Haven, Yale University Press.

Lash, S. and Urry, J. (1987) *The End of Organised Capitalism*. Cambridge, Polity.

Laslett, P. (1988) 'Social structural time: an attempt at classifying types of social change by their characteristic pace', in M. Young and T. Schuller (eds) *The Rhythms of Society*. London, Routledge, 17–36.

Latour, B. (1988) *The Pasteurization of France*. Cambridge, MA, Harvard University Press.

—— (1993) *We Have Never Been Modern*. Harvard University Press.

—— (1997a) 'Trains of thought, Piaget formalism and the fifth dimension', *Common Knowledge* 6: 170–91.

—— (1997b) 'On actor-network theory: a few clarifications', CSTT, Keele univerity. Available at: ,http//www.keele.ac.uk/depts/stt/staff/jl/pubs-jl2.htm..

—— (1998) *Artefaktens återkomst: ett möte mellan organisationsteori och tingens sociologi.* Nerenius & Santérus förlag.

—— (1999) 'On recalling ANT', in J. Law and J. Hassard (eds) *Actor Network Theory and After*. Oxford, Blackwell, 15–25.

Lauer, R.H. (1981) *Temporal Man: The Meaning and Uses of Social Time*. New York, Praeger.

Law, J. (1994) *Organizing Modernity*. Oxford, Blackwell.

Law, J. and Benschop, R. (1997) 'Resisting pictures: representation, distribution and ontological politics', in K. Hetherington and R. Munro (eds) *Ideas of Difference: Social Spaces and the Labour of Division*, Oxford, Blackwell.

Le Guin, U. (1990) *Dancing at the Edge of the World*. New York, Perennial.

Leach, W.R. (1984) 'Transformations in a culture of consumption: women and department stores, 1890–1925', *The Journal of American History* 71(2): 319–42.

Leccardi, C. (1996) 'Rethinking social time: feminist perspectives', *Time and Society* 5(2): 151–79.

Leccardi, C. and Rampazi, M. (1993) 'Past and future in young women's experience of time', *Time and Society* 2(3): 353–79.

Lee, N. (1998) 'Two speeds: how are real stabilities possible', in R. Chia (ed.) *Organized Worlds*. Oxford, Blackwell, 39–66

Lee, W. (1994) 'Restructuring the local welfare state: a case study of Los Angeles', doctoral dissertation, University of Southern California.

—— (1997) 'Poverty and welfare dependency: The case of Los Angeles County in the 1980s', *Environment and Planning A* 29: 443–58.

Lefebvre, H. (1991a) *The Production of Space*. Oxford, Blackwell.

—— (1991b) *A Critique of Everyday Life*. London, Verso.

—— (1992) *Elements de Rhythmanalyse: Introduction a la Connaissance des Rhythmes*. Paris, Editions Syllepie.

—— (1995) 'Elements of rhythmanalysis', in E. Kofman and E. Lebas (eds) *Henri Lefebvre: Writings on Cities*. Oxford, Blackwell.

—— (1996) *Writings on Cities*. Oxford, Blackwell.

Leira, A. (1983) 'Women's work strategies. An analysis of the organisation of everyday life in an urban neighbourhood', in L. Arnlaug (ed.) *Work and Womanhood*, report 8/83. Oslo, Institute for Social Research, 125–60.

Lenntorp, B. (1998) 'Orienteringsanvisning i ett forskningslandskap', in M. Gren and P.O. Hallin (eds) *Svensk kulturgeografi: en exkursion inför 2000-talet*. Karlstad, Student Litteratur.

Lerman, P. (1982) *Deinstitutionalization and the Welfare State*. New Brunswick, NJ, Rutgers University Press.

Lestienne, R. (1998) *The Creative Power of Chance*. Urbana, University of Illinois Press.

Levinas, E. (1985) *Ethics and Infinity*, tr. R.A. Cohen. Pittsburgh, Duquesne University Press.

——— (1989) *The Levinas Reader*, ed. S. Hand. Oxford, Blackwell.

Levine, R. (1989) 'The pace of life', *Psychology Today* October: 42–6.

——— (1997) *A Geography of Time*. New York, Basic Books.

Lévi-Strauss, C. (1978) *Structural Anthropology*. Harmondsworth, Penguin.

Lewis, C.S. (1991) *The Lion, the Witch and the Wardrobe*. London, HarperCollins.

——— (1997) [1950] *The Chronicles of Narnia: The Lion, the Witch and the Wardrobe*. London, Collins.

Lewis, R. (ed.) (1969) *Staffordshire Pottery Industry*. Staffordshire County Council Education Department, Local History Source Books, No. 4.

Ley, D. (1988) 'From urban structure to urban landscape'. *Urban Geography* 9(1): 98–105.

Leyshon, A. (1995) 'Annihilating space?: the speed-up of communications', in J. Allen and C. Hamnett (eds) *A Shrinking World? Global Unevenness and Inequality*. Milton Keynes, Open University Press.

Liljeström, R. and Jarup, B. (1983) *Vardagsvett och vetenskap i vårdarbete*. Stockholm, Svenska kommunalarbetareförbundet.

Lindström, M. Nilsson (1998) 'Tradition och överskridande', Lund, Lund Dissertations in Sociology 21.

Lloyd, G. (1993) *Being in Time. Scales of Narrativism, Philosophy and Literature*. London, Routledge.

Lobel, A. (1979) *Days with Frog and Toad*. New York, Scholastic.

Longhurst, R. (1997) '(Dis)embodied geographies', *Progress in Human Geography* 21(4): 486–501.

Loy, D. (1996) *Lack and Transcendence: The Problem of Death and Life in Psychotherapy, Existentialism and Buddhism*. Atlantic-Highlands, New Jersey, Humanities Press.

——— (1997) *Nonduality: A Study in Comparative Philosophy*. New York, Humanity Books.

Luhmann, N. (1971a) *Politische Planung. Aufsätze zur Soziologie von Politik und Verwaltung*. Opladen, Westdeutscher Verlag.

——— (1971b) 'Die Knappheit der Zeit und die Vordringlichkeit des Befristeten', in *Politische Planung*. Opladen, Westdeutscher Verlag, 143–64.

——— (1978) 'Temporalisation of complexity', in R.F. Geyer and J. van der Zouwen (eds) *Sociocybernetics: An Actor-Orientated Social Systems Approach*. London, Martinus Nijhoff, 92–113.

—— (1979) 'Zeit und Handlung – eine vergessene Theorie', *Zeitschrift für Soziologie* 8: 63–81.

—— (1980) *Gesellschaftstruktur und Semantik. Studien zur Wissenssoziologie der modernen Gesellschaft*. Frankfurt a. M., Suhrkamp.

—— (1982a) *The Differentiation of Society*, tr. S. Holmes and C. Larmore. New York, Columbia University Press.

—— (1982b) 'The future cannot begin: Temporal structures in modern society', in *The Differentiation of Society*. New York, Columbia University Press, 271–89.

—— (1982c) 'World-time and system history', in *The Differentiation of Society*. New York, Columbia Universtiy Press, 289–324.

—— (1995) *Social Systems*. Stanford University Press.

Luke, A. (1998) 'The material effects of the word: apologies, "stolen children" and public discourse', *Discourse: Studies in the Cultural Politics of Education* 19(3): 343–68.

Lynch, K. (1972) *What Time is this Place*. Cambridge, Cambridge University Press.

—— (1989) 'Solitary labour: its nature and marginalisation', *The Sociological Review* 37(1): 1–14.

Lyotard, J.-F. (1984) *The Postmodern Condition*. Manchester, Manchester University Press.

MacIntyre, A. (1990) *Three Rival Versions of Moral Inquiry*. London, Gerald Duckworth.

Maffesoli, M. (1996) *The Time of the Tribes*. London, Sage.

—— (1998) 'Presentism – or the value of the cycle', in S. Lash, A. Quick and R. Roberts (eds) *Time and Value*. London, Sage, 103–12.

Mantoux, P. (1961) *The Industrial Revolution in the Eighteenth Century*. London, Jonathan Cape.

Marin, L. (1984) *Utopics: Spatial Play*. London, Macmillan.

—— (1993) 'Frontiers of Utopia: past and present', *Critical Inquiry* 19(3): 397–420.

Markus, T. (1993) *Buildings and Power*. London, Routledge.

Marvin, C. (1988) *When Old Technologies Were New: Thinking About Electric Communication in the Late Nineteenth Century*. New York, Oxford University Press.

Marx, K. (1938) *Capital, Volume 1*. London, George Allen and Unwin.

—— (1973) *Grundrisse: Foundations of the Critique of Political Economy*. Harmondsworth, Middlesex, Penguin Books.

—— (1987) *The Communist Manifesto*. Toronto, Canadian Scholars Press.

Marx, L. (1997a) 'Technology: the emergence of a hazardous concept', *Social Research* 64(3): 965–88.

—— (1997b) 'In the driving-seat? The nagging ambiguity in historians' attitudes to the rise of technology', *Times Literary Supplement* 29 August: 3–4.

Mason, S. (1944) 'The scientific revolution and the Protestant Reformation – I', *Annals of Science* 9(1).

Massey, D. (1992) 'Politics and space/time', *New Left Review* 196: 65–84.

—— (1993) 'Power-geometry and a progressive sense of place', in J. Bird, B. Curtis, T. Putnam, G. Roberston and L. Tickner (eds) *Mapping the Futures: Local Cultures, Global Change*. London, Routledge, 59–69.

—— (1994) *Space, Place and Gender*. Cambridge, Polity Press.

—— (1995) 'Masculinity, dualisms and high technology', *Transactions of the Institute of British Geographers* 20(4): 487–99.

—— (1998) 'Philosophy and the politics of spatiality', in *Power-Geometries and the Politics of Space-Time*, Hettner Lecture, University of Heidelberg, 25–42.

—— (2000) 'Travelling thoughts', in P. Gilroy, L. Grossberg and A. McRobbie (eds) *Without Guarantees: In Honour of Stuart Hall*. London, Verso, 225–32.

Massumi, B. (1992) *A Users Guide to Capitalism and Schizophrenia: Deviations from Deleuze and Guattari*. Cambridge, MA, MIT Press.

Matthews, G.B. (1994) *The Philosophy of Childhood*. Cambridge, MA, Harvard University Press.

Maturana, H.R. (1988) 'Reality: the search for objectivity or the quest for a compelling argument', *The Irish Journal of Psychology* 9(1): 25–82.

Maturana, H.R. and Varela, F.J. (1980) *Autopoiesis and Cognition: The Realization of the Living*. Boston, Riedel.

—— (1987) *The Tree of Knowledge: The Biological Roots of Human Understanding*. New Science Library.

Maturana, H.R., Coddou, F. and Méndez, C.L. (1988) 'The bringing forth of pathology', *The Irish Journal of Psychology* 9(1): 144–72.

May, J. (1991) 'Putting some space back in to time: the spatio-temporal self and the turn to the postmodern', in C. Philo (ed.) *New Words, New Worlds: Reconceptualising Social and Cultural Geography*. Aberystwyth, Cambrian Printers, 168–72.

—— (1994) 'Towards a sociology of time-space compression', unpublished Ph.D. thesis. University College London.

—— (1996a) 'A little taste of something more exotic: the imaginative geographies of everyday life', *Geography* 81(1): 57–64.

—— (1996b) 'In search of authenticity off and on the beaten track', *Environment and Planning D: Society and Space* 14: 709–36.

McAdams, D. (1990) 'Unity and purpose in human lives: the emergence of identity as a life story', in M. Riley, B. Huber and B. Hess (eds) *Social Change and the Life Course*. Newbury Park, Sage.

McGregor, R. (1997) *Imagined Destinies: Aboriginal Australians and the Doomed Race Theory, 1880–1939*. Melbourne, University of Melbourne Press.

McKendrick, N. (1961) 'Josiah Wedgwood and factory discipline', *The Historical Journal* 4(1): 30–55.

—— (1982) 'Josiah Wedgwood and the commercialization of the potteries', in N. McKendrick, J. Brewer and J.H. Plumb (eds) *The Birth of a Consumer Society*. London, Europa, 100–45.

McTaggart, J.M.E. (1927) *The Nature of Existence*, volume 11, Book V. Cambridge, Cambridge University Press.

Mead, G.H. (1934) *Mind, Self, and Society*. Chicago, University of Chicago Press.

—— (1959) *The Philosophy of the Present*, first published 1932, ed. A.E. Murphy, Preface by J. Dewey. La Salle, IL, Open Court.

Meares, R. (1992) *The Metaphor of Play*. Melbourne, Hill of Content.

Mehrotra, R. (1999) 'Working in Bombay: the city as generator of practice', in C. Davidson (ed.) *Anytime*. Cambridge, MA, MIT Press, 64–9.

Melbin, M. (1987) *Night as Frontier: Colonizing the World After Dark*. London, Macmillan.

Mellencamp, P. (1992) *High Anxiety: Catastrophe, Scandal, Age and Comedy*. Bloomington, University of Indiana Press.

Merleau-Ponty, M. (1962) *Phenomenology of Perception*, tr. C. Smith. London, Routledge and Kegan Paul.

Meteyard, E. (1865, 1866) *Life of Josiah Wedgwood*, two volumes. London, Hurst and Blackett.

Michael, M. (1996) *Constructing Identities: The Social, the Nonhuman and Change*. Sage Publications.

Michie, R.C. (1997) 'Friend or foe? Information technology and the London Stock Exchange since 1700', *Journal of Historical Geography* 23(3): 304–26.

Miller, H. (1991) 'Modeling accessibility using space-time prism concepts within geographical information systems', *International Journal of Geographical Information Systems* 5(3): 287–301.

Miller, J.H. (1995) *Topographies*. Stanford, CA, Stanford University Press.

Milner, M. (1987) *The Suppressed Madness of Sane Men*. London, Tavistock.

Mitchell, D. (1997) 'The annihilation of space by law: The roots and implications of anti-homeless laws in the United States', *Antipode* 29(3): 303–35.

Montreal Gazette, 4 April 1883.

Montreal Star, 19 November 1883.

Moore, R.O. (2000) *Savage Theory: Cinema as Modern Magic*. Durham, NC, Duke University Press.

Morelli, G. (1997) 'Sicilian time', *Time and Society* 6(1): 55–70.

Morgan, D.H.J. (1981) 'Men, masculinity and the process of sociological enquiry', in H. Roberts (ed.) *Doing Feminist Research*. London, Routledge and Kegan Paul, 83–113.

Morris, M. (1988) *The Pirate's Fiancée: Feminism, Reading, Postmodernism*. London, Verso.

—— (forthcoming) ' "The fairies of the day": modernity and enlightenment in the work of Ernestine Hill', in P. Hamilton and L. Johnson (eds) *Localising Modernity*.

Motevasel, I. (1996) 'Omsorg, ansvar och service. Exempel från manligt yrkesarbete', in R. Eliasson (ed.) *Omsorgens skiftningar. Begreppet, vardagen, politiken, forskningen*. Lund, Studentlitteratur, 52–70.

Mullarkey, J. (1999) *Bergson and Philosophy*. Edinburgh, Edinburgh University Press.

Mumford, L. (1934) *Technics and Civilization*. New York, Harcourt Brace.

Munn, N.D. (1992) 'The anthropology of time: a critical essay', *Annual Review of Anthropology* 21: 93–123.

Murdoch, J. (1997a) 'Towards a geography of heterogenous associations', *Progress in Human Geography* 23(1): 321–37.

—— (1997b) 'Inhuman/nonhuman/human: actor-network theory and the prospects for a nondualistic and symmetrical perspective on nature and society', *Environment and Planning D: Society & Space* 15: 731–56.

Närvänen, A.-L. (1994) *Temporalitet och social ordning. En tidssociologisk diskussion utifrån vårdpersonalens uppfattningar om handlingsmöjligheter i arbetet*, Linköping Studies in Arts and Science. 117.

Natter, W. (1993) 'The city as cinematic space: modernism and place in Berlin, symphony of a city', in S. Aitken and L. Zonn (eds) *Place, Power, Situation and Spectacle: A Geography of Film*. Lanham, MD, Rowman & Litlefield Publishers, 203–28.

Negrey, C. (1993) *Gender, Time and Reduced Work*. New York, State University of New York.

Nietzsche, F. (1986) 'The wanderer and his shadow', in *Human, All Too Human*, tr. R.J. Hollingdale, Cambridge, Cambridge University Press.

Nowotny, H. (1982) 'Nie von Zeit allein…', *Feministische Studien* 1(1): 9–18.

—— (1992) 'Time and social theory: towards a social theory of time', *Time and Society* 1(3): 421–54.

—— (1994) *Time: The Modern and Postmodern Experience*. Polity Press, Cambridge.

Nussbaum, M.C. (1999) 'The professor of parody', *New Republic* 22 February: 37–45.

Nye, D.E. (1994) *American Technological Sublime*. Cambridge, MA, MIT Press.

Oakley, A. (1974) *Housewife*. London, Allen Lane.

O'Brien, R. (1991) *Global Financial Integration: The End of Geography*. London, Pinter.

O'Connor, J. (1973) *The Fiscal Crisis of the State*. New York, St Martin's Press.

Ó hÁilín, T. (1972) 'Irish revival movements', in S. Ó Túama (ed.) *The Gaelic League Idea* Cork, Mercier Press.

Olsson, G. (1980) *Birds in Egg / Eggs in Bird*. London, Pion Limited.

—— (1991) *Lines of Power / Limits of Language*. Minneapolis, University of Minnesota Press.

—— (1993) 'Chiasm of thought-and-action', *Environment and Planning D: Society & Space* 11: 279–94.

—— (1998) 'Horror vacui', in M. Gren and P.O. Hallin (eds) *Svensk kulturgeografi: en exkursion inför 2000-talet*. Karlstad, Student Litteratur.

O'Malley, M. (1992) 'Time, work and task orientation: a critique of American historiography', *Time and Society* 1: 341–8.

Osborne, B. and Pike, R. (1984) 'Lowering "the walls of oblivion": the revolution in postal communications in central Canada, 1851–1911', in *Canadian Papers in Rural History* 4: 200–25.

Ottawa Daily Citizen, 5 April and 19 November 1883.

Owen, R. (1970) *A New View of Society*. Harmondsworth, Penguin.

Paolucci, G. (1998) 'Time shattered: the post-industrial city and women's temporal experience', *Time and Society* 7(2) 265–82.

Parkes, D. and Thrift, N. (1978) 'Putting time in its place', in T. Carlstein, D. Parkes and N. Thrift (eds) *Spacing Time and Timing Space*. London, Arnold, 119–29.

—— (1980) *Themes, Spaces, Places: A Chronogeographic Perspective*. Chichester, John Wiley.

Parr, H. and Philo, C. (1996) *A Forbidding Fortress of Locks, Bars and Padded Cells: The Locational History of Mental Health Care in Nottingham* Historical Geography Research Series, Number 32, Department of Geography, University of Glasgow.

Parr, J. (1990) *The Gender of Breadwinners: Women, Men and Change in Two Industrial Towns, 1880–1950*. Toronto, Toronto University Press.

Parris, H. (1965) *Government and the Railways in Nineteenth-century Britain*. London, Routledge & Kegan Paul.

Pateman, C. (1988) *The Sexual Contract*. Cambridge, Polity.

Paterson, K. (1998) *Parzival*. New York, Lodestar Books.

Pawson, E. (1992) 'Local times and Standard Time in New Zealand', *Journal of Historical Geography* 18(3): 278–87.

People to People, Nation to Nation: Highlights from the Report of the Royal Commission on Aboriginal Peoples (1996) ,http://www.inac.gc.ca/rcap/report/..

Perkin, H. (1970) *The Age of the Railway*. London, Panther.

Petersilia, J. (1997) 'Justice for all? Offenders with mental retardation and the California correctional system', *The Prison Journal* 77(4): 355–81.

Peterson, P. (1981) *City Limits*. Chicago, University of Chicago Press.

Phillips, A. (1988) *Winnicott*. London, Fontana.

—— (1993) *On Kissing, Tickling and Being Bored*. London, Faber and Faber.

Philo, C. (1989) ' "Enough to drive one mad": the organization of space in 19th-century lunatic asylums,' in M. Dear and J. Wolch (eds) *The Power of Geography*. Boston, Unwin, 258–90.

—— (1992) 'Foucault's geography', *Environment and Planning D: Society and Space* 10.

Piaget, J. (1973) *The Child's Conception of the World*, tr. J. and A. Tomlinson. Frogmore, Paladin.

Pile, S. (1996) *The Body and the City: Psychoanalysis, Space and Subjectivity*. London, Routledge.

Pile, S. and Thrift, N. (1995) *Mapping the Subject: Geographies of Cultural Transformation*. London, Routledge.

Pitt, W. (1808) *General View of the Agriculture of the County of Stafford*, 2nd edition. London, Richard Phillips.

Plot, R. (1686) *The Natural History of Staffordshire*. Oxford.

Pococke, R. (1888) *The Travels Through England of Dr. Richard Pococke*, ed. J. Cartwright for The Camden Society, N.S. 42 (Nichols and Sons, Westminster).

Pollard, S. (1965) *The Genesis of Modern Management*. London, Edward Arnold.

Poulet, G. (1966) *The Metamorphoses of the Circle*, tr. C. Dawson and E. Coleman. Baltimore, The John Hopkins Press.

Pred, A. (1977) 'The choreography of existence: comments on Hägerstrand's time-geography and its usefulness', *Economic Geography* 53: 207–21.

—— (1981a) 'Production, family and free-time projects: a time-geographic perspective on the individual and societal change in nineteenth-century U.S. Cities', *Journal of Historical Geography* 7(1): 3–36.

—— (1981b) 'Power, everyday practice and the discipline of human geography', in A. Pred (ed.) *Space and Time in Geography*. Lund Studies in Geography series B, No. 48: 30–55.

—— (1981c) 'Social reproduction and the time-geography of everyday life', *Geografiska Annaler series B* 63B(1): 5–22.

—— (ed.) (1981d) *Space and Time in Geography: Essays Dedicated to Torsten Hägerstrand*. Lund Studies in Geography, series B, no. 48.

—— (1984) 'Place as historically contingent process: structuration and the time-geography of becoming places', *Annals of the Association of American Geographers* 74(2): 279–97.

—— (1990a) 'In other wor(l)ds: fragmented and integrated observations on gendered languages, gendered spaces and local transformations', *Antipode* 22(1): 33–52.

—— (1990b) *Making Histories and Constructing Human Geographies*. Boulder, CO, Westview Press.

—— (1991a) 'Vega symposium introduction: everyday articulations of modernity'. *Geografiska Annaler*, 73 B: 3–5.

—— (1991b) 'Spectacular articulations of modernity: the Stockholm exhibition of 1897'. *Geografiska Annaler*, 73 B: 45–84.

—— (1992) 'Pure and simple lines, future lines of vision: The Stockholm exhibition of 1930', *Nordisk Samhällsgeografisk Tidskrift* 15: 3–61.

—— (1995) *Recognizing European Modernities: A Montage of the Present*. London, Routledge.

—— (1996) 'The choreography of existence: comments on Hägerstrand's time-geography and its usefulness,' in J. Agnew, D. Livingstone and A. Rogers (eds) *Human Geography, An Essential Anthology*. London, Blackwell, 636–49.

Prigogine, 1. and Stengers, 1. (1984) *Order out of Chaos: Man's New Dialogue with Nature*. London, Heinemann.

Pringle, J.F. (1972) *Lunenburgh or the Old Eastern District*. Cornwall, Ontario, Standard Printing House, repr. Belville, Ontario, Mika Silk Screening Ltd.

Probyn, E. (1996) *Outside Belongings*. London and New York, Routledge.

Probyn, E. (forthcoming) *Visceral Proximities: Eating, Sex & Ethics*. London, Sage.

Proust, M. (1983) *Remembrance of Things Past*, volume 3, tr. C.K. S. Moncrieff, T. Kilmartin and A. Mayor. Harmondsworth, Penguin.

Prude, J. (1983) *The Coming of Industrial Order: Town and Factory Life in Rural Masachusetts 1810–1860*. Cambridge, Cambridge University Press.

Pulido, L. (1996) 'A critical review of the methodology of environmental racism research', *Antipode* 28(2): 142–59.

Quick, A. (1998) 'Time and the event', in S. Lash, A. Quick A. and R. Roberts (eds) *Time and Value*. London, Sage, 103–12.

Rabin, J. (1999) 'Probe of police demanded in fatal shooting', *Los Angeles Times* 24 May: A1.

Rabinow, P. (1997) (ed.) 'Introduction', *Michel Foucault: Ethics, Subjectivity and Truth*. New York, The New Press.

Rajchman, J. (1997) *Constructions*. Cambridge, MA, MIT Press.

Rammstedt, 0. (1975) 'Alltagsbewusstsein von Zeit', *Kölner Zeitschrift für Soziologie und Sozialpsychologie* 27: 47–63.

Rämö, H. (1999) 'An Aristotelian human time-space manifold: from *chronochora* to *Kairotopos*', *Time and Society* 8(2): 309–28.

Réda, J. (1997) *The Ruins of Paris*. London, Reaktion Books.

Reiho, M. (158) *The Soto Approach to Zen*. Tokyo, Layman Buddhist Society Press.

Reilly, R. (1992) *Josiah Wedgwood*. London, Macmillan.

Relph, E. (1976) *Place and Placelessness*. London, Pion.

Ricoeur, P. (1984) *Time and Narrative*, volume 1. Chicago, University of Chicago Press.

—— (1988) *Time and Narrative*, volume 3. Chicago, University of Chicago Press.

Rivera, C. (1998) 'Skid Row's changing face', *Los Angeles Times* 28 July: A1.

Robins, K. (1991) 'Prisoners of the city: whatever could a postmodern city be?', *New Formations* 15: 1–22.

—— (1999) 'Foreclosing the city? The bad idea of virtual urbanism', in J. Downey and J. McGuigan (eds) *Technocities*. London, Sage, 34–59.

Robinson, J.P. and Godbey, G. (1997) *Time for Life*. Pennsylvania State University Press.

Robson, B. (1973) *Urban Growth: An Approach*. London, Methuen.

Rodowick, D.N. (1997) *Gilles Deleuze's Time Machine*. Durham, NC, Duke University Press.

Rorty, R. (1989) *Contingency, Irony, and Solidarity*. Cambridge, Cambridge University Press.

Rose, G. (1993) *Feminism and Geography*. Polity Press.

Rose, G. (1997) 'Situating knowledges: positionality, reflexivities and other tactics', *Progress in Human Geography* 21(3): 305–20.

Rose, H. (1994) *Love, Power and Knowledge*. Cambridge, Polity Press.

Rose, J. (1993) *The Case of Peter Pan*. Philadelphia, University of Pennsylvannia.

Rose, N. (1996) *Inventing Ourselves: Pyschology, Power and Personhood*. Cambridge, Cambridge University Press.

Ross, E. (1991) *Full of Hope and Promise: The Canadas in 1841*. Montreal and Kingston, McGill-Queen's University Press.

Rostow, W. (1960) *The Stages of Economic Growth*. Cambridge, Cambridge University Press.

Rothman, D. (1971) *The Discovery of the Asylum: Social Order and Disorder in the New Republic*. Boston, Little, Brown and Company.

—— (1980) *Conscience and Convenience: The Asylum and its Alternatives in Progressive America*. Boston, Little, Brown and Company.

Rowe, S. and Wolch, J. (1990) 'Social networks in time and space: homeless women in Skid Row, Los Angeles', *Annals of the Association of American Geographers* 80: 184–204.

Royal Commission on the Relations of Capital and Labor in Canada.(1889) Evidence. Ontario, Ottawa, Queen's Printer and Controller of Stationary.

Sassen, S. (1999) 'Juxtaposed temporalities: producing a new zone', in C. Davidson (ed.) *Anytime*. Cambridge, MA, MIT Press, 121–42.

Sayer, A. (1985) 'The difference that space makes', in D. Gregory and D. Urry (eds) *Social Relations and Spatial Structures*. London, Macmillan.

Schein, R.H. (1997) 'The place of landscape: A conceptual framework for an American scene', *Annals of the Association of American Geographers* 87(4): 660–80.

Schivelbusch, W. (1977) *The Railway Journey: Trains and Travel in the 19th Century*. New York, Urizen Books.

—— (1986) *The Railway Journey: The Industrialisation of Time and Space in the Nineteenth Century*. Berkeley, University of California Press.

—— (1988) *Disenchanted Night: The Industrialisation of Light in the Nineteenth Century*. Berkeley, University of California Press.

Schlör, J. (1998) *Nights in the Big City: Paris, Berlin, London, 1840–1930*. London, Reaktion.

306

Schön, D.A. (1983) *The Reflective Practitioner: How Professionals Think in Action*. New York, Basic Books.

Schöps, M. (1980) *Zeit und Gesellschaft*. Stuttgart, Ferdinand Enke Verlag.

Schor, J. (1992) *The Overworked American*. New York, Basic Books.

Schrift, A.D. (1995) 'Refiguring the subject as a process of self', *New Formations* 25: 28–39.

Schumpeter, J. (1964 [1939]) *Business Cycles: A Theoretical, Historical, and Statistical Analysis of the Capitalist Process*. New York, McGraw Hill.

Schutz, A. (1971) *Collected Papers*, volume I, *The Problem of Social Reality*, ed. M. Natanson. The Hague, Martinus Nijhoff.

Schutz, A. and Luckmann, T. (1973) *The Structures of the Life-World*, tr. R.M. Zaner and H.T. Engelhardt Jr. London, Heinemann.

Scranton, P. (1983) *Proprietary Capitalism: the Textile Manufacture at Philadelphia, 1800–1885*. Cambridge, Cambridge University Press.

Scull, A. (1982) *Museums of Madness*. London, Allen Lane.

—— (1984) *Decarceration: Community Treatment and the Deviant – A Radical View*, 2nd edition. New Brunswick NJ, Rutgers University Press.

Segel, J. (1998) *Body Ascendant: Modernism and the Physical Imperative*. Baltimore, MD, Johns Hopkins University Press.

Seigworth, G.J. (2000) 'Banality for cultural studies', *Cultural Studies*, 14: 227–68.

Semple, J. (1993) *Bentham's Prison*. Oxford, Clarendon Press.

Sendak, M. (1967) *Where the Wild Things Are*. London, The Bodley Head.

Senior, E.K. (1984) *From Royal Township to Industrial City: Cornwall 1784–1984*. Belleville, Ontario, Mika Publishing Company.

Serres, M. (1995) *Angels: A Modern Myth*, tr. F. Cowper. Paris, Flammarion.

Serres, M. and Latour, B. (1995) *Conversations on Science, Culture, and Time*, tr. R. Lapidus. Ann Arbor, University of Michigan Press.

Shapcott, M. and Steadman, P. (1978) 'Rhythms of urban activity', in T. Carlstein, D. Parkes and N. Thrift (eds) *Timing Space and Spacing Time (vol. 2): Human Activity and Time Geography*. London, Arnold, 49–74.

Shaw, S. (1970) *A History of the Staffordshire Potteries*. Newton Abbott, David and Charles.

Sheedy, J. (1981) *The Rediscovery of Ireland's Past: The Celtic Revival 1830–1930*. London, Thames and Hudson.

Shields, R. (1992) 'A truant proximity: presence and absence in the space of modernity', *Environment and Planning D: Society and Space* 10:181–98.

Showalter, E. (1986) 'Shooting the rapids', *Oxford Literary Review*: 218–24.

Simmel, G. (1994) 'Bridge and door', *Theory, Culture & Society*, 11(1): 5–10.

Skowronek, S. (1995) 'Order and change'. *Polity* 28(1): 91–6.

Smart, C. and Neale, B. (1999) *Family Fragments?*. Cambridge, Polity Press.

Smiles, S. (1905) *Josiah Wedgwood: His Personal History*. London, John Murray.

—— (1968) *Lives of the Engineers*. Newton Abbott, David and Charles.

Smith, A. (1922) *The Wealth of Nations*. London, Methuen.

Smith, D.M. (1998) 'How far should we care? On the spatial scope of beneficence', *Progress in Human Geography* 22 (1):1 5–38.

Smith, K. and Ferri, E. (1996) *Parenting in the 1990s: Family and Parenthood, Policies and Practice*. London, Family Policy Studies Unit.

Smith, N. (1996) *The Revanchist City*. New York, Blackwell Press.

Smith, N. and Katz, C. (1993) 'Grounding metaphor: towards a spatialised politics', in M. Keith and S. Pile (eds) *Place and the Politics of Identity*. London, Routledge, 67–83.

Smith, S. and M. Lipsky (1993) *Nonprofits for Hire: The Welfare State in the Age of Contracting*. Cambridge MA, Harvard University Press.

Smith, W.H. (1851) *Canada: Past, Present and Future, Being an Historical, Geographical, Geological and Statistical Account of Canada West*, volume 2. Toronto, Thomas Maclear.

Snow, R.P. and Brissett, D. (1986) 'Pauses: explorations in social rhythm', *Symbolic Interaction* 9(1): 1–18.

Sobel, D. (1998) *Longitude*. London, Fourth Estate.

Soja, E. (1980) 'The socio-spatial dialectic', *Annals of the Association of American Geographers* 70: 207–25.

—— (1989) *Postmodern Geographies: The Reassertion of Space in Social Theory*. London, Verso.

Sørensen, B.A. (1982) 'Ansvarsrasjonalitet: om mål – middeltenkning blant kvinner', in H. Holter (ed.) *Kvinner i Felleskap*. Oslo, Universitetsforlaget, 392–402.

St Augustine (1978) 'Some questions about time' in R.M. Gale (ed.) *The Philosophy of Time*. Brighton, Harvester Press, 38–55.

Stafford, B.M. (1991) *Body Criticism*. Cambridge, MA, MIT Press.

Staples, W.G. (1990) *Castles of Our Conscience: Social Control and the American State, 1800–1930*. New Brunswick NJ, Rutgers University Press.

Starr, P. (1988) 'The meaning of privatization', *Yale Law and Policy Review* 6: 6–41.

Stegmüller, W. (1969) *Hauptströmungen der Gegenwart*. Stuttgart, Alfred Körner Verlag.

Stein, J. (1992) 'Industrialising Cornwall: time, space and the pace of change in a nineteenth-century Ontario town', unpublished MA thesis, Queen's University.

—— (1995) 'Time, space and social discipline: factory life in Cornwall, Ontario, 1867–1893', *Journal of Historical Geography* 21(3): 278–99.

—— (1996) 'Annihilating space and time: the modernisation of fire-fighting in late nineteenth-century Cornwall, Ontario', *Urban History Review* 24(2): 3–11.

—— (1999) 'The telephone: its social shaping and public negotiation in late nineteenth- and early twentieth-century London', in M. Crang, P. Crang and J. May (eds) *Virtual Geographies: Bodies, Space and Relations*. London and New York, Routledge, 44–62.

Steiner, G. (1989) *Real Presences*. London, Faber and Faber.

Stekete, M. (1997) 'Sorry seems to be the hardest word', *Australian*, 24 May.,'Focus': 22.

Stephens, C. (1989) ' "The most reliable time": William Bond, the New England rail-roads, and time awareness in 19th-century America', *Technology and Culture* 30(1): 1–24.

Stewart, N. (1998) 'Re-languaging the body: phenomenological description and the dance image', *Performance Research* 3(2) 42–53.

Stewart, S. (1984) *On Longing: Narratives of the Miniature, the Gigantic, the Souvenir and the Collection*. Baltimore, Johns Hopkins University Press.

Stoner, M. (1995) *The Civil Rights of Homeless People: Law, Social Policy, and Social Work Practice*. New York, Aldine de Gruyter.

308

Storper, M. and Walker, R. (1989) *The Capitalist Imperative: Territory, Technology, and Industrial Growth*. New York, Blackwell.

Strathern, M. (1992) 'Foreword: the mirror of technology', in R. Silverstone and E. Hirsch (eds) *Consuming Technologies: Media and Information in Domestic Spaces*. London, Routledge.

Suderburg, E. (ed.) (2000) *Space, Site, Intervention: Situating Installation Art*. Minneapolis, University of Minnesota Press.

Takahashi, L. (1996) 'A decade of understanding homelessness in the USA: from characterization to representation', *Progress in Human Geography* 20(3): 291–310.

Takahashi, L. and Dear, M. (1997) 'The changing dynamics of community opposition to human service facilities', *Journal of the American Planning Association* 63(1): 79–93.

Tann, J. (1970) *The Development of the Factory*. London, Cornmarket Press.

Tarr, J.A. and Dupuy, G. (eds) (1988) *Technology and the Rise of the Networked City in Europe and North America*. Philadelphia, Temple University Press.

Tavuchis, N. (1991) *Mea Culpa: A Sociology of Apology and Reconciliation*. Palo Alto, Stanford University Press.

Taylor, M. and Bassi, A. (1998) 'Unpacking the state: The implications for the third sector of changing relationships between national and local government', *Voluntas: International Journal of Voluntary and Nonprofit Organizations* 9(2): 113–36.

Terdiman, R. (1992) 'The response of the other', *Diacritics* 22(2): 2–10.

Thompson, D. (1996) *The End of Time*. London, Minerva.

Thompson, E.P. (1967) 'Time, work-discipline and industrial capitalism', *Past and Present* 38: 56–97.

Thompson, W. (1967) *The Imagination of an Insurrection: Dublin Easter 1916*. Oxford, Oxford University Press.

Thrift, N. (1981) 'Owners' time and own time: the making of a capitalist time consciousness, 1300–1880', in A. Pred (ed.) *Space and Time in Geography*, Lund Studies in Geography, Series B, Lund, University of Lund, 56–84.

—— (1983) 'On the determination of social action in space and time', *Society and Space* 1: 23–57.

—— (1988) ' "Vivos Voco": ringing the changes in the historical geography time consciousness', in M. Young and T. Schuller (eds) *The Rhymns of Society*. London, Routledge, 53–94.

—— (1989) 'Images of social change', in C. Hamnett, L. McDowell and P. Sarre (eds) *The Changing Social Structure*. London, Sage.

—— (1990a) 'Transport and communication', in R.A. Dodgson and R.A. Butlin (eds) *An Historical Geography of England and Wales*. London, Academic Press, 453–86.

—— (1990b) 'For a new regional geography 1', *Progress in Human Geography*, 4(2).

—— (1991) 'For a new regional geography 2', *Progress in Human Geography*, 15(4): 456–65.

—— (1993) 'For a new regional geography 3', *Progress in Human Geography* 17(1): 92–100.

—— (1994) 'Inhuman geographies: landscapes of speed, light and power', in P. Cloke, M. Doel, D. Matless, M. Phillips and N. Thrift *Writing the Rural: Five Cultural Geographies*. London, Paul Chapman, 191–248.

—— (1995) 'A hyperactive world', in R.J. Johnston, P.J. Taylor and M.J. Watts (eds) *Geographies of Global Change: Remapping the World in the Late Twentieth Century*. Oxford, Blackwell, 18–35.

—— (1996) *Spatial Formations*. London, Sage.

—— (1997) 'Cities without modernity, cities without magic', *Scottish Geographical Magazine* 113(3): 138–50.

—— (1998) 'Re-imagining places, re-imagining identities', in H. Mackay (ed.) *Consumption and Everyday Life*. London, Sage, 159–212.

—— (2000) 'Everyday life in the city', in G. Bridge and S. Watson (eds) *The Blackwell Companion to the City*. Oxford, Blackwell.

Thrift, N. and Pred, A. (1981) 'Time-geography: a new beginning', *Progress in Human Geography* 5: 227–86.

Tobar, H. (1991) 'A dumping place for the mentally ill', *Los Angeles Times*, 25 August: A1, A4, A27.

Tong, R. (1993) *Feminine and Feminist Ethics*. Belmont, Wadsworth.

Torrey, E.F. *et al.* (1992) *Criminalizing the Seriously Mentally Ill: The Abuse of Jails as Mental Hospitals*. Washington DC, National Alliance for the Mentally Ill and Public Citizen's Health Research Group.

Toulmin, S. (1992) *Cosmopolis: The Hidden Agenda of Modernity*. Chicago, University of Chicago Press.

Tournier, M. (1983) *The Fetishist*. London, Methuen.

Travers, P.L. (1993) *What the Bee Knows*. Harmondsworth, Penguin.

Tredinnick, M. (1997) 'Review of P. Miller: *The Last One Who Remembers*', *Australian Book Review* 195: 48.

Tucker, B.M. (1984) *Samuel Slater and the Origins of the American Textile Industry, 1790–1860*. Ithaca, Cornell University Press.

Tunnicliff, W. (1787) *A Topographical Survey of the Counties of Stafford, Chester, and Lancaster*. Nantwich, E. Snelson.

Turner, F.J. (1894) 'The significance of the frontier in American History', Annual Report of the American Historical Association for 1893, Washington: US Government Printing Office. Reprinted in F.J. Turner (1962) *The Frontier in American History*. New York, Holt, Rinehart and Wilson, 1–38.

—— (1896) 'The problem of the West', *Atlantic Monthly*, September. Reprinted in F.J. Turner (1962) *The Frontier in American History*. New York, Holt, Rinhart and Wilson, 205–21.

Turner, V. (1969) *The Ritual Process*. Ringwood, Penguin.

Ure, A. (1967) *The Philosophy of Manufactures*. London, Frank Cass.

Urry, J. (1990) *The Tourist Gaze: Leisure and Travel in Contemporary Societies*. London, Sage.

—— (1992) 'Time and space in Giddens' social theory', in C. Bryant and D. Jary (eds) *Giddens' Theory of Structuration: A Critical Appreciation*. London, Routledge, 160–75.

—— (1995) *Consuming Places*. London, Routledge.

—— (1999) *Sociology Beyond Societies: Mobility for the Twenty-First Century*. London, Routledge.

Valentine, G. (1992) 'Images of danger: women's sources of information about the spatial distribution of male violence', *Area* 24: 22–9.

—— (1997) ' "My son's a bit dizzy." "My wife's a bit soft": Gender, children and cultures of parenting,' *Gender, Place and Culture* 4: 37–62.

Van Gennep, A. (1960) *Rites of Passage*, tr. M.B. Vizedom. London, Routledge & Kegan Paul.

Varela, F.J., Thompson, E. and Rosch, E. (1991) *The Embodied Mind*. Massachusetts Institute of Technology.

Varela, F.J. and Dupuy, G. (eds) (1992) *Understanding Origins*. Netherlands, Kluwer Academic Publishers.

Verdier, Y. (1981) *Tvätterskan, sömmerskan, kokerskan*. Stockholm, Atlantis.

Virilio, P. (1993) 'The third interval: a critical transition', in V.A. Conley (ed.) *Rethinking Technologies*. Minneapolis, University of Minnesota Press, 3–12.

—— (1997) *Open Sky*. London, Verso.

von Herder, J. G. (1968 [1783]) *Reflections on the Philosophy of the History of Mankind*, abridged and translated by T. Manuel. Chicago, Chicago University Press.

Wærness, K. (1984) 'The rationality of caring', *Economic and Industrial Bureaucracy* 5(2): 185–211.

Wallin, E. (1980) *Vardagslivets generativa grammatik*. CWK, Gleerup.

Ward, J. (1984) *The Borough of Stoke-Upon-Trent*. Stoke-On-Trent, Webberly.

Ward, J.T. (1970) *The Factory System*, 2 vols. Newton Abbott, David and Charles.

Ware, C.F. (1931) *The Early New England Cotton Manufacture: A Study in Industrial Beginnings*. Boston and New York, Houghton Mifflin Co.

Watson, S. (1998) 'The new Bergsonism', *Radical Philosophy* 97.

Watts-Miller, W. (2000) 'Ritual and time in Durkheim', *Time and Society* 9(1): 5–20.

Weatherill, L. (1971) *The Pottery Trade in North Staffordshire, 1660–1760*. Manchester, Manchester University Press.

Wedgwood, J. (1969) 'An address to the young people of The Pottery', ed. R. Lewis, Staffordshire Pottery Industry, Staffordshire County Council Education Department, Local History Source Books, No. 4. p. 17.

Wedgwood, J.C. (1913) *Staffordshire Pottery and Its History*. London, Sampson Low.

—— (1915) *The Personal Life of Josiah Wedgwood The Potter*. London, Macmillan.

Weinberg, S. (1988) *The First Three Minutes*. New York, Harper.

Weir, L. (1993) 'Limitations of New Social Movement Analysis', *Studies in Political Economy*, 40: 73–101.

Weisbrod, B. (1998) 'Institutional form and organizational behavior,' in W. Powell and E. Clemens (eds) *Private Action and the Public Good*. New Haven, CT, Yale University Press, 69–84.

Weiss, B. (1996) *The Making and Unmaking of the Haya Lived World*. Durham, NC, Duke University Press.

Welch, D. and Tanahashi, K. (1985) *Moon in a Dewdrop: Writings of Zen Master Dogen*. San Francisco, North Point Press.

Whipp, R. (1981) 'A time to every purpose: an essay on time and work', in P. Joyce (ed.) *The Historical Meanings of Work*. Cambridge, Cambridge University Press, 210–36.

Whorf, B.L. (1956) *Language, Thought and Reality*. Cambridge, Mass., MIT Press.

Wilshire, B. (1991) *Role Playing and Identity*. Bloomington, Indiana University Press.

Wilson, E. (1991) *The Sphinx in the City: Urban Life, the Control of Disorder, and Women*. London, Virago Press.

Wilson, J. (1987) *The Truly Disadvantaged*. Chicago, University of Chicago Press.

Wilson, D. and Huff, J.O. (1994) 'Contemporary human geography – the emergence of structuration in inequality research', in D. Wilson and J.O. Huff (eds) *Marginalized Places and Populations*. Westport CT, Praeger Press, xiii–xxv.

Winnicott, D.W. (1965a) *The Family and Individual Development*. London, Tavistock.

—— (1965b) *The Maturational Processes and the Facilitating Environment*. London, Hogarth Press.

—— (1991) *Playing and Reality*. London, Routledge.

Wolch, J. (1989) 'The shadow state: transformations in the voluntary sector,' in M. Dear and J. Wolch (eds) *The Power of Geography*. Boston, Unwin, 197–221.

—— (1990) *The Shadow State: Government and Voluntary Sector in Transition*. New York, The Foundation Center.

—— (1996) 'From global to local: the rise of homelessness in Los Angeles during the 1980s,' in A. Scott and E. Soja (eds) *The City: Los Angeles and Urban Theory at the End of the Twentieth Century*, eds. Berkeley, University of California Press, 390–424.

—— (1998) 'America's new urban policy: welfare reform and the fate of American cities', *Journal of the American Planning Association* 64(1): 8–10.

Wolch, J. and Dear, M. (1993) *Malign Neglect: Homelessness in an American City*. San Francisco, Jossey-Bass.

Wolch, J. and Sommer, H. (1997) *Los Angeles in an Era of Welfare Reform: Implications for Poor People and Community Well-Being*. Los Angeles: Southern California Inter-University Consortium on Homelessness and Poverty.

Wolch, J., Rahimian, A. and Koegel, P. (1993) 'Daily and periodic mobility patterns of the urban homeless', *The Professional Geographer* 45(2): 159–69.

Wolpert, E. and Wolpert, H. (1974) 'From asylum to ghetto', *Antipode* 6: 63–76.

Wolpert, J., Dear, M., and Crawford, R. (1975) 'Satellite mental health facilities', *Annals of the Association of American Geographers* 65: 24–35.

Wood, D. (1990) *Philosophy and the Limit*. London, Unwin Hyman.

Wright, T. (1997) *Out of Place: Homeless Mobilizations, Subcities, and Contested Landscapes*. Albany, SUNY Press.

Yaghmaian, B. (1998) 'Globalization and the state: The political economy of global accumulation and its emerging mode of regulation', *Science and Society* 62(2): 241–65.

Young, A. (1770) *A Six Months Tour Through the North of England*, volume 3. London.

Young, M. (1988) *The Metronomic Society*. London, Thames and Hudson.

Zelinsky, W. (1973) *The Cultural Geography of the United States*, New Jersey, Prentice-Hall.

Zerubavel, E. (1981) *Hidden Rythms: Schedules and Calendars in Social Life*, Chicago, University of Chicago Press.

Zimring, F. and Hawkins, G. (1991) *The Scale of Imprisonment*. Chicago, University of Chicago Press.

Zukin, S. (1995) *The Cultures of Cities*. New York, Blackwell.

INDEX

313